Cyber Forensics

Cyber Forensics

Examining Emerging and Hybrid Technologies

Edited by
Albert J. Marcella

CRC Press
Taylor & Francis Group
Boca Raton London New York

CRC Press is an imprint of the
Taylor & Francis Group, an **informa** business

Dedication
To Our Families....
Thank you for your
Strength, Support, Love

Contents

Preface

A global dependency on accurate and timely information is being demonstrated daily across world markets. Data is an international commodity of tremendous value as well as a necessity for organizations of all sizes, financial strength, and geographic footprint.

Threat actors, be they cyber criminals, terrorists, hacktivists, or disgruntled employees, are employing sophisticated attack techniques and anti-forensic tools to mask their attacks.

As the influence of emerging and hybrid technologies continues to grow in daily business decisions, the proactive use of cyber forensics to better assess the risks that the exploitation of these technologies pose to enterprise-wide operations is rapidly becoming a strategic business necessity.

This book moves beyond the typical, technical approach to discussing cyber forensic processes and procedures; instead, the authors examine how cyber forensics can be applied to identifying, collecting, and examining evidential data from emerging and hybrid technologies. Each author examines the process of cyber forensic investigation on an emerging or hybrid technology, while mindful of the influence, affect, and impact of the technology on general business operations.

Beyond the cyber forensic practitioner, this book is an essential resource for both the technical and non-technical executive, manager, attorney, auditor, information security professional, and general interest reader who is seeking an authoritative source on how cyber forensics may be applied to both evidential data collection and to proactively managing today's and tomorrow's emerging and hybrid technologies.

The authors, who have contributed their expertise to the chapters embodied in this book, have stepped beyond the typical, technical approach to discussing cyber forensic processes and procedures, instead, explore how cyber forensic tools and techniques can be proactively applied to examining and managing emerging and hybrid technologies.

Written by professionals responsible for routinely performing forensic investigations, presenting legal arguments and evidence in court, along with information security, privacy, and information technology (IT) audit professionals, this book examines the role which cyber forensics plays in such critical business areas as the Internet of Things (IoT); cloud computing; risk mitigation and management; fraud; operational technologies and Supervisory Control and Data Acquisition (SCADA) systems; mobile technologies; and emerging fields, such as unmanned aircraft systems and social network forensics.

The reader is presented with an enlightening discussion on cyber forensics applied to compliance and auditing in Chapter 1, written by Douglas Menendez, CIA, CISA. Douglas also examines the application of risk management to cyber forensics in Chapter 7 and provides the reader with a survey and review of cyber forensic tools in Chapter 10.

In Chapter 2, The Internet of Things (IoT) and the interrelationship with cyber forensics is expertly addressed by Detective Patrick Wilds, CISSP, CFCE, a 25-year police force veteran and digital forensic examiner.

Albert J. Marcella, wearing both the hats of editor and author, tackles the emerging field of applying cyber forensics to examining unmanned aircraft systems (UASs) and unmanned aerial vehicles (UAVs, aka drones) in Chapter 3 and explores the role of cyber forensics in investigating breach and security activities involving operational technology and industrial control systems in Chapter 6.

In Chapter 4, Ronald L. Krutz, Ph.D., P.E., CISSP, ISSEP, an accomplished author, security expert, and founder of the Carnegie Mellon Research Institute Cybersecurity Center, analyzes and provides a comprehensive examination of cloud forensics.

Dr. James Curtis, Ph.D., PMP, a 24-year career Air Force officer and former Presidential Communications Officer for the President of the United States, discusses the structure of social media, networks, and engineering and the unique threats and challenges these techniques pose within the realm of cybersecurity forensic analysis, in Chapter 5 "Forensics of the Digital Social Triangle with an Emphasis on DeepFakes".

Detective Andy Hrenak, a 30-year police force veteran and digital forensic examiner, shares his extensive knowledge and breadth of hands-on examination experience by providing the reader with an introduction to mobile device forensics in Chapter 8.

While identifying, collecting, and analyzing electronic data are essential steps in performing a cyber forensic investigation, doing so in a manner that, if required, the data can be submitted as evidential matter in a court of law is paramount. Given the reality that cyber forensics has become increasingly important to the field of forensic accounting, it is inevitable that cyber forensic professionals will likely be called upon to assist the forensic accountant in the performance of their investigatory responsibilities.

Chapter 9, Forensic Accounting and the Use of E-Discovery and Cyber Forensics, written by Richard Dippel, JD, MBA, CPA, provides the reader with an examination of the interrelationship that exists between the legal system, cyber forensic investigations, and the forensic accounting profession.

The use of proven cyber forensic techniques, when applied to existing, emergent, and hybrid technologies, is shown by the authors to foster a greater cybersecurity awareness and posture for the organization, and a heightened, proactive response to threat actors who seek to exploit these technologies.

The reader of this text will be pleased to learn that accompanying the extensive body of material presented in this book, the Publisher has established an eResources companion site. Appendix material, including recommended additional reading resources and publications along with a complete and comprehensive glossary of terms used in this book, has been provided as a downloadable eResource, available through the Publisher's website at https://routledge.com/9780367524180. Readers are encouraged to access and download these additional, value-added resources, as you look further into the role of cyber forensic examination of emerging and hybrid technologies.

It has been both my honor and privilege to work with a world-class team of cyber forensic researchers, investigators, authors, practitioners, law enforcement, and cybersecurity professionals, each dedicated to sharing their knowledge, experience, and insights into the application of cyber forensic methodologies and procedures to both emerging and hybrid technologies.

Thank you, James Curtis, Richard Dippel, Andrew Hrenak, Ronald L. Krutz, Douglas Menendez and Patrick Wilds.

Albert J. Marcella, Ph.D., CISA, CISM
Editor

Acknowledgements

We wish to thank the anonymous reviewers identified by Taylor & Francis Group, LLC, our Publisher, who initially reviewed our proposed book and provided feedback and comments on the proposed content when the authors initially brought the idea for this book to CRC Press.

Our thanks go to the anonymous online contributors, drone pilots, technicians, users, and guests on the following forums: Airdata UAV Forum, DJI Phantom Drone, DJI UAV Online, Pixhawk, and UAV Coach, who posted countless questions and responses, provided technical assistance and insights, uploaded flight logs for analysis, and who make the UAS/UAV forums a viable resource for everyone.

We would also like to thank:

- Desert Rotor of Scottsdale, AZ, for permission to use a photo of the company's 12PCX HOTAS HD Portable GCS, Dual Screen UAV Ground Control Station
- Paul Thelu and Drotek Electronics, Avignonet-Lauragais France, for permission to use a photo of the company's The Pixhawk® 3 Pro Autopilot
- Elliot Webb, founder and chief pilot at High Line Drones LLC, of Quinebaug, CT, for providing the Putnam Connecticut Town-Scape image and associated EXIF data
- Holybro organization for permission to include photos of the company's Kakute F7 All-In-One flight controller
- UAV Navigation S. L., of Madrid, España, for permission use a photo of the company's Flight Data Recorder
- Dr. Fahad E. Salamh, at Purdue University and Digital Forensics and Incident Response consultant, for his excellent and detailed review of materials for the chapter Cyber Forensics: Examining Commercial Unmanned Aircraft Systems (UASs) and Unmanned Aerial Vehicles (UAVs). Dr. Salamh's research 'A 3-Dimensional UAS Forensic Intelligence-Led Taxonomy (U-FIT),' will soon be independently published.
- Myles Levin, CEO and Founder, Daubert Tracker, LLC and Principal and Co-founder, Expert Witness Profiler, LLC, for use of the Daubert Tracker graphic and personal information provided related to the Expert Witness Profiler
- Robert White, Bradley Smith, Russ Curry, Alex De Looze (Aerialtronics), David Webster (Autel Robotics), and Justin Slater (Phantom Pilot), who responded to specific UAS/UAV research questions and provided flight logs for analysis: thank you for your contributions and for so graciously responding to requests for further information
- Dr. Keyun Ruan for permission to use data from her research in her publication 'Cloud forensics definitions and critical criteria for cloud forensic capability: an overview of survey results,' published in Digital Investigation

Editor

Dr. Albert J. Marcella is a Professor of Management in the Management Department, Walker School of Business and Technology, at Webster University in Saint Louis, Missouri. Dr. Marcella is also President of Business Automation Consultants, (BAC) LLC, an international IT consultancy, which he founded in 1984.

Dr. Marcella is an internationally recognized public speaker, researcher, IT consultant, and workshop and seminar leader with 40 years of experience in IT audit, risk management, IT security, and assessing internal controls.

Dr. Marcella has authored numerous articles and 28 books on various IT-, audit-, and security-related subjects. Dr. Marcella's clients include organizations in financial services, IT, banking, petrolchemical, transportation, services industry, domestic, and international public utilities, and departments of government, nonprofits, and international agencies.

Dr. Marcella is the 2016 recipient of the Information Systems Security Association's Security Professional of the Year award and recipient of the Institute of Internal Auditors Leon R. Radde Educator of the Year 2000 award.

Contributors

Dr. James Curtis has over 40 years of leadership, information technology, cybersecurity, project management, and logistics experience. His expertise encompasses systems analysis; software development lifecycle; encryption systems; cybersecurity critical infrastructures and risk analysis and management; and IT strategic planning.

Highlights of a distinguished 24-year Air Force career include being the first Cybersecurity (electronic security) Officer for a Major Air Force Command (Strategic Air Command) leading a team of cyber professionals in establishing and operating an information security/TEMPEST program for military installations, weapons systems, and IT systems.

Dr. Curtis also developed the cyber security assessment program for new weapon systems. From 1989 to 1993, he served in the White House, where he supported the President of the United States of America as the 'President's Communicator.' In this role, he managed all presidential communication systems for the President, White House staff, and US Secret Service.

Applicable publications include 'Ensuring Resilience and Prosperity in a Digital Economy,' Participant for National Council on Competitiveness Cyber Committee Report, 2018, and 'A Critique and Policy Analysis of the United States Air Force Computer Security Program,' University of Oklahoma, 1985.

Richard Dippel is an Associate Professor of Accounting in the Business Department, Chair of the Business Department, and Program Director of the Master of Science in Forensic Accounting graduate degree program in the George Herbert Walker School of Business and Technology at Webster University (Saint Louis, Missouri, USA).

Rich has represented clients in a variety of business civil litigation matters, including disputes involving corporate, commercial, and bankruptcy issues. He also has managed the financial aspects of various businesses and advised businesses on a variety of legal and business topics.

Rich earned his B.S. in Business Administration, Master of Business Administration, and Juris doctor degrees from Saint Louis University.

He is a Certified Public Accountant (CPA) and is licensed to practice law in the state courts of Missouri and Illinois and several federal trial and appellate courts, including the U.S. Supreme Court.

Andrew Hrenak is a former active duty Marine and thirty-year veteran of the Hazelwood, Missouri, Police Department, serving at the rank of Detective. Throughout his career, Andy participated in the initiation of small unit assets to address current and future needs of the organizational mission he is a component of.

Andy's career as a Police Officer spans from general patrol, investigations of crimes against property and persons, to his current position as a Digital Forensic Examiner within the Regional Computer Crimes and Enforcement Group (RCCEEG), Saint Louis, MO.

In his current position, Andy supports local, state, and various federal investigative agencies in the recovery and analysis of information that is useful in the investigative process. Those cases involve online investigative techniques, network acquisitions, personal computer and mobile device repair, and electronic memory module data extraction at a physical level.

Andy actively participates as a Peer Review Coach for the International Association of Computer Investigative Specialists (IACIS) in the Certified Forensic Computer Examiner (CFCE) process.

Dr. Ronald L. Krutz is an Adjunct Professor for Webster University. He has over 30 years of experience in industrial automation and control systems, distributed computing systems, computer architectures, information assurance methodologies, and information security training.

Dr. Krutz has been a Senior Information Security Consultant at Security Research Solutions, Lockheed Martin, BAE Systems, and REALTECH Systems Corporation; an Associate Director of the Carnegie Mellon Research Institute (CMRI); and a professor in the Carnegie Mellon University Department of Electrical and Computer Engineering.

He was also a lead instructor for (ISC)2 Inc. in their Certified Information Systems Security Professionals (CISSP) training seminars. Dr. Krutz founded the CMRI Cybersecurity Center and was founder and Director of the CMRI Computer, Automation and Robotics Group.

As a published author, Dr. Krutz has authored the *CISSP Prep Guide* and 11 other cybersecurity texts for John Wiley and Sons and ISA, holds seven patents in digital systems and has published over 40 technical papers. He has also authored two university texts on microprocessors and logic design and a text on digital interfacing techniques. He also served as a guided missile officer in the U.S. Army Ordnance Corps.

Douglas Menendez is a talented and seasoned executive with over 30 years of experience in Risk Management, Auditing, and Compliance. He brings a wealth of expertise with particular strengths in areas such as Information Technology Controls/Security/Audit, Enterprise & Operational Risk Management, Process Improvement, and Regulatory Compliance. Doug has worked across multiple industry-leading companies, including Mastercard, Graybar Electric, Enterprise Holdings, and Express Scripts.

In addition to his executive career, Doug is past president of both the IIA (Institute of Internal Auditors) and ISACA (Information Systems Audit & Control Association) St. Louis Chapters and has held many committee and Board positions with both professional associations. Doug is also a member of InfraGard (FBI's public–private security partnership), the Information Systems Security Association (ISSA), local Data Analytics User Groups, and is on the SecureWorld Expo Advisory Council.

Doug co-authored the book *Cyber Forensics: A Field Manual for Collecting, Examining, and Preserving Evidence of Computer Crimes, 2nd Edition*. He has also written articles for authoritative journals and contributed to research publications.

Patrick Wilds is a 25-year veteran police Detective, having spent the last 14 years as a digital forensic examiner. Pat conducts forensic examinations for local, state, federal and military law enforcement agencies. He is also the supervisor of the digital evidence section of his department's crime laboratory.

Pat provides training to examiners, law enforcement, prosecutors, and private sector entities on forensics and information security. He has instructed and proctored courses for the National Computer Forensics Institute and U.S. Department of Justice.

He has over 1500 hours of professional training in forensics, information security and internet-based investigations.

Chapter 1

Cyber forensics
Compliance and auditing

Douglas Menendez

CONTENTS

INTRODUCTION

The constructs of compliance and auditing may vary depending upon industry and application. To begin this chapter on common ground, we will first take a brief look at exactly what compliance and auditing is, from a broad, more global perspective. In many instances throughout this book, the reader will encounter terms such as examiner and investigator. While auditing involves both the process of examination and investigation, there is both an operational as well as functional difference between the two processes.

Let's start with some definitions of compliance and auditing.

> *The definition of compliance is:* 'the action of complying with a command," or "the state of meeting rules or standards.' *In the corporate world, it's defined as the process of making sure your company and employees follow all laws, regulations, standards, and ethical practices that apply to your organization and industry*[1]

The definition of an *audit* is the process of evaluation or analysis of something to determine its accuracy. In the business world, *auditing* can be focused on financial, operational, or information technology:

Financial Auditing:

> The process of verifying a company's financial information. An auditor examines a company's accounting books and records in order to determine whether the company is following appropriate accounting procedures. An auditor issues an opinion in a report that says whether the financial statements "present fairly" the company's financial position and its operational results in accordance with Generally Accepted Accounting Principles (GAAP).[2]

Operational Auditing:

> An independent review and examination of records and activities to assess the adequacy of operational controls, to ensure compliance with established policies and operational procedures, and to recommend necessary changes in controls, policies, or procedures.[3]

Information Technology Auditing:

> An independent review and examination of system records and activities in order to test the adequacy and effectiveness of data security and data integrity procedures, to ensure compliance with established policy and operational procedures, and to recommend any necessary changes.[4]

There are also two main categories of auditing: internal and external.

Definition of Internal Auditing:

Internal auditing is an independent, objective assurance and consulting activity designed to add value and improve an organization's operations. It helps an organization accomplish its objectives by bringing a systematic, disciplined approach to evaluate and improve the effectiveness of risk management, control, and governance processes.[5]

Definition of External Auditing:

External auditing is an independent function outside of the organization that assesses the financial and risk associated aspects in order to comply with statutory audit requirements. The main role of external audit is to provide an opinion whether the company financial statements present a true and fair view of the company's financial results. The external audit function is managed by the external auditor, who in the United States is typically a Certified Public Accountant.[6]

The audit work performed by an auditor is different from the investigation work performed by cyber forensic professionals (see Table 1.1).

The remainder of this chapter will focus the reader's attention on a review and examination of auditing and compliance and the rapidly growing field of cyber forensics.

As defined by UpGuard, Cyber forensics is a branch of forensic science focused on the recovery and investigation of material found in digital devices and cybercrimes. Throughout this book, cyber forensics, digital forensics, and computer forensics are used interchangeably.

As society increases reliance on computer systems and cloud computing, cyber forensics becomes a crucial aspect of law enforcement agencies and businesses. The reader interested in a deeper review of cyber forensics and cloud computer is directed to Chapter 4 and Ronald L. Krutz's examination of the subject.

Table 1.1 Auditing vs. investigation comparison[7]

Basis for comparison	Auditing	Investigation
Meaning	The process of inspecting the books of accounts of an entity and reporting on it is known as Auditing.	An inquiry conducted for establishing a specific fact or truth is known as Investigation.
Nature	General Examination	Critical and in-depth examination
Evidences	The evidences are persuasive in nature.	The evidences are unquestionable; therefore, its nature is decisive.
Time Horizon	Annually	As per requirement
Performed by	Certified Public Accountant or Chartered Accountant	Experts
Reporting	General Purpose	Confidential
Obligatory	Yes	No
Appointment	An auditor is appointed by the shareholders of the company.	The management or shareholders or a third party can appoint investigator.
Scope	Seeks to form an opinion on financial statements.	Seeks to answer the questions that are asked in the engagement letter.

Cyber forensics is concerned with the identification, preservation, examination, and analysis of digital evidence, using scientifically accepted and validated processes to be used in and outside of a court of law.

While its roots stretch back to the personal computing revolution in the late 1970s, cyber forensics began to take shape in the 1990s, and it wasn't until the early 21st century that countries like the United States began rolling out nation-wide policies.[8]

Addressed throughout this chapter will be a discussion of a cyber forensics event timeline, relevant laws, and regulations along with applicable cyber forensic policies and procedures. Equally important will be a review and discussion of best practices for cyber forensics compliance, along with cyber forensic certifications.

By the end of this chapter, we will examine the role of audit in cyber forensics and using cyber forensics proactively to mitigate fraud.

CYBER FORENSICS EVENT TIMELINE

Before we look at a cyber forensics event timeline, it is important to first understand some of the major milestones in cybersecurity breaches.

Cybersecurity is an evolving field that is in a constant state of flux (see Table 1.2). Hackers are unrelenting in their search for vulnerabilities to exploit, while information security professionals try to assure that information and assets are properly protected. By understanding the cyber events of the past, we can hopefully learn and improve our future cybersecurity policies, processes, and procedures.

WHY IS CYBER FORENSICS IMPORTANT?

Cyber forensics is important because it is used in both criminal and private investigations. Traditionally, it is associated with criminal law where evidence is collected to support or negate a hypothesis before the court. Collected evidence may be used as part of intelligence gathering or to locate, identify, or halt other crimes. As a result, data gathered may be held to a less strict standard than traditional forensics.

In civil cases, cyber forensics may help with electronic discovery (eDiscovery). A common example is following an unauthorized network intrusion. A forensic examiner will attempt to understand the nature and extent of the attack, as well as try to identify the attacker.

The most common use of cyber forensics is to support or refute a hypothesis in a criminal or civil court:

- **Criminal cases:** These involve the alleged breaking of laws and law enforcement agencies and their cyber forensic examiners.
- **Civil cases:** These involve the protection of rights and property of individuals or contractual disputes between commercial entities where a form of cyber forensics called electronic discovery (eDiscovery) may be involved.

Cyber forensic experts are also hired by the private sector as part of cybersecurity and information security teams to identify the cause of data breaches, data leaks, cyber-attacks and

Table 1.2 Cybersecurity breach milestones[9]

Date	Threat Actor	Description
Early 1970s	Bob Thomas	Thomas wrote the 'Creeper,' a self-replicating program that used ARAPNET to infect DEC PDP-10 computer and display the message, 'I'm the creeper, catch me if you can!'
1976–2006	Greg Chung Boeing Corporation	Chung stole $2 billion (US) worth of aerospace docs and gave them to China. 225,000 pages of sensitive material were recovered in his home. This was one of the largest insider attacks in history with malicious intent to supply China with proprietary military and spacecraft intel.
2013	Edward Snowden	Former CIA employee and contractor for the US government copied and leaked classified information from the National Security Agency.
2013–2014	Unknown	Largest Data Breach. Yahoo reported a breach by a group of hackers that jeopardized the accounts of all 3 billion users. Everything from names to passwords and security question answers were compromised. Yahoo failed to report this breach until 2016 and was fined $35 million by the SEC for failure to disclose the breach in a timely manner.
2015	Unknown	The US Office of Personnel Management fell victim to an attack that stole 4.2 million personnel files of former and current government employees. This included 21.5 million security clearance background investigation files and 5.6 million fingerprints.
2016	Unknown	Panamanian law firm Mossack Fonseca suffered a data breach in April 2016 that exposed 2.6 terabytes of sensitive data totaling 11.5 million files. The leaked data included 4.8 million emails, 2.2 million PDF documents, 1.1 million image files, 3 million database records, and 320,000 other text files.
2017	Unknown	The First Ransomeworm, WannaCry, a ransomware cryptoworm, targeted computers running the Microsoft Windows operating system and demanded ransom payments in the Bitcoin cryptocurrency.
2018	Unknown	An 'unauthorized party' acquired data associated with 150 million of Under Armour's MyFitnessPal user accounts.
2019	Unknown	Mobile game producer Zynga announced that a hacker had accessed account log-in information for 218 million customers. Hackers took log-in credentials, usernames, email addresses, log-in IDs, some Facebook IDs, some phone numbers, and Zynga account IDs.
2020	Unknown	Mathway, a popular website for helping students and children learn mathematics suffered from a data breach, resulting in more than 25 million records being exposed.

other cyber threats. Cyber forensic analysis may also be part of incident response to help recover or identify any sensitive data or personally identifiable information (PII) that was lost or stolen in a cybercrime.[10]

CYBER FORENSICS AND TODAY'S AUDITING PROFESSION

Cyber forensic professionals and auditing professionals frequently work together when there is a suspicion of fraud that is detected as part of an internal or external audit. A cyber forensic professional who has special training in forensic audit techniques will be used by the audit team to perform additional steps beyond the regular audit procedures.

The cyber forensic professional will work to:

- Identify what fraud, if any, is being carried out;
- Determine the time period during which the fraud has occurred;
- Discover how the fraud was concealed;
- Identify the perpetrators of the fraud;
- Quantify the loss suffered due to the fraud;
- Gather relevant evidence that is admissible in the court;
- Suggest measures that can prevent such frauds in the company in future.

The cyber forensic professional will start by collecting evidence. By the conclusion of the audit, the cyber forensic professional is required to understand the possible type of fraud that has been carried out and how it has been committed. The evidence collected should be adequate enough to prove the identity of the fraudster(s) in court, reveal the details of the fraud scheme, and document the amount of financial loss suffered and the parties affected by the fraud.

A logical flow of evidence will help the court in understanding the fraud and the evidence presented. Forensic auditors are required to take precautions to ensure that documents and other evidence collected are not damaged or altered by anyone.

Common techniques used for collecting evidence in a forensic audit include the following:

- Substantive techniques – For example, doing a reconciliation, review of documents, etc.
- Analytical procedures – Used to compare trends over a certain time period or to get comparative data from different segments
- Computer-assisted audit techniques – Computer software programs that can be used to identify fraud
- Understanding internal controls and testing them so as to understand the loopholes which allowed the fraud to be perpetrated
- Interviewing the suspect(s)

Once the evidence is analyzed, a report is required so that it can be presented to a client about the fraud. The report should include the findings of the investigation, a summary of the evidence, an explanation of how the fraud was perpetrated, and suggestions on how internal controls can be improved to prevent such frauds in the future. The report needs to be presented to a client so that they can proceed to file a legal case if they so desire.

Lastly, the cyber forensic professional needs to be present during court proceedings to explain the evidence collected and how the suspect was identified. They should simplify the complex accounting issues and explain in layman's language so that people who have no understanding of the accounting terms can still understand the fraud that was carried out.

To summarize, a forensic audit is a detailed engagement that requires the expertise of not only accounting and auditing procedures but also expert knowledge regarding the legal framework. A forensic auditor is required to understand various frauds that can be carried out and of how evidence needs to be collected.[11]

CYBER FORENSICS: A TIMELINE OF SIGNIFICANT CONTRIBUTIONS

The following table provides the reader with a concise review of the more significant events that paved the way for the field today, which we call cyber forensics (see Table 1.3).

Table 1.3 History of cyber forensics: A timeline[12]

Date	Event
1966	The first federally prosecuted case of computer crime in the United States. It involved a consultant who programmed and maintained the computer system of a Minneapolis bank.
1973	The Equity Funding Insurance Company fraud leads to an increased interest in digital forensics.
1988	The court decided that if an individual independently searches a computer, finds alarming results, and reports it to law enforcement, then the action doesn't violate the Fourth Amendment.
2000	The 'I Love You' virus case demonstrates that crime that effects the globe yet originated in an unrestricted country would have complicated legal effects.
2001	The U.S. Department of Justice released the Technical Working Group for Electronic Crime Scene Investigation's report, Electronic Crime Scene Investigation: A Guide for First Responders.
2002	The U.S. Department of Justice's Computer Crime and Intellectual Property section suggests that computer files are just as private as information stored in something like a filing cabinet. This also prevented law enforcement from searching electronic devices at their leisure.
2003	In the Jessica Chapman murder case, cell phones led investigators to yet another way to track down suspects.
2005	Because some digital evidence is very work-intensive or costly to produce, courts start using a system of 'burden and cost' to determine whether it would be possible and worthwhile to try and get certain evidence.
2005	During the Zubulake case, three categories are made for reasonably accessible data: active, online data; near-line data; and offline storage. Backup tapes and erased or fragmented tapes are categorized as hard to recover.
2006	Amendments to the U.S. Federal Rules of Civil Procedure clearly define e-discovery data as Electronically Stored Information (ESI), which cleared up confusion in courts.
2008	Criminals begin to create anti-forensic tools to destroy evidence of their activities and distract investigators.
2008	George Socha and Thomas Gelbman create the Electronic Discovery Reference Model, a six-step, widely accepted framework for e-discovery.
2008	The Good Practice Guide for Computer-Based Electronic Evidence is published by the Association of Police Officers and acknowledges that the traditional 'pull the plug' technique leads to loss of data.
2013	US President Obama issued Executive Order (EO) 13636, *Improving Critical Infrastructure Cybersecurity*, which calls for a voluntary risk-based cybersecurity framework (the Cybersecurity Framework, or CSF) that is 'prioritized, flexible, repeatable, performance-based, and cost-effective.'
2014	Five laws were passed in the United States, including the National Cybersecurity Protection Act (NCPA) and the Cybersecurity Enhancement Act.
2016	Obama developed a Cybersecurity National Security Action Plan (CNAP).
2018	The EU General Data Protection Regulation (GDPR) was set into place on 14 April 2016, with a date of enforcement of 25 May 2018. The GDPR aims to bring a single standard for data protection among all member states in the EU.
2020	Some 92% of United Nations Member States have developed reforms of legislation on cybercrime and electronic evidence.

CYBER FORENSICS: SOLVING DIGITAL CRIMES ONE BYTE AT A TIME

Cyber forensics is rapidly becoming a way and a means for not only law enforcement but for organizations to proactively address unauthorized computer activities, cyber events, and the misuse or inappropriate use of computer hardware and software.

In a 24/7, always on, digitally connected, global, mobile society, cyber forensics, the ability to search for and acquire digital evidence has assisted in solving crime and mitigating risks – risks to organizations and to individuals.

The reader may find the following examples of applied cyber forensics of interest, prior to moving onto the next section of this chapter, where we will discuss laws and regulations relevant to cyber forensics.

CASE: MATT BAKER – IMPRISONMENT FOR ALLEGED MURDER

Synopsis: In 2006, a case caught public attention, against Matt Baker for allegedly murdering his wife, which according to the news and evidence found, later on, reported to be an apparent suicide.

Cyber Forensic Analysis: Computer forensic analysts investigated Matt Baker's background – then Baptist Texas preacher – and sought out the data on his laptop, despite having a suicide note from the crime scene. After going through his search history, they found out that not only did Matt Baker enter a query related to 'overdosing on sleeping pills,' but he also ventured through different pharmaceutical operations to gain access to the drugs. The case was closed in 2010 when the court of law sentenced Matt Baker to life imprisonment or 65 years for murdering his wife.[13]

CASE: KRENAR LUSHA – INVESTIGATION OF ALLEGED TERRORISM

Synopsis: Krenar Lusha, an alleged terrorist of British nationality, was held accountable for his acts on terrorism after a ton of ammunition was found in his apartment in 2009.

Cyber Forensic Analysis: The police raided his apartment after computer forensic analysts investigated his search history and pattern on the Internet and hacked his MSN account only to find out that he has not only been searching for tutorials on making explosive devices, but also introducing himself as a sniper. The computer forensic analysts managed to retrieve all deleted conversations through their hacking skills and tools.[13]

CASE: NATHANIEL SOLON – CHILD PORNOGRAPHY

Synopsis: Downloading media from the Internet arguably became most popular with the arrival of Napster, where one could download completely free music from any person connected to the service who was willing to share. Other person-to-person (p2p) programs soon followed, such as LimeWire and Share Bear. Nathaniel Solon apparently used this p2p network to illegally download music, video games, and later, child pornography.

Cyber Forensic Analysis: He was discovered by the Internet Crimes Against Children Agency, who found that not only was he downloading such files, but distributing them as well.[14]

CASE: HASSAN ABU-JIHAAD – TERRORISM AND ESPIONAGE

Synopsis: Hassan was serving as a signalman aboard the USS *Benfold*. Little did anyone know at the time, he was also a homegrown radical who was secretly in touch with al Qaeda financiers, sharing classified details about the vulnerabilities and movements of the ships just six months after al Qaeda operatives had killed 17 Americans aboard the USS *Cole* in the port of Yemen.

Cyber Forensic Analysis: When British authorities raided the apartment of Babar Ahmad, a Briton later charged with raising money for al Qaeda through a London-based organization called Azzam Publications. Its former website, www.azzam.com, was hosted on servers in Connecticut.

In Ahmad's flat was a floppy disk with a password-protected document detailing what was then classified information about the travel and security weaknesses of the USS Benfold and the sister ships in its convoy. That document, it was proved at trial, was sent by Abu-Jihaad while aboard the Benfold, endangering the lives of his own shipmates and countless others.[15]

FUTURE CHALLENGES FOR CYBER FORENSICS

So, what lies ahead in the future for Cyber Forensics? Below are some of the future challenges for Cyber Forensics:

- Cyber forensics is a critical aspect of modern law enforcement investigations and deals with how data is gathered, studied, analyzed, and stored. This includes the recovery and investigation of data found in electronic devices. Due to the nature of flash memory, and a lack of sufficient protocols in place to outline effective data-retrieval techniques for solid state discs (SSDs) and universal serial bus (USB) flash drives, data forensic examiners face many challenges that sometimes impede their ability to operate successfully.

  In addition to the numerous technical complications that investigators face, there are also many legal matters to consider. These legal issues are not secondary considerations whereas having valid search authority is a primary requirement. It is important not to overlook or minimize the importance of the legal difficulties surrounding digital forensic investigations.[16]

- Data Volume and Velocity. Nowadays, most information is created, stored, modified, and accessed purely in digital form. This knowledge highlights the importance of digital investigations, because most of our daily activities and interactions are digitally recorded in some form, meaning that critical evidence in criminal investigations must be extracted from an electronic device. Organizations now capture and process greater volumes of data than ever before. Only a few years ago, working with a 100-megabyte file was considered a lot of data. Today, data can be measured in zettabytes, or ZBs, which is equal to 1 trillion megabytes.

  Many organizations find themselves 'drowning in data.' Beyond the vast amount of data collected, today's globalization and connectivity result in data produced at incredible and increasing speeds. IBM estimates that approximately 90% of all the data in the world was created in the past two years alone.

  In 2012, 2.8 ZBs were created; in 2020, the total data generated annually is forecasted to reach 40 ZBs. User-generated content such as photos and videos and devices with sensors that constantly generate data—commonly referred to as the Internet of Things (IoT)—contribute significantly to the mountain of digital information.[17]

- Though there are many issues that law enforcement officers encounter when attempting to retrieve digital data, the two that will currently present the biggest challenges are cloud computing and encryption. Cloud computing has changed the way that data is stored. It is possible to store data blocks in different jurisdictions, meaning officials in the United States could be faced with trying to retrieve cloud stored data that is in another country.[18]
- The use of encryption technology to protect computer data is growing—and that fact presents a challenge for forensic investigators. Without a decryption key, forensic tools cannot be used to find digital evidence. Even with the key, searching encrypted data can be tricky and time consuming. The move to encryption is coming from hardware and software companies who are embedding encryption technology into their products. Cyber forensic investigators are limited to the information on the device that they can access. If a hard drive is fully encrypted, they have no easy access to the stored data and the investigative options become limited.[19]

In short, cyber forensics is, and will continue to be, a highly valuable tool in criminal investigations. Law enforcement agencies need to be equipped with the proper people, tools, and resources to legally conduct these types of investigations. As society becomes increasingly reliant on various communication technologies, more evidence will be found digitally. This area poses significant challenges for investigators, due to rapidly changing technologies, accessibility, retrieval, and legal issues.[20]

As we continue our review into compliance, auditing and cyber forensics, we next examine the various and relevant laws and regulations related to the field of cyber forensics.

CYBER FORENSICS RELEVANT LAWS AND REGULATIONS

In the previous section, we introduced some of the key definitions and distinguished between auditing and investigations. We looked back at some cybersecurity breach milestones, discussed why cyber forensics is important, and reviewed a timeline of significant cyber forensic contributions. We noted some significant crimes that have been solved through the use of cyber forensics and listed some future challenges for cyber forensics.

In the following section, we will explore laws and regulations that are relevant to cyber forensics, including the Computer Fraud and Abuse Act (CFAA), Hacking laws and Internet laws. For an in-depth look at forensic accounting and the use of e-discovery and cyber forensics (see Chapter 9, authored by Richard Dippel).

It is important for the cyber forensic professional to understand the relevant laws and regulations at the national and state level, in order to operate within the aspects of those laws while preforming any cyber forensics investigation.

The following is a review and overview of significant cyber laws.

COMPUTER FRAUD AND ABUSE ACT (CFAA)

The CFAA was enacted in 1986, as an amendment to the first federal computer fraud law, to address hacking. Details of the offenses addressed by the CFAA can be found in Table 1.4. Over the years, it has been amended several times, most recently in 2008, to cover a broad range of conduct far beyond its original intent. The CFAA prohibits intentionally accessing a computer without authorization or in excess of authorization, but fails to define what 'without authorization' means. With harsh penalty schemes and malleable provisions, it has become a tool ripe for abuse and use against nearly every aspect of computer activity.

Table 1.4 Provisions of the Computer Fraud and Abuse Act[22]

Offense	Section	Sentence*
Obtaining National Security Information	(a)(1)	10 years (20)
Accessing a Computer and Obtaining Information	(a)(2)	1 or 5 years (10)
Trespassing in a Government Computer	(a)(3)	1 year (10)
Accessing a Computer to Defraud and Obtain Value	(a)(4)	5 years (10)
Intentionally Damaging by Knowing Transmission	(a)(5)(A)	1 or 10 years (20)
Recklessly Damaging by Intentional Access	(a)(5)(B)	1 or 5 years (20)
Negligently Causing Damage and Loss by Intentional Access	(a)(5)(C)	1 year (10)
Trafficking in Passwords	(a)(6)	1 year (10)
Extortion Involving Computers	(a)(7)	5 years (10)
Attempt and Conspiracy to Commit such an Offense	(b)	10 years for attempt but no penalty specified for conspiracy in section (c)

*The maximum prison sentences for second convictions are noted in parentheses.

As technology advances, the use of the criminal law to regulate conduct using such technology also advances. Perceptions concerning the role of technology in both traditional and high-tech criminal conduct prompted Congress to enact the first federal computer crime law 30 years ago. Increases in computer availability and mainstream usage, however, have propelled government regulation of computer conduct into overdrive.

Over the course of 30 years, federal computer crimes went from non-existent to touching on every aspect of computer activity for intensive and occasional users alike.

The CFAA is not without its critics, however. The National Association of Criminal Defense Lawyers (NACDL) states that the breadth and ambiguity of the CFAA are deeply troubling. NACDL supports wholesale reform of the CFAA and, in particular, believes violations of website terms of services should not be federal crimes.

NACDL opposes any additional expansion of the CFAA and is actively working to reform the CFAA through *amicus* support, coalition building, and legislative advocacy.[21]

Cybercrime federal legislation – evolution

The Computer Fraud and Abuse Act (CFAA), 18 U.S.C. § 1030, is a civil and criminal cybercrime law prohibiting a variety of computer-related conduct and has seen an evolution since its inception, responding proactively to the increase in digital crime.

Since the original enactment of the CFAA in 1984, technology and the human relationship to it have continued to evolve. Although Congress has amended the CFAA on numerous occasions to respond to new conditions (see Table 1.5), the rapid pace of technological advancement continues to present novel legal issues under the statute.

Although sometimes described as an anti-hacking law, the CFAA is much broader in scope. Indeed, it prohibits seven categories of conduct, including certain exceptions and conditions:

1. Obtaining national security information through unauthorized computer access and sharing or retaining it;

Table 1.5 Changes to the computer fraud and abuse act[23]

Year	Computer Fraud and Abuse Act (CFAA)
1984	Congress passes the Comprehensive Crime Control Act (CCCA), which included the first federal computer crime statute, later codified at 18 U.S.C. § 1030
1986	Congress passes the Computer Fraud and Abuse Act (CFAA)
1994	CFAA amended to cover several other computer-related acts including: • Theft of property via computer that occurs as part of a scheme to defraud • Intentional alteration, damage, or destruction of data belonging to others • Distribution of malicious code and denial of service • Trafficking in passwords and similar items.
2001	Congress expands the CFAA through the USA Patriot Act. The most significant change was the expanded definition of 'protected computer' to include computers located outside the United States; specifically, those computers 'located outside the United States that [are] used in a manner that affects interstate or foreign commerce or communications of the United States.'
2008	CFAA (18 U.S.C. § 1030(a)(7)) expanded to criminalize not only explicit threats to cause damage to a computer but also threats to (1) steal data on a victim's computer, (2) publicly disclose stolen data, or (3) not repair damage the offender already caused to the computer. Created a criminal offense for conspiring to commit a computer hacking offense under section 1030. Established a mechanism for civil and criminal forfeiture of property used in or derived from § 1030 violations. Broadened the definition of 'protected computer' in 18 U.S.C. § 1030(e)(2) to the full extent of Congress's commerce power by including those computers used in or affecting interstate or foreign commerce or communication.

2. Obtaining certain types of information through unauthorized computer access;
3. Trespassing in a government computer;
4. Engaging in computer-based frauds through unauthorized computer access;
5. Knowingly causing damage to certain computers by transmission of a program, information, code, or command;
6. Trafficking in passwords or other means of unauthorized access to a computer;
7. Making extortionate threats to harm a computer or based on information obtained through unauthorized access to a computer.[24]

STATE LEGISLATION

There are many states that are passing their own legislation on computer crimes. It is important for the cyber forensic professional to monitor these pending legislations, which when they become law, could potentially impact a cyber forensic investigation.

Computer Crimes Legislation can be found on each respective State's Legislation website. Below are 12 bills that were active in October 2020 (see Table 1.6).

In addition to monitoring pending state cybercrimes legislation, both the auditor and cyber forensic professional should understand law and protocols for prosecuting computer crimes.

This U.S. Department of Justice (DOJ) has published a manual that examines the federal laws that relate to computer crimes. The focus is on those crimes that use or target computer networks, which are interchangeably referred to as 'computer crime,' 'cybercrime,' and 'network crime.' Examples of computer crime include computer intrusions, denial of service (DoS) attacks, viruses, and worms. The DOJ does not attempt to cover issues of state law and do not cover every type of crime related to computers, such as child pornography or phishing.

Table 1.6 Pending state legislature cybersecurity bills

State	Bill number	Title
Connecticut	HB 5511	An Act Concerning an Analysis Of Municipal Cybersecurity[25]
Illinois	HB 5204	Cybersecurity Legal Defense[26]
Maryland	HB 635	Criminal Law – Crimes Involving Computers - Malware and Ransomware[27]
Minnesota	HF 4085	Unauthorized access of critical state information technology system crime established.[28]
	SF 4297	Critical state information technology systems unauthorized access crime establishment.[29]
New Jersey	A 3984	Creates affirmative defense for certain breaches of security.[30]
	A 4518	Increases penalty for 'bombing' online meeting or teleconference under certain circumstances.[31]
	S 1374	Establishes 'Internet Predator Investigation and Prosecution Fund' with $200 assessment on persons convicted of certain offenses.[32]
New York	AB 2124	Creates specific computer crimes as well as increasing penalties for crimes committed with the aid of a computer.[33]
Ohio	HB 368	To enact the Ohio Computer Crimes Act.[34]
Virginia	SB 378	Computer trespass; expands the crime.[35]
	SB 844	Computer trespass; expands the crime.[36]

This manual is intended as assistance, not authority. The research, analysis, and conclusions therein reflect current thinking on difficult and dynamic areas of the law; they do not represent the official position of the Department of Justice or any other agency. This manual has no regulatory effect, confers no rights or remedies, and does not have the force of law or a U.S. Department of Justice directive (see United States v. Caceres, 440 U.S. 741 (1979)).

Electronic copies of this document are available from the DOJ website, www.cybercrime. gov. The DOJ may update the electronic version periodically and we advise cyber forensic professionals and others interested to check the website's version for the latest developments.

The Table of Contents of the manual is as follows:

Chapter 1: Computer Fraud and Abuse Act
Chapter 2: Wiretap Act
Chapter 3: Other Network Crime Statutes
Chapter 4: Special Considerations
Chapter 5: Sentencing
There are also several useful Appendices. The cyber forensic professional will find this to be a valuable resource.[37]

In addition to the CFAA and various state regulations, the cyber forensic professional and auditor should understand the various hacking laws.

HACKING LAWS AND PUNISHMENTS

While there are many types of crimes that can be committed using a computer, some of the most prominent cases involve hacking. Hardly a day goes by when there is not some sort of data breach being covered in the news and on social media. These hackers have infiltrated businesses in all industries, non-profit organizations, and government agencies.

However, not all hacking is a criminal event. There are numerous types of 'hacking' and 'hackers.' So, let's first cover some definitions to better understand the differences.

Definition of hacking and types of hackers

Hacking can be defined as the act of breaking into a computer system. Hacking can be categorized as either authorized or unauthorized. On the authorized side, organizations can hire hackers (ethical or white-hat hackers) to test existing infrastructures for bugs and loopholes so that these weaknesses can be fixed before being exploited by bad or black-hat hackers. The black-hat hackers are a type of malicious hackers which accesses a computer system without prior consent or authorization. These malicious hackers can include cyber criminals, spammers, hacktivists, and disgruntled insiders.

Federal hacking laws

There are several federal laws that address hacking, including:

- The Computer Fraud and Abuse Act (CFAA) (covered earlier in this section)
- The Stored Communications Act (SCA)
- The Electronic Communications Privacy Act (ECPA)
- The Defend Trade Secrets Act (DTSA)

The Stored Communications Act mirrors the prohibitions of the CFAA and protects stored electronic communications and data or data at rest (including email, texts, instant messages, social media accounts, cloud computing and storage, and blogs/microblogs).

The Electronic Communications Privacy Act (EPCA), a counterpart law to the SCA forbids intentional interception of electronic communications in transit or 'data in motion,' rather than 'data at rest.'

The DTSA allows an owner of a trade secret to sue in federal court when its trade secrets have been misappropriated.

Hacking laws: State laws

Although much of the focus is on federal laws, states have enacted hacking laws as well. While every state has computer crime laws, some states address hacking more specifically with laws that prohibit unauthorized access, computer trespass, and the use of viruses and malware.

For example, approximately half of the states in the country have laws that target the use of DoS attacks. In this form of hacking, an intruder floods the system or servers with traffic, denying access to legitimate users.

Ransomware occurs when malware is installed on someone's computer, denying access to the computer unless a ransom is paid. Several states have laws that specifically criminalize ransomware.[38]

Having presented relevant discussion addressing cyber forensic laws and regulations, we next move onto cyber forensic policies and controls.

CYBER FORENSICS POLICIES AND CONTROLS

In the previous section, we reviewed the relevant cyber forensics laws and regulations that are important to the cyber forensic professional. In this section, we will discuss the key cyber forensics policies and controls. Following a defined process is necessary to have a defendable and repeatable procedure that, if necessary, will stand up in a court of law.

Because different organizations are subject to different laws and regulations, this section should not be used as a guide to executing a digital forensic investigation, construed as legal advice, or used as the basis for investigations of criminal activity.

Organizations should use this guide as a starting point for developing a forensic capability in conjunction with extensive guidance provided by legal advisors, law enforcement officials, and management.

For the purposes of this section, we will highlight the process as defined in NIST Special Publication 800-86, Guide to Integrating Forensic Techniques into Incident Response.

The process for performing cyber forensics comprises the following basic phases:

- **Collection:** Identifying, labeling, recording, and acquiring data from the possible sources of relevant data, while following procedures that preserve the integrity of the data.
- **Examination:** Forensically processing collected data using a combination of automated and manual methods, and assessing and extracting data of particular interest, while preserving the integrity of the data.
- **Analysis:** Analyzing the results of the examination, using legally justifiable methods and techniques, to derive useful information that addresses the questions that were the impetus for performing the collection and examination.
- **Reporting:** Reporting the results of the analysis, which may include describing the actions used, explaining how tools and procedures were selected, determining what other actions need to be performed (e.g., forensic examination of additional data sources, securing identified vulnerabilities, improving existing security controls), and providing recommendations for improvement to policies, procedures, tools, and other aspects of the forensic process.[39]

This section provides general recommendations for performing the forensic process. To start with, organizations should define a set of policies to address forensic considerations.

Policies

Organizations should ensure that their policies include specific statements outlining all important forensic considerations, such as working with law enforcement, tracking activity and ongoing examination of forensic policies, protocols, and procedures. At a high level, policies should allow approved staff to track system and network activity and, under appropriate circumstances, carry out investigations for legitimate reasons.

Policies should be reviewed regularly, especially for organizations that have national or international operations, due to new laws, updates to regulations, and new rulings from the courts. Policies should also provide guidance on the appropriate use of forensic tools. This would include who is authorized to use forensic tools and under what circumstances. A more detailed review of cyber forensics tools is provided in Chapter 10.

Guidelines and procedures

Organizations should have forensic guidelines and procedures that include a protocol for investigating a cyber forensic event. This should include step-by-step procedures for performing essential tasks, such as imaging a hard disk, capturing volatile information, and maintaining a documented chain of custody.

The purpose of the guidelines and procedures is to promote consistent, reliable, and precise forensic actions that are especially relevant for events that could lead to the documented evidence being used in criminal prosecution or internal disciplinary action. This is especially true with electronic evidence that can be easily created, altered, or manipulated.[40]

Now that we have outlined the need for an organization to establish their policies, guidelines, and procedures, we will go into some additional detail for performing the forensic process.

PERFORMING THE FORENSIC PROCESS

The most common goal of performing forensics is to gain a better understanding of an event of interest by finding and analyzing the facts related to that event. As described above, forensics may be needed in many different situations, such as evidence collection for legal proceedings and internal disciplinary actions and handling of malware incidents and unusual operational problems. Regardless of the need, forensics should be performed using the four-phase process shown in Figure 1.1.

This section describes the basic phases of the forensic process: collection, examination, analysis, and reporting. During collection, data related to a specific event is identified, labeled, recorded, and collected, and its integrity is preserved. In the second phase, examination, forensic tools and techniques appropriate to the types of data that were collected are executed to identify and extract the relevant information from the collected data while protecting its integrity. Examination may use a combination of automated tools and manual processes.

The next phase, analysis, involves analyzing the results of the examination to derive useful information that addresses the questions that were the impetus for performing the collection and examination. The final phase involves reporting the results of the analysis, which may include describing the actions performed, determining what other actions need to be performed, and recommending improvements to policies, guidelines, procedures, tools, and other aspects of the forensic process.

As shown at the bottom of Figure 1.1, the forensic process transforms media into evidence, whether evidence is needed for law enforcement or for an organization's internal usage. Specifically, the first transformation occurs when collected data is examined, which extracts data from media and transforms it into a format that can be processed by forensic tools. Second, data is transformed into information through analysis.

Finally, the information transformation into evidence is analogous to transferring knowledge into action using the information produced by the analysis in one or more ways during the reporting phase. For example, it could be used as evidence to help prosecute a specific individual, actionable information to help stop or mitigate some activity, or knowledge in the generation of new leads for a case.

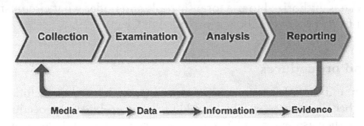

Figure 1.1 Forensic process[41]

Phase 1 – Data collection

The first step in the forensic process is to identify potential sources of data and acquire data from them. This includes desktop computers, servers, laptop computers, internal drives, and external storage devices. Also, cell phones, digital cameras, video recorders, and other network devices and logs can contain evidence. Table 1.7 shows the related control steps for this phase.

Phase 2 – Examination

After data has been collected, the next phase is to examine the data, which involves assessing and extracting the relevant pieces of information from the collected data. This phase may also involve bypassing or mitigating operating system (OS) or application features that obscure data and code, such as data compression, encryption, and access control mechanisms. An acquired hard drive may contain hundreds of thousands of data files; identifying the data files that contain information of interest, including information concealed through file compression and access control, can be a daunting task. Table 1.8 shows the related control steps for this phase.

Phase 3 – Analysis

Once the relevant information has been extracted, the analyst should study and analyze the data to draw conclusions from it. The foundation of forensics is using a methodical approach

Table 1.7 Steps in the data collection process

Control Steps – Safeguarding Digital Evidence
- o Secure the physical area of the scene
- o Take possession of all hardware and other storage devices
- o Ensure to protect all volatile digital evidence
- o Obtain all relevant logs and data
- o Document all evidence that will be removed from the scene
- o Preserve the chain of custody

Control Steps – Transferring Digital Evidence
- o Securely transport evidence from scene to storage
- o Preserve chain of custody during transportation

Control Steps – Storing Digital Evidence
- o Store evidence in a controlled access area
- o Restrict and log all access to the evidence area
- o Preserve chain of custody while in secure storage

Table 1.8 Steps in the examination process

Control Steps – Ensure Integrity of Digital Evidence
- o Follow established digital forensic investigation methodology
- o Ensure to write-protect all digital evidence source media

Control Steps – Extract Digital Evidence
- o Extract digital evidence in order of volatility
- o Extract non-volatile digital evidence

Control Steps – Copy Digital Evidence
- o Make forensic copies of all digital evidence

Control Steps – Authenticate Digital Evidence
- o Authenticate all digital evidence as identical to the original
- o Time stamp all copies of the authenticated digital evidence

Control Steps – Documenting Acquisition Process
- o Document all actions through chain of custody documentation

to reach appropriate conclusions based on the available data or determine that no conclusion can yet be drawn. The analysis should include identifying people, places, items, and events, and determining how these elements are related so that a conclusion can be reached. Table 1.9 shows the related control steps for this phase.

Phase 4 – Reporting

The final phase is reporting, which is the process of preparing and presenting the information resulting from the analysis phase. Table 1.10 shows the related control steps for this phase.

The key take-aways from this four-phase forensic process are as follows:

- Organizations should perform forensics using a consistent process.
- Analysts should be aware of the range of possible data sources.
- Organizations should be proactive in collecting useful data.

Table 1.9 Steps in the analysis process

Control Steps – Plan Analysis
- o Evaluate all available evidence about the event (digital and physical)
- o Ascertain if additional specialized forensic expertise is required
- o Determine which forensic tools are most appropriate to use

Control Steps – Analyze Evidence
- o Analyze evidence using the most appropriate forensic tools available
- o Follow the requirements of the 'best evidence rule' applicable in your jurisdiction

Control Steps – Create Timeline
- o Reconstruct sequence of events
- o Match digital evidence with other known facts about the event

Control Steps – Formulate Conclusions
- o Draw results based on the evidence reviewed
- o Document the finding

Control Steps – Document Results
- o Document all steps of the analysis of the evidence
- o Continue to maintain the chain of custody

Table 1.10 Steps in the reporting process

Control Steps – Prepare Report
- o Define the target audience (law enforcement, senior management, etc.)
- o Gather and organize all evidence needed for the report
- o Organize any extra displays needed for report presentation
- o Continue to maintain the chain of custody

Control Steps – Present Report
- o Present the report and supporting evidence in a consistent, clear way to ensure that the audience understand the examination results
- o If necessary, use displays and charts to help explain the more technical areas of the report

Control Steps – Preserve Report and Supporting Evidence
- o After the report has been presented, ensure the preservation of all evidence, in the event that any follow-up is required

Control Steps – Post Report
- o Conduct a thorough post-examination review to help identify lessons learned and ways to improve the forensic process
- o Identify any updates needed to policies or procedures and communicate these to all appropriate individuals and teams

- Analysts should perform data collection using a standard process.
- Analysts should use a methodical approach to studying the data.
- Analysts should review their processes and practices.[42]

Now that we have presented general recommendations for performing the forensic process, in the next section we will review the best practices for cyber forensics compliance.

QUALITY STANDARDS FOR DIGITAL FORENSICS

In June of 2019, the Council of the Inspectors General on Integrity and Efficiency developed a set of Quality Standards for Digital Forensics. As dependence on computers, tablets, and mobile devices increases and the cost of digital storage decreases, the amount of Electronically Sored Information (ESI) continues to increase rapidly. If accessed correctly and legally, this digital information can be extremely valuable for investigative use. This section outlines standards in two areas: management and personnel. Management standards pertain to the organization and the environment in which digital forensics are performed. Personnel standards pertain to the qualifications and proficiency of individuals conducting digital forensics.[43]

MANAGEMENT STANDARDS

A. Digital Forensic Competency

Not all organizations need to be capable of executing a digital forensics examination. If an organization does not have the capability, then it must have a policy to direct how an event requiring digital forensics would be handled. If the organization performs its own digital forensic investigation, it must follow its own documented methodology.

Guidelines

1. **Overview** – Digital devices are everywhere today. The volume of data created every day is growing at an exponential rate. Some of this data may eventually become evidence that could be used to convict or absolve in a court of law. So, it is important that digital forensic examinations are conducted by experienced forensic professionals.
2. **Jurisprudence** – Before starting a digital forensic investigation, consideration must be given to the legal precedence concerning the authority to gather and analyze data. This is usually in the form of a search warrant or legal consent. Because of the large volume of data that is involved, the digital forensic professional needs to understand the scope of these legal documents to ensure the evidence obtained is pertinent to the authorization. It is best to work with legal counsel in all cases.
3. **Data Integrity** – Because digital data can be fragile and is subject to manipulation, it is important to ensure that data is handled in strict accordance with evidence handling procedures. This includes the steps of collecting, transporting, and storing digital evidence. Throughout all steps, the chain of custody must be preserved and documented.
4. **Documentation of Forensic Activities** – It is imperative that all phases of digital forensic activities be meticulously documented. From collection, examination, analysis, and reporting, the evidence chain of custody must be preserved. For the final report, containing a conclusion or opinion needs to be directly supported by the documented digital evidence examined.

5. **External Forensic Expertise** – If your organization does not have the digital forensic expertise or experience, then it is advisable to enter into an agreement with an outside professional digital forensic examiner. Before selecting and contracting with a digital forensic provider, it is important to perform adequate due diligence to help select the firm or individual that would best meet the needs of your organization. Be sure to check references and validate any certifications of the proposed vendor.

B. Quality Assurance

1. **Overview** – Organizations that perform their own digital forensic activities should establish a quality assurance function to ensure that all digital forensic examinations conform to established policies and procedures that support high quality, consistent results.
2. **Independent Reviews** – All digital forensics examination reports should be reviewed by another qualified individual to ensure compliance with the organization's forensic policy. Also, a sample of digital forensics documentation should be independently reviewed by a qualified individual to ensure consistency and completeness of the examination documentation.
3. **Tool Validation** – Since the use of digital forensic tools is a critical component of any examination, organizations should carefully evaluate digital forensic tools before purchasing. To ensure that any tool functions correctly and as intended, validation by an outside digital forensic authority is appropriate. Refer to Chapter 1 for further information on digital forensic tools and utilities.
4. **Review of Quality Assurance Process** – Every organization should review its own digital forensics examination policies and procedures at least once per year. This includes review of the quality assurance process. The field of digital forensics is rapidly evolving and organizations want to ensure that they keep up with the latest tools and methodologies.

WORKFORCE STANDARDS

A. Criteria

Workforce standards are applicable to all individuals who perform digital forensic activities within an organization. Organizations may differ in the job titles for those individuals who perform digital forensics, i.e., examiner, analyst, specialist, etc.

1. **Competency** – The organization is responsible for ensuring that the cyber forensics tasks and activities are performed only by individuals who have the experience and technical competency to perform those tasks.
2. **Sourcing** – The Human Resources and Recruiting functions of an organization should help set the minimum job requirements, provide formal job descriptions, and assist hiring managers in finding and interviewing prospective candidates. This also includes setting reasonable and market-appropriate salary and compensation packages to attract and retain top talent. Cyber-positions are in high demand and are commanding top dollar in many markets.
3. **Education** – For most of today's cyber forensic positions, a degree from an accredited four-year college is a minimum requirement. Many universities are offering cyber-related master's programs and accelerated degree programs to help fill the abundance of open cybersecurity and forensics positions. A bachelor's or master's degree will provide the student with an exposure to both the technical and management.

4. **Experience** – Depending on the level of the position, organizations may allow candidates to substitute on the job experience for some or all of the educational requirements. Of course, this would be experience that provided the relevant knowledge, skills and abilities required.

5. **Intangibles** – Digital forensics professionals are held to the highest standard of conduct, ethics, honesty, and integrity. Throughout the course of their work they may encounter sensitive and private data that demands the utmost confidentiality. Candidates should expect to go through criminal background checks, drug testing, previous employment and reference checks, and other background investigations.

6. **Training** – At the entry-level, candidates may be expected to demonstrate some basic cyber-skills as part of the job screening process. Once hired all candidates should expect to go through orientation to understand the benefits, policies, and expectations of the organization.

B. Expertise

1. **Certification** – There are many certification programs that the cyber forensic professional can pursue. Studying for and taking any exams requires an effort and helps demonstrate a professional's dedication to increasing their competency in the field, with the reward of a recognized certification to show for their hard work. Information about several cyber forensic certifications is provided in the next section.

2. **Continuing Professional Education** – Once certified, all cyber forensic professionals are expected to stay current in their field through annual continuing professional education (CPE) requirements. This can be through formal training classes, conference and seminars, online training, and self study.[11]

In the previous section, we have covered a set of quality standards for digital forensics and a taxonomy for cyber forensics compliance. These can be used by organizations as 'best practices,' to assist with continuous improvement of their cyber forensics program. In the next section, we will review various cyber forensic certifications.

CYBER FORENSIC CERTIFICATIONS

Digital forensics or computer forensics certifications have experienced an incredible growth and market appeal over the past several years. The continued increase in cyber-criminal activities and the need to identify digital evidence in cases ranging from divorces, medical malpractice suits, civil disputes to breaches of industrial control systems, have fueled this growth.

Attaining certification as a cyber/computer forensic examiner/investigator denotes a level of competency in digital forensic techniques, methods, policies, procedures, and required standards of practice.

In addition, the application of both legal and ethical principles to warrant accurate, comprehensive, and reliable digital evidence obtained in a manner that such evidence is permissible, in a law court.

In the following section, we will review several cyber forensic certifications. The cyber forensic professional can choose from vendor/product specific certifications, or certifications offered by various organizations/associations.

Obtaining a certification in cyber forensic examination is a plus for today's cyber forensic professionals. Certifications can provide the following benefits:

1. Expand your knowledge and skills – giving you the key tools and methods to draw upon when needed, including:
   a. ensuring that all digital evidence recovered during an investigation will be accepted in a court of law,
   b. tracing back the digital trail to identify the cyber-criminal after a breach or loss of data has occurred.
2. Provides recognition – of your forensic capabilities, knowledge, and the ability to apply forensics processes.
3. Builds professional credibility – shows your commitment to professionalism and cyber forensic standards.
4. Gives you a competitive advantage – against other candidates during the interview process.
5. Establishes you as a continuous learner – staying up to date in the cyber forensic field is essential.
6. Increase your earning potential – many salary surveys indicate certifications can lead to increased compensation.

There are several options for cyber forensic certifications. Some are product specific and others are product agnostic and focus more on methodology, regardless of the product used. Below is some information regarding several of the more popular cyber forensic certifications.

CFCE – CERTIFIED FORENSIC COMPUTER EXAMINER

The Certified Forensic Computer Examiner (CFCE) certification program is based on a series of core competencies in the field of computer/digital forensics. IACIS offers the CFCE certification program to prospective candidates who wish to attain the CFCE certification. The program is comprised of two phases:

- Peer review phase – Candidates complete four scenario-based problems guided by a forensic professional through a mentored process whereby candidates are able to submit reports or assessment documents after completing each practical exercise.
- Certification Phase – An independent exercise wherein the candidate must complete a practical exercise and written final examination. Upon successful completion, the candidate will be awarded the CFCE certification.

Each certified CFCE member must satisfy recertification requirements every three years. IACIS offers proficiency tests on a regular basis for organizations or laboratories that require frequent proficiency tests. Likewise, proficiency tests are offered to certified CFCE members in the third year from the initial date of certification for the sole purpose of recertification.[45] Table 1.11 provides an overview of the CFCE certification.

Table 1.11 CFCE overview

Certification	Certified Forensic Computer Examiner (CFCE)
Prerequisites	72 hours of training in computer/digital forensics comparable to CFCE core competencies; BCFE training course meets training requirement
Exam	Two-part process: Peer review (must pass to proceed to subsequent phase) and certification phase (includes hard-drive practical and written examination)
Cost	The fee to take the CFCE exam is $750.00.
Website	www.iacis.com/certification/cfce/

CHFI – COMPUTER HACKING FORENSIC INVESTIGATOR

Computer hacking forensic investigation is the process of detecting hacking attacks and properly extracting evidence to report the crime and conduct audits to prevent future attacks.

Computer crime in today's cyber world is on the rise. Computer investigation techniques are being used by police, government, and corporate entities globally and many of them turn to the EC-Council for the Digital Forensic Investigator CHFI Certification Program.

Computer Security and Computer investigations are changing terms. More tools are invented daily for conducting computer investigations, be it computer crime, digital forensics, computer investigations, or even standard computer data recovery. The tools and techniques covered in EC-Council's CHFI program will prepare the student to conduct computer investigations using ground-breaking digital forensics technologies.

Computer forensics is simply the application of computer investigation and analysis techniques in the interests of determining potential legal evidence. Evidence might be sought in a wide range of computer crime or misuse, including but not limited to theft of trade secrets, theft of or destruction of intellectual property, and fraud. CHFI investigators can draw on an array of methods for discovering data that resides in a computer system, or recovering deleted, encrypted, or damaged file information known as computer data recovery.[46] Table 1.12 provides an overview of the CHFI certification.

GCFA – GIAC CERTIFIED FORENSIC ANALYST

The GCFA certifies that candidates have the knowledge, skills, and ability to conduct formal incident investigations and handle advanced incident handling scenarios, including internal and external data breach intrusions, advanced persistent threats, anti-forensic techniques used by attackers, and complex digital forensic cases. The GCFA certification focuses on core skills required to collect and analyze data computer systems. Table 1.13 provides an overview of the GCFA certification.

Table 1.12 CHFI Overview

Certification	Computer Hacking Forensic Investigator (CHFI)
Prerequisites	EC-Council training recommended but not required. Without training you must have two years information security work experience.
Exam	One exam (150 questions, 4 hours, passing score 70%)
Cost	The fee to take the CHFI exam is $500.00, plus $100.00 application fee.
Website	www.eccouncil.org/programs/computer-hacking-forensic-investigator-chfi/

Table 1.13 GCFA overview

Certification	GIAC Certified Forensic Analyst
Prerequisites	None. However, SANS course FOR508 is recommended.
Exam	One exam (115 questions, 3 hours, passing score 71%)
Cost	The fee to take the GCFA exam is $1,899.00 (without training)
Website	www.giac.org/certification/certified-forensic-analyst-gcfa

Areas Covered:

- Advanced Incident Response and Digital Forensics
- Memory Forensics, Timeline Analysis, and Anti-Forensics Detection
- Threat Hunting and APT Intrusion Incident Response[47]

GCFE – GIAC CERTIFIED FORENSIC EXAMINER

The GIAC Certified Forensic Examiner (GCFE) certification validates a practitioner's knowledge of computer forensic analysis, with an emphasis on core skills required to collect and analyze data from Windows computer systems. GCFE certification holders have the knowledge, skills, and ability to conduct typical incident investigations including e-Discovery, forensic analysis and reporting, evidence acquisition, browser forensics, and tracing user and application activities on Windows systems. An overview of the GCFE certification is provided in Table 1.14.

Areas Covered

- Windows Forensics and Data Triage
- Windows Registry Forensics, USB Devices, Shell Items, Key Word Searching, Email, and Event Logs
- Web Browser Forensics (Firefox, IE, and Chrome) and Tools (Nirsoft, Woanware, SQLite, ESEDatabaseView, and Hindsight)[48]

CCE – CERTIFIED COMPUTER EXAMINER

The International Society of Forensic Computer Examiners (ISFCE) principal certification is the Certified Computer Examiner (CCE)®.

The goal of the CCE competencies is to outline the necessary level of proficiency required for a CCE test candidate.

The CCE testing process is designed to test an applicant's proficiency in several areas pertinent to digital forensics. The applicant is required to complete an online test and forensically examine three pieces of media, submitting a report after each examination.[49] See Table 1.15 for an overview of the CCE certification.

CERTIFICATIONS COMPARED: GCFE VS. CFCE VS. CCE

As cybercrimes grow in terms of number of attacks and cost to organizations and businesses, it is obvious that concentrating not only on the prevention but also on the investigation of cases is paramount.

Table 1.14 GCFE overview

Certification	GIAC Certified Forensic Examiner
Prerequisites	None. However, SANS course FOR500 is recommended.
Exam	One exam (115 questions, 3 hours, passing score 71%)
Cost	The fee to take the GCFA exam is $1,899.00 (without training).
Website	www.giac.org/certification/certified-forensic-examiner-gcfe

Table 1.15 CCE overview

Certification	ISFCE Certified Computer Examiner
Prerequisites	Complete training at a CCE Bootcamp Authorized Training Center or possess a minimum of 18 months of verifiable professional experience conducting digital forensics examinations.
Exam	One exam, four parts (on-line written exam and practical exam, passing score 70%)
Cost	The fee to take the GCFA exam is $485.00
Website	www.isfce.com/certification.htm

Digital forensics, then, is playing a growing role and companies are more and more on the lookout for knowledgeable professionals, including investigators and examiners. This branch of forensic science encompasses the collection, preservation, analysis, and reporting of evidence for many purposes, including legal proceedings. The investigator/examiner will be involved in the recovery and scrutiny of material found in electronic systems or digital devices to identify the cause of data breaches or leaks.

Considering computers as a crime scene, a digital forensic examiner will move just like any other criminal investigator to understand the nature and extent of an incident. They will use analysis techniques, reconstructing the events relating to an intrusion or extracting data needed for a case.

Forensic examiners have the task of collecting data and information from electronic systems and are responsible for independently analyzing evidence from hardware or files located on a computer. They are also responsible for the proper handling and examination of digital evidence. Then they'll produce written analysis of their findings and may be called to testify in court as an expert witness.

The field is quickly evolving and examiners' techniques are becoming more sophisticated, which requires them to have specialized, up-to-date knowledge. An investigation requires examiners to use computer forensic methods to determine the source, cause, and scope of the incident as quickly as possible. So, in addition to them needing a solid knowledge of IT hardware and software concepts, it is crucial for a professional to know how to use the latest forensic tools to find data, anomalies, and malicious activity in digital media.

IT professionals can prepare themselves to assist the cyber forensic professional, or to potentially obtain a position as an examiner or cyber forensics investigator, by earning professional computer forensics certifications such as the GIAC® Certified Forensic Examiner (GCFE), IACIS's CFCE, and ISFCE's CCE. Any of these qualifications can be a great asset to demonstrate a competency in this profession.[50]

VENDOR-SPECIFIC CERTIFICATIONS

AccessData FTK Certification
AccessData Certified Examiner (ACE)

The ACE® credential demonstrates your proficiency with Forensic Toolkit® technology. Although there are no prerequisites, ACE candidates will benefit from taking the FTK® BootCamp and FTK® Intermediate courses as a foundation. See Table 1.16 for the ACE certification requirements.[51]

Table 1.16 AccessData Certified Examiner

Certification	AccessData Certified Examiner (ACE)
Prerequisites	None. Candidates will benefit from the FTK BootCamp and FTK Intermediate courses as a foundation.
Exam	One written exam. 88 questions, minimum passing score is 80%.
Cost	The fee to take the ACE exam is $100.00.
Website	https://training.accessdata.com/exam/accessdata-certified-examiner

ENCASE CERTIFIED EXAMINER (ENCE) CERTIFICATION PROGRAM

The EnCase™ Certified Examiner (EnCE) program certifies both public and private sector professionals in the use of Opentext™ EnCase™ Forensic. EnCE certification acknowledges that professionals have mastered computer investigation methodology as well as the use of EnCase software during complex computer examinations.[52] See Table 1.17 for the EnCase certification overview.

BEST DIGITAL FORENSICS CERTIFICATIONS

There are an appreciable number of available, high-quality certification programs that focus on digital investigations and forensics. However, there are also many certifications and programs in this area that are far less transparent and widely known.

There's been a steady demand for digital forensics certifications for the past several years, mainly owing to the following:

- Computer crime continues to escalate. As more cybercrimes are reported, more investigations and qualified investigators are needed. This is good news for law enforcement and private investigators who specialize in digital forensics.
- There's high demand for qualified digital forensics professionals because nearly every police department needs trained candidates with suitable credentials.

Table 1.17 EnCE Overview

Certification	EnCase Certified Examiner (EnCE)
Prerequisites	64 hours of authorized computer forensic training (online or classroom) or 12 months' work experience in computer forensics.
Exam	Phase I – Written exam taken with ExamBuilder. 180 questions, 2-hour time limit, minimum passing score is 80%. Phase II – Practical exam, 18 questions, minimum passing score is 85%
Cost	The fee to take the EnCE exam with ExamBuilder is $200.00 or International $300.00.
Website	www.opentext.com/products-and-solutions/services/training-and-learning-services/ encase-training/examiner-certification

Table 1.18 Job board search results[53]

	SimplyHired	Indeed	LinkedIn Jobs	LinkUp	Total
Vendor Neutral					
CFCE (IACIS)	63	82	117	46	308
CHFI (EC-Council)	106	140	253	68	567
GCFA (SANS GIAC)	422	489	857	294	2,062
GCFE (SANS GIAC)	203	226	433	143	1,005
Vendor Specific					
ACE (AccessData)	25	29	31	12	97
EnCE (EnCase)	110	154	237	114	615

- IT professionals interested in working for the federal government (either as full-time employees or private contractors) must meet certain minimum training standards in information security. Digital forensics qualifies as part of the mix needed to meet them, which further adds to the demand for certified digital forensics professionals.

As a result, there is a continuing rise of companies that offer digital forensics training and certifications. Alas, many of these are "private label" credentials that are not well recognized. Making sense of all options and finding the right certification for you may be trickier than it seems.

To help choose the top five certifications for 2019, the Business News Daily looked at several popular online job boards to determine the number of advertised positions that require these certifications. While the actual results vary from day to day and by job board, Table 1.18 will provide the reader with an idea of the number of digital forensic jobs with specific certification requirements.

If you look around online, you will find numerous other forensics hardware and software vendors that offer certifications. Prior to investing in a certification, you might want to research the sponsoring organization's history and the number of people who've earned its credentials, and then determine whether the sponsor not only requires training but stands to profit from its purchase.

You might also want to ask a practicing digital forensics professional if they've heard of the certifications you found on your own and, if so, what that professional thinks of those offerings.[54]

THE ROLE OF AUDIT IN CYBER FORENSICS

In this section, we will explore the role of audit in cyber forensics. As defined in the first section of this chapter, we will consider both the external audit and internal audit perspectives.

External audit's role in cyber forensics

To begin this section, we have summarized comments from Ms. Kathleen M. Hamm, Board Member of the Public Company Accounting Oversight Board (PCAOB). Her speech was given on May 2, 2019 at the Baruch College 18th Annual Financial Reporting Conference. The PCAOB oversees the audits of public companies and Securities and Exchange Commission (SEC)-registered brokers and dealers in order to protect investors and further the public interest in the preparation of informative, accurate, and independent audit reports. In other words, they audit the auditors, specifically the public accounting firms.[55] In other words, they audit the auditors, specifically the public accounting firms.

The larger public accounting firms are known as the 'Big 4' – PricewaterhouseCoopers (PwC), Ernst & Young (EY), Deloitte, and KPMG. While much of the work performed by these accounting firms is focused on auditing financial statements, they also provide a wide array of tax, advisory, assurance, forensics, and consulting services to their global client base.

In her speech, Ms. Hamm highlighted the promise and the threat of technology. Technology is a key strategic imperative for today's auditors. Emerging technology is also helping to make auditors more efficient and effective through the use of data analytics, artificial intelligence, and robotic process automation.

However, with this emerging technology also come real risks: program coding errors, unauthorized access, and the growing interconnectivity of the 'Internet of Things,' just to name a few. Data breaches are on the rise. With significant data breaches, organizations have obligations to provide required disclosures to shareholders and the general public. In many cases, cyber forensic experts are called in to identify the cause and extent of the breach.

The next area is cyber-enabled fraud. One prominent type of cyber-enabled fraud involves criminals masquerading as company executives, sending emails to finance, and accounting employees requesting a transfer of funds. 'The FBI estimates that business email compromises have cost companies more than $5 billion over the past five years.' Again, cyber forensic experts assist with tracing the fraud trail, and determining the cause, which usually includes a breakdown in a company's internal controls.

So, what is the role of auditors as it relates to these and other threats facing our financial reporting system? Today, based on the current standards, an auditor of public company financial statements plays an important, but limited, role with respect to cyber security. The auditor focuses on the information technology that the public company uses to prepare its financial statements. These methods and procedures are known as internal controls over financial reporting (ICFR).

Can auditors do more? Hamm states that it should start with risk assessments performed at the beginning of financial statement audits. When auditors perform their risk assessments they should consider any cybersecurity risks that could have a material effect on a company's financial statements. As part of those risk assessments, auditors should understand the controls in place and methods used by the company to prevent and detect cyber-incidents that may have a material effect on the company's financial reporting.

What is the auditor's responsibility if a company experiences a cyber-incident? The audit firm would obtain information on the breach from the company's cyber forensic team (if they have one), or they would bring in their own cyber forensic professionals, or both depending on the significance of the incident. They would assess the nature and extent of the breach, including what was stolen, altered, or destroyed. The auditors would then determine the expected effect on the company's operations and financial statement implications.

As Ms. Hamm states the financial effect could include:

- Loss of revenue from disrupted operations
- Costs associated with securing, reconfiguring, and replacing systems
- Costs of conducting forensic inquiries
- Costs of defending against enforcement investigations and civil actions
- Payment of regulatory fines and monetary penalties to harmed private parties

Also, there is the negative reputational and possible stock prices impacts.[56]

The America Institute of Certified Public Accountants (AICPA) develops standards for audit of private companies and other services provided by CPAs. These standards are reflected in the AICPA's Statement on Auditing Standards (SAS) No. 99, *Consideration of Fraud in a Financial Statement Audit*, codified as AU Section 316. It is this standard that directs auditors, if a risk of material misstatement due to fraud is identified, to assign persons with specialized skills, such as forensic specialists to the engagement.

Internal audit

Internal auditors are more accountable today, more than ever before, for detecting fraud. Often fraudulent transactions are concealed in computer files and networks. So, it is necessary for internal auditors to understand the forensic steps necessary to preserve data for forensic examination.

Whether it be financial fraud, a data breach, or other cybercrime, there is a high probability that there will be electronic evidence. This evidence may reside on company-owned devices, employee-owned devices (if BYOD – Bring Your Own Device is allowed), or on an array of removable media. Digital evidence is extremely fragile, and unless properly handled by those who first find the evidence, the data could be made inadmissible in a court of law.

If a computer is compromised, with malware or ransomware, that computer contains evidence of the attack. A computer can also be a tool used to commit a crime, like financial fraud. The auditor might be the first to discover such evidence through audit testing or use of data analytics. It is essential that the auditor be familiar with the procedures to follow when working with electronic records, as these records may eventually be needed as evidence. The use of digital data extraction tools for data analytics, prior to using forensic tools can result in a permanent loss of evidence.

Internal auditors and cyber forensic professionals have several skill sets in common. Both involve critical thought, attention to detail, and comprehension of cause and effect. Essentially both professions need to be able to 'think like a criminal' in order to catch the criminal. It is likely that the cyber forensic professional will have a greater understanding of electronic evidence than most internal auditors.

The internal auditor should have certain technical knowledge and skills to be able to properly work with digital files that may be evidence of a fraud or cybercrime. Before calling in a computer forensic team, the internal auditor has the responsibility to secure digital evidence. The related data needs to be preserved as quickly as possible without disrupting the business.

The technical skills needed for working with digital evidence collection are based on the following seven requirements, each calling for a forensic skill set.

1. Familiarity with cyber forensic laws and regulations. The auditor should have a basic familiarity with the various cyber forensics laws and regulations, including the Computer Fraud and Abuse Act (CFAA) and other state and local regulations.
2. Understanding of different operating systems (MS-Windows, MAC OS, Linux, Android, iOS, etc.). The auditor will have to perform a preliminary analysis of electronic data. The auditor requires a basic knowledge of the various operating systems in order to locate the proper files.
3. Quickly identifying relevant digital data. The auditor needs to know how to conduct a read-only search that does not change the data and will not alert any fraud suspicions. If a read-only search does not follow clear guidelines, it could preclude the data from being used as evidence in a court of law.
4. Correctly preserving data. The auditor needs to know how to maintain the date and timestamps in any file that is being reviewed for potential fraud. Date and timestamp data display when files changes have been made and help determine who made the changes.
5. Properly protecting data. The auditor should be able to use hashes to determine if sensitive files have been changed. Hashes easily determine whether the integrity of the file has been compromised.

6. Properly collecting data. When an initial analysis of data is conducted, the auditor should use data collection methods, such as imaging software to locate and obtain electronic evidence by making a bit-stream image.
7. Ensuring the chain of custody. The process for managing electronic data should be part of the audit plan. Carefully documenting the steps throughout the investigation will help ensure the irrefutable accuracy of the evidence.

The technical skills needed to perform these activities are already partially embedded into the auditing profession as information systems security auditors may already use software tools in their audits. If necessary, the computer forensic expert will take the investigation onward, but if an auditor is suspicious that fraud is present, the evidence must be securely collected before the electronic data vanishes.

Auditors and cyber forensic professionals need to be certain that they are working together to capture the digital fraudsters who themselves may only be present in a digital form when they carry out the financial fraud. Any delay in collecting digital evidence means it is likely to disappear. Any mistakes in collecting digital evidence means it is legally unusable.

It is recommended that specific training in digital investigative techniques be used to help auditors understand how to successfully work with computer forensic specialists in collecting digital information.[57]

In the previous section, we covered the role of external and internal auditors in cyber forensics. To wrap up this chapter, the final section will highlight several cyber forensics case studies.

CYBER FORENSICS CASE STUDIES

'Crime Scene!' Does this word bring an image of blood splattered on the floor, bullet holes, chalk outline, and an investigator mimicking Sherlock Homes in his overcoat and tweed hat? But this is not exactly what we are addressing in this chapter. We are referring to a cyber-crime where there is no blood spatter, no fingerprint, no bullet holes, and no misplacement of things. Finding evidence in cybercrime is an entirely different story, one where the protagonist is a detective behind a screen.

With 95% of the Americans owning mobile phones today, the existence of data is staggering. But it is not just mobile phones that forms a part of investigation, but other devices like laptop, desktop, tab, juke box, play station, smart watches, and everything under the IoT family are responsible for exchange of data. The advancement of technology adds more to the volume of data, and therefore, digital forensics should be expanded to adapt to meet the needs of the users. The emergence of higher sophisticated devices has stressed on the importance of digital forensics too.

For the reader interested in a further discussion on IoT and the role of cyber forensics, see Patrick Wilds examination of the topic in Chapter 3. For more on mobile forensics, Andrew Hrenak provides the reader with an introduction to the subject in Chapter 8.

EMINENT CASES SOLVED WITH DIGITAL FORENSICS

Listed below are several high-profile criminal cases that were solved with the help of digital forensics.[58]

Larry Jo Thomas—2016

Larry Joe Thomas of Indianapolis was convicted of two counts of murder and a count of attempted robbery in the death of Rita Llamas-Juarez on February 29, 2016, in a parking lot in the 3900 block of North Post Road, in Indianapolis. Indianapolis Metropolitan Police Department (IMPD) detectives interviewed witnesses who had accompanied Llamas-Juarez to meet the seller of an iPhone 6 posted on the app OfferUp. Witnesses told police that Llamas-Juarez was sitting in the passenger seat of a vehicle when three men in their early 20s approached. After talking for a few moments, witnesses said, one of the men pulled an automatic rifle out of his jacket, pointed it at Llamas-Juarez and fired once into his chest. Electronic and cellphone records were used to help identify Thomas as a suspect in the murder. The IMPD digital forensics unit provided data that linked Thomas to the OfferUp post through his Facebook account.[59]

Mikayla Munn—2016

A Manchester University student, Mikayla Munn, gave birth to a baby in her dorm room bathtub. She immediately drowned her new born in the bath tub but covered it up stating that she was not aware of her pregnancy and labor pains were felt while taking a bath, followed by the baby's arrival. On verifying her digital assets, investigators found that she had searched on Google for 'at home abortions' and 'ways to cut the umbilical cord of a baby.' Munn pleaded guilty to neglect and was imprisoned for nine years.[60]

Ross Comptown—2017

Ross Compton from Middletown, Ohio, was convicted on the grounds of aggravated arson and insurance fraud of his Court Donegal house. The incident costed him $4 million in damage. When Ross submitted fake medical certificates describing his heart illness, the data from his pacemaker served as evidence before the court of law. The data collected from pacemaker included his heart rate, pacer demand, and heart rhythms which helped prove arson and insurance fraud.[61]

There are many cases of criminal and civil types where the gathered digital evidence has helped uncover hidden scams.[62]

SUMMARY

This initial chapter provided some definitions of compliance and auditing and how to differentiate between auditing and a cyber forensic investigation. We then introduced some of the key events over the years for both cybersecurity and cyber forensics to provide some perspective on how much arena for cyber forensics has expanded. Existing and proposed laws at both the federal and state level were presented, as it is important for the cyber forensic professional needs to understand and operate within the context of these regulations.

To ensure a consistent and sustainable cyber forensics examination approach, we included a policy and controls section for the cyber forensic process and highlighted some important cyber forensic quality standards. We presented some of the key certifications that are available for cyber forensic professionals and the benefits that obtaining and maintaining a certification provides. We wrapped up the chapter describing the difference between internal and external audit and their roles in cyber forensics, and a description of a few prominent legal cases that were solved using cyber forensics.

Continuing on, in Chapter 2, Patrick Wilds presents both a timely and critical topic 'IoT and the Role of Cyber Forensics.'

NOTES

1   Gasior, M., (April 8, 2020), "What Is Corporate Compliance and Why It's Important," PowerDMS, www.powerdms.com/blog/what-corporate-compliance-is-why-compliance-is-important/#:~:

text=The%20definition%20of%20compliance%20is,apply%20to%20your%20organiza-tion%20and, retrieved October 1, 2020.

2 (n.a.), (n.d.) "Audit definitions" www.yourdictionary.com/audit, retrieved October 1, 2020.

3 Ibid.

4 Ibid.

5 (n.a.), (n.d.) "Definition of Internal Auditing" https://na.theiia.org/standards-guidance/manda-tory-guidance/Pages/Definition-of-Internal-Auditing.aspx, retrieved October 1, 2020.

6 (n.a.), (n.d.) "What is External Audit?" www.differencebetween.com/difference-between-internal-and-vs-external-audit/, retrieved October 7, 2020.

7 Surbhi, S., (June 6, 2015), "Difference Between Auditing and Investigation", https://keydiffer-ences.com/difference-between-auditing-and-investigation.html, retrieved October 7, 2020.

8 Tunggal, A., (October 2, 2020), "What is Digital Forensics?", www.upguard.com/blog/digital-forensics, retrieved October 7, 2020.

9 Hospelhorn, S., (March 29, 2020), "8 Events That Changed Cybersecurity Forever", www.varo-nis.com/blog/events-that-changed-cybersecurity/, retrieved October 7, 2020.

10 Tunggal, A., (October 2, 2020), "What is Digital Forensics?", www.upguard.com/blog/digital-forensics, retrieved October 7, 2020.

11 Corporate Finance Institute (n.d.), https://corporatefinanceinstitute.com/resources/knowledge/accounting/what-is-a-forensic-audit, used with permission, retrieved October 8, 2020.

12 (n.a.), (n.d.), "History of Digital Forensics: A Timeline," http://digitalforensicsncssmo.weebly.com/timeline.html, retrieved October 8, 2020.

13 Lee, B., (October 4, 2019), "Examples of Digital Crimes Solved with Cyber Forensics," www.inboundwriter.com/technology/computer-forensics-is-the-answer-to-digital-crimes/, retrieved October 8, 2020.

14 The Editors, (n.d.), "15 Criminal Cases Solved with Digital Evidence", www.brainz.org/15-crim-inal-cases-solved-digital-evidence/, retrieved October 13, 2020.

15 (n.a.), (March 10, 2008), "Passing Secrets at Sea", https://archives.fbi.gov/archives/news/sto-ries/2008/march/secrets031008, retrieved October 13, 2020.

16 Lytle, A., Stephens, N., Conner, J., Bashiri, S., and Jones, S., (March 2018), "Digital Forensics and Enforcement of the Law," Ball State University, Center for Information and Communication Sciences, IEEE Internet Policy Newsletter, https://internetinitiative.ieee.org/newsletter/march-2018/digital-forensics-and-enforcement-of-the-law, retrieved October 13, 2020.

17 Stippich, W., Preber, B., (2016), "Data Analytics: Elevating Internal Audit's Value", Institute of Internal Auditors Publication. The Four V's of Data: Volume, Velocity, Variety and Veracity, p. 10., retrieved October 15, 2020.

18 Lytle, A., Stephens, N., Conner, J., Bashiri, S., and Jones, S., (March 2018), "Digital Forensics and Enforcement of the Law," Ball State University, Center for Information and Communication Sciences, IEEE Internet Policy Newsletter, https://internetinitiative.ieee.org/newsletter/march-2018/digital-forensics-and-enforcement-of-the-law, retrieved October 15, 2020.

19 Spruill, A., (September–October 2011), "Digital Forensics & Encryption", Evidence Technology Magazine, www.evidencemagazine.com/v9n5.htm, retrieved October 15, 2020.

20 Lytle, A., Stephens, N., Conner, J., Bashiri, S., and Jones, S., (March 2018), "Digital Forensics and Enforcement of the Law," Ball State University, Center for Information and Communication Sciences, IEEE Internet Policy Newsletter, https://internetinitiative.ieee.org/newsletter/march-2018/digital-forensics-and-enforcement-of-the-law, retrieved October 15, 2020.

21 (n.a.), (n.d.), "Computer Fraud and Abuse Act (CFAA), National Association of Criminal Defense Lawyers, 1660 L St. NW, 12th Floor, Washington, DC 20036, (202) 872-8600, www.nacdl.org/Landing/ComputerFraudandAbuseAct, retrieved October 1, 2020, used with permission.

22 (n.a.), (n.d.), "Computer Fraud and Abuse Act (CFAA)", www.nacdl.org/Landing/ComputerFraudandAbuseAct, retrieved October 1, 2020, used with permission.

23 (n.a.), (n.d.), "CFAA Background", National Association of Criminal Defense Lawyers (NACDL) www.nacdl.org/Content/CFAABackground, retrieved October 1, 2020, used with permission.

24 Berris, P., (September 21, 2020), "Cybercrime and the Law: Computer Fraud and Abuse Act (CFAA) and the 116th Congress," R46536, Congressional Research Service, https://crsreports.congress.gov/product/pdf/R/R46536, retrieved November 30, 2020.

25 (n.a.), (n.d.), "Connecticut General Assembly", "Raised H.B. No. 5511 Session Year 2020", www. cga.ct.gov/asp/cgabillstatus/cgabillstatus.asp?selBillType=Bill&bill_num=HB5511&which_ year=2020, retrieved October 19, 2020.

26 (n.a.), (n.d.), "Illinois General Assembly", "Bill Status of HB5204 101st General Assembly", www. ilga.gov/legislation/BillStatus.asp?DocNum=5204&GAID=15&DocTypeID=HB&LegId=12562 5&SessionID=108&GA=101, retrieved October 19, 2020.

27 (n.a.), (n.d.), "House Bill 635", http://mgaleg.maryland.gov/2020RS/bills/hb/hb0635f.pdf, retrieved October 19, 2020.

28 (n.a.), (n.d.), "Minnesota Legislature", "Office of the Revisor of Statutes" "HF 4085), www.revi-sor.mn.gov/bills/bill.php?b=house&f=HF4085&ssn=0&y=2020, retrieved October 20, 2020.

29 (n.a.), (n.d.), "Minnesota Legislature", "Office of the Revisor of Statutes", "SF 4297", www. revisor.mn.gov/bills/text.php?number=sf4297&version=latest&session_number=0&session_ year=2020, retrieved October 20, 2020.

30 (n.a.), (n.d.), "Assembly, No. 3984", "State of New Jersey", "219th Legislature", www.njleg.state. nj.us/2020/Bills/A4000/3984_I1.HTM, retrieved October 21, 2020.

31 (n.a.), (n.d.), "Assembly, No. 4518", "State of New Jersey", "219th Legislature", www.njleg.state. nj.us/2020/Bills/A5000/4518_I1.HTM, retrieved October 21, 2020.

32 (n.a.), (n.d.), "Senate, No. 1374", "State of New Jersey", "219th Legislature", www.njleg.state. nj.us/2020/Bills/S1500/1374_I1.HTM, retrieved October 21, 2020.

33 (n.a.), (n.d.), "New York State Assembly", "AB2124 Summary", https://nyassembly.gov/ leg/?default_fld=%0D%0At&leg_video=&bn=AB2124&term=&Summary=Y, retrieved October 22, 2020.

34 (n.a.), (n.d.), "The Ohio Legislature", "House Bill 368", www.legislature.ohio.gov/legislation/ legislation-summary?id=GA133-HB-368, retrieved October 22, 2020.

35 (n.a.), (n.d.), "LIS Virginia Law", "Code of Virginia", "18.2-152.4 Computer trespass: penalty", https://law.lis.virginia.gov/vacode/18.2-152.4/, retrieved October 22, 2020.

36 Ibid.

37 Office of Legal Education Executive Office for United States Attorneys, (2015), "Prosecuting Computer Crimes", www.justice.gov/sites/default/files/criminal-ccips/legacy/2015/01/14/ccman-ual.pdf, retrieved October 27, 2020.

38 Legal writers and editors, (May 2, 2019), FindLaw, "Hacking Laws and Punishments", https:// criminal.findlaw.com/criminal-charges/hacking-laws-and-punishments.html, retrieved October 27, 2020.

39 K, Kent, Chevalier, S., Grance, T., Dang, H., (August, 2006), "Guide to Integrating Forensic Techniques into Incident Response, NIST 800-86, https://nvlpubs.nist.gov/nistpubs/Legacy/SP/ nistspecialpublication800-86.pdf, retrieved October 27, 2020.

40 Ibid.

41 Ibid.

42 von Solms, S., Louwrens. C., Reekie, C., Grobler, T., Chapter 27. A Control Framework for Cyber Forensics. 2006 in International Federation for Information Processing, Volume 222, Advances in Cyber Forensics II. eds Olivier. M., Shenoi, S., (Boston: Springer). pp. 343–355. https://link. springer.com/search?query=A+Control+Framework+for+Cyber+Forensics%2C+solms, Retrieved October 1, 2020, used with permission.

43 Whitcomb, T., Missal, M., (June 18, 2019), "Quality Standards for Digital Forensics", www.ignet. gov/content/quality-standards, retrieved November 25, 2020.

44 Ibid.

45 (n.a.), (n.d.), "Certification", www.iacis.com/certification/, retrieved October 27, 2020.

46 (n.a.), (n.d.), "Computer Hacking Forensic Investigator Certification", www.eccouncil.org/pro-grams/computer-hacking-forensic-investigator-chfi/, retrieved October 27, 2020.

47 (n.a.), (n.d.), "Cyber Security Certification: GCFA", www.giac.org/certification/certified-forensic-analyst-gcfa, retrieved October 28, 2020.

48 (n.a.), (n.d.), "Cyber Security Certification: GCFE", www.giac.org/certification/certified-forensic-examiner-gcfe, retrieved October 28 2020.

49 (n.a.), (n.d.), "Certified Computer Examiner", www.isfce.com/index.html, retrieved October 28, 2020.

50 Brecht, D., (February 26, 2020), "Certifications compared: GCFE vs. CFCE vs. CCE", https://resources.infosecinstitute.com/topic/gcfe-vs-cfce-vs-cce/, retrieved October 28, 2020, Reprinted with permission from resources.infosecinstitute.com.

51 (n.a.), (n.d.), "AccessData Certified Examiner", https://training.accessdata.com/exam/accessdata-certified-examiner, retrieved October 28, 2020.

52 (n.a.), (n.d.), "EnCase Certified Examiner (EnCE)", www.opentext.com/products-and-solutions/services/training-and-learning-services/encase-training/examiner-certification, retrieved October 28, 2020.

53 (Tittel, E., Lindros, K., Kyle, M.), (n.d.), "Best Digital Forensics Certifications," Business News Daily, www.businessnewsdaily.com/10755-best-digital-forensics-certifications.html, retrieved November 30, 2020.

54 Ibid.

55 (n.a.), (n.d.), "PCAOB Mission, Vision and Values", https://pcaobus.org/about/mission-vision-values, retrieved November 25, 2020.

56 Hamm, K., (May 2, 2019), "Cybersecurity: Where We Are; What More Can be Done? A call for Auditors to Lean In", https://pcaobus.org/News/Speech/Pages/hamm-cybersecurity-where-we-are-what-more-can-be-done.aspx, retrieved October 29, 2020, used with permission.

57 Smith, G. Stevenson. (January 2005), Computer Forensics: Helping to Achieve the Auditor's Fraud Mission? *Journal of Forensic Accounting*, 6, 119–134, www.researchgate.net/publication/280578933_Computer_Forensics_Helping_to_Achieve_the_Auditor%27s_Fraud_Mission, Reprinted with permission from the National Association of Certified Valuators and Analysts® (NACVA®) and the Journal of Forensic and Investigative Accounting (JFIA). All rights reserved. To learn more, please visit https://www.nacva.com/jfia, retrieved October 29, 2020.

58 EC-Council, March 24, 2020, "5 Cases Solved Using Extensive Digital Forensic Evidence", https://blog.eccouncil.org/5-cases-solved-using-extensive-digital-forensic-evidence/, retrieved October 30, 2020.

59 Staff Reports, (May 17, 2018), "Jury finds 22-year-old guilty of murdering man at cellphone sale", www.wishtv.com/news/jury-finds-22-year-old-guilty-of-murdering-man-at-cellphone-sale/, retrieved November 29, 2020.

60 (n.a.), (July 25, 2018), "Ex-Manchester University student sentenced in newborn's dorm death," www.indystar.com/story/news/crime/2018/07/25/mikayla-munn-sentenced-death-her-newborn-baby-dorm-manchester-university/832684002/, retrieved November 22, 2020.

61 Pack, L. (July 24, 2018), "Arson suspect in unique case featuring pacemaker data is back in custody," www.journal-news.com/news/arson-suspect-unique-case-featuring-pacemaker-data-back-custody/dn6JyzsOemZovpayJMZLNJ/, retrieved November 22, 2020.

62 EC-Council, March 24, 2020, "5 Cases Solved Using Extensive Digital Forensic Evidence", https://blog.eccouncil.org/5-cases-solved-using-extensive-digital-forensic-evidence/, retrieved October 30, 2020.

Chapter 2

IoT and the role of cyber forensics[1]

Patrick Wilds

CONTENTS

THE INTERNET OF THINGS (IOT) – BEGINNINGS

Describing the IoT

Definitions

> The Internet of Things ("IoT") refers to the ability of everyday objects to connect to the Internet and to send and receive data. It includes, for example, Internet-connected cameras that allow you to post pictures online with a single click; home automation systems that turn on your front porch light when you leave work; and bracelets that share with your friends how far you have biked or run during the day.[2]

> Over the past decade, there has been a surge in the development of new 'smart' devices that can connect to the internet and be controlled using applications remotely. This network of devices and other items embedded with sensors, electronics, software and connectivity is called the Internet of Things (IoT).[3]

> The IoT represents a technologically optimistic future, where the objects will be able to utilize the Internet and make intelligent collaborations with each other anywhere and anytime. In particular, the IoT combines a wide range of technologies, such as sensors, actuators, Internet, cloud computing as well as many communication infrastructures.[4]

There have been numerous definitions offered by academia, government, and industry that attempt to define the IoT. Many definitions include functionality examples to better clarify how the systems operate and the interrelation between devices and entities within and outside the IoT. There are characteristics to many of the descriptions that appear common within the understanding of what the IoT is and how it functions. They are 'smart,' can operate independently and are becoming, or have become, ubiquitous. The next sections describe the IoT in more detail by discussing the devices themselves, how they communicate and what capabilities they have, or can have.

Purpose

The things, or devices, are designed to operate independent of the need for user input to start a process or function. Using an array of sensors, the devices are designed by their manufacturers to perform specific functions autonomously and communicate data without intervention. As a result, the growing number of objects and offered functionalities have created a ubiquitous computing landscape. At the same time the independent operating capability and small size of the devices often makes them unnoticed by humans.

Development

Though not the inventor of the IoT, Kevin Ashton is often credited with coining the term Internet of Things during a presentation to Procter and Gamble in 1999.[5] Before being called the IoT there were devices in operation that fit into the general definitions discussed above.

In the early 1980s several graduate students and an engineer at Carnegie Mellon University set about the task of modifying a Coke machine with sensors that would monitor the stock and temperature of soda. This information was transmitted to a server connected to the university network as well as ARPANET. Before walking to the machine for a Coke, anyone with access to the network could check to see if the machine had soda available, and if it was cold. Sometime later another graduate student modified an M&M machine close by with a similar system.[6]

A toaster is more widely believed to be the first IoT device. In 1989, Interop President Dan Lynch challenged John Romkey to connect a toaster to the Internet. Working with Simon Hackett, Romkey connected a Sunbeam toaster to the Internet and in 1990 presented it at the Interop Internet networking show. There was only one control; turn the toaster on. The toaster still had the shortcoming of needing a person to put the bread in, so in 1991 a small robotic crane was added to the system that would pick the bread up and place it in the toaster.[7]

CHARACTERISTICS OF IOT DEVICES

Typically, when people think of computers they think of PCs, laptops, servers and mobile devices such as smart phones and tablets. These generally have moderate to high capacity processing ability and storage. They are able to perform a multitude of functions represented by the wide variety of applications available to the user. In the case of computers, user input is generally needed to begin the performance or execution of a task. Though mobile devices also require user initiation of many functions, they have autonomous capabilities giving them characteristics found in IoT devices.

IoT devices change this paradigm by offering an ever-growing list of functions that can be performed outside the need of user intervention to start, or continue, an operation. What is built into these devices to allow such functionality is discussed below.

Sensors

The purpose of an IoT device is to produce data from input created by actions or conditions that activate the sensor(s) embedded in the device. An example would be a potentiometer used to detect position. An IoT device with a potentiometer could be placed on a door to report when it is opened. The sensor facilitates the connection of the IoT device to the world around it as the sensor reacts to its surroundings based on preset conditions or activities.

The resulting data is then collected and communicated. In some cases, the reported data can result in the execution of an actuator to perform another function.[8] Opening the door could activate an alarm state, triggering other devices to automatically lock other doors within a structure. Table 2.1 provides examples of sensor capabilities and types.

Table 2.1 Types of sensors[9]

Sensor types	Sensor description	Examples
Position	A position sensor measures the position of an object; the position measurement can be either in absolute terms (absolute position sensor) or in relative terms (displacement sensor). Position sensors can be linear, angular, or multi-axis.	Potentiometer, inclinometer, proximity sensor
Occupancy and motion	Occupancy sensors detect the presence of people and animals in a surveillance area, while motion sensors detect movement of people and objects. The difference between the two is that occupancy sensors will generate a signal even when a person is stationary, while a motion sensor will not.	Electric eye, RADAR
Velocity and acceleration	Velocity (speed of motion) sensors may be linear or angular, indicating how fast an object moves along a straight line or how fast it rotates. Acceleration sensors measure changes in velocity.	Accelerometer, gyroscope
Force	Force sensors detect whether a physical force is applied and whether the magnitude of force is beyond a threshold.	Force gauge, viscometer, tactile sensor (touch sensor)
Pressure	Pressure sensors are related to force sensors and measure the force applied by liquids or gases. Pressure is measured in terms of force per unit area.	Barometer, bourdon gauge, piezometer
Flow	Flow sensors detect the rate of fluid flow. They measure the volume (mass flow) or rate (flow velocity) of fluid that has passed through a system in a given period of time.	Anemometer, mass flow sensor, water meter
Acoustic	Acoustic sensors measure sound levels and convert that information into digital or analog data signals.	Microphone, geophone, hydrophone
Humidity	Humidity sensors detect humidity (amount of water vapor) in the air or a mass. Humidity levels can be measured in various ways: absolute humidity, relative humidity, mass ratio, and so on.	Hygrometer, humistor, soil moisture sensor
Light	Light sensors detect the presence of light (visible or invisible).	Infrared sensor, photodetector, flame detector
Radiation	Radiation sensors detect radiations in the environment. Radiation can be sensed by scintillating or ionization detection.	Geiger–Müller counter, scintillator, neutron detector
Temperature	Temperature sensors measure the amount of heat or cold that is present in a system. They can be broadly of two types: contact and non-contact. Contact temperature sensors need to be in physical contact with the object being sensed. Non-contact sensors do not need physical contact, as they measure temperature through convection and radiation.	Thermometer, calorimeter, temperature gauge

(Continued)

Table 2.1 (Continued) Types of sensors

Sensor types	Sensor description	Examples
Chemical	Chemical sensors measure the concentration of chemicals in a system. When subjected to a mix of chemicals, chemical sensors are typically selective for a target type of chemical (e.g., a CO2 sensor senses only carbon dioxide).	Breathalyzer, olfactometer, smoke detector
Biosensors	Biosensors detect various biological elements such as organisms, tissues, cells, enzymes, antibodies, and nucleic acids.	Blood glucose biosensor, pulse oximetry, electrocardiograph

It is possible for one device to contain more than one sensor, allowing for the customization of services offered to the consumer. For example, smart home devices could be equipped with a position sensor, humidity sensor, and acoustic sensor. During the process of setting up the service, the homeowner can choose to place the device on the front door to sense when it is opened, in the basement to detect high levels of moisture, or near to a smoke detector to detect when it is sounding.

Information generated as a result of sensor activation would be transmitted to a device controller that would then pass the data to a wireless access point in the home and on to the homeowners' smart phone. In this example, the homeowner would be notified in near real time of events occurring at home regardless of his or her location.

There are some limitations that prevent IoT devices from offering a level of functionality often found in personal computers, laptops, servers, and even smart phones.

Memory and processing

IoT devices, which also perform calculations, are characterized as having low memory capacity and processing power.[10] The small size and intended functionality of many devices are factors physically restricting the size of the components that can be included in the design. Think of an activity tracker worn on a wrist. This IoT device is smaller than the CPU and attendant cooling fan found in a typical desktop computer. The CPU in the activity tracker needs to be much smaller to exist in the same chassis as a circuit board, communication module, power supply, etc. As a result, it does not have the processing capability of the desktop CPU.

This holds true for memory as well. Devices have a very small storage capacity, if any at all. Many devices have no storage and will immediately send data to the Fog or Cloud when generated. Memory in the form of Read Only Memory (ROM) and Random Access Memory (RAM) are limited due to the factors mentioned above and this has an effect on the operating system utilized in the device.[11]

Power capacity

Lower power capacity also contributes to the description of IoT devices as resource constrained.[12] Though some devices can be wired to power sources, many are designed to rely on internal power for function. The activity tracker is just one example of a device that cannot be connected to an external power source during use.

Some devices are designed to run for years on battery power before it needs to be replaced or the device discarded. Others are designed to harvest energy from the environment. Some examples include the collection and conversion of solar energy, or the conversion of movement into energy.

Operating systems

As devices have become more complex, the need for operating systems led to the development of many options for use in the IoT. However, as device resources are still limited, the deployment of Real-Time Operating Systems (RTOS) is usually required. RTOS are able to function in real time, processing data and application commands with little or no delay.[13] Such functionality allows the resource-constrained devices to operate with little or no data storage capacity. It also provides the ability to report sensor-generated data in real time as required by the design of numerous devices.

There are several factors taken into consideration when manufacturers and developers choose which operating system to use.

- Scalability – the operating system needs to work with a variety of processors as there are often more than one in a device. Additionally, the processors may be a mix of 8, 16, and/or 32bit.
- Modularity – the operating system as a module providing minimal necessary service will allow the use of other modules specific to the intended functionality of the device. Modularity also requires less memory.
- Connectivity – the operating system should support a variety of wired and wireless protocols and standards.[14]

Other factors include:

- Footprint – a small operating system should have a minimal requirement for resources from the device
- Portability – the operating system should have the ability to work on multiple hardware platforms
- Security – security can be added to the operating system
- Reliability – the operating system should be able to run for extended periods of time without human intervention to correct errors.[15]

Table 2.2 provides examples of operating systems, includes memory requirements and whether it is an RTOS. This table is illustrative of the wide variety of operating systems available.

Hardware

Hardware architecture is also extremely varied. There are a large number of companies that produce, or have produced for them, IoT devices to fit business functions or sell to other markets. Smart homes, personally worn IoT devices, business-oriented devices, and health care solutions are a few examples of sectors and customers impacted by the growing IoT ecosystem.

Many of the device designs are proprietary and contribute to the heterogeneous nature of the IoT. This heterogeneity is reflected in a number of different data extraction methods that complicate cyber forensic procedures and efforts to obtain that data.[17]

Communications

There are a wide variety of communication protocols that exist to enable device connectivity. Which to use is often determined by the intended function of the device. Short-range connectivity may be desirable for smart home devices, or even devices worn on the body. Long-range connectivity would be more useful for soil sensors spread over a large geographic

Table 2.2 IoT operating systems[16]

OS name	Min. RAM	Min. ROM	Real Time
Contiki	10 KB	30 KB	Partial
TinyOS	1 KB	4 KB	No
RIOT	1.5 KB	5 KB	Yes
Mantis	14 KB	50 KB	Partial
FreeRTOS	1 KB	10 KB	Yes
Nano-RK	2 KB	18 KB	Yes
LiteOS	4 KB	128 KB	Yes
Apache Mynewt	16 KB	128 KB	Yes
Zephyr OS	8 KB	128 KB	Yes
Ubuntu Core Snappy	128 MB	350 MB	No
Android Things	512 MB	4 GB	No
Windows 10 IoT	256 MB	200 MB	Partial
WindRiver VxWorks	1 MB	128 KB	Yes
Micrium μC/OS	1 KB	6 KB	Yes
MicroEJ OS	32 KB	128 KB	Yes
Express Logic ThreadX	1 KB	2 KB	Yes
Nucleus RTOS	2 KB	12 KB	Yes

area. Protocols that can operate on low power are more useful for devices that run on battery power while devices connected to a power source could enjoy greater data transmission speed from communication protocols that require high power.

This section, while not exhaustive, provides examples of several communication protocols used in the IoT.

- 6LoWPAN operates on the IEEE 802.15.4 standard which allows communication on the Internet using IPv6 addressing. Devices utilizing the protocol can be connected directly to the Internet without the need for intermediate gateways for IP addressing. The standard was created for low power consumption enabling operation on battery powered devices. This is a short-range protocol.
- Zigbee is another protocol based on the IEEE 802.15.4 standard. It has a short transmission range and low power consumption, meaning it can run on battery powered devices.
- Bluetooth Low Energy (BLE) based on Bluetooth was designed to provide short-range data transfer with low power consumption. It can communicate with existing technology that already provides Bluetooth support.
- Z-Wave was designed for low power consumption in smart home and small business applications. The design also included small data packets and low speed transmission. Z-Wave utilizes controlling devices and slave nodes. Controlling devices send commands to the nodes while the nodes can only reply and execute the commands. Nodes cannot initiate communication.
- RFID (Radio Frequency Identification) uses two devices. The first is the radio frequency (RF) tag and the second is a reader device, or just reader. Data programmed into the tag is static. Two approaches to tag-reader interaction are active reader tag system, in which the tag contains a battery for power, and passive reader tag system, where there is no tag power source. There is also an active reader active tag system in which both items have a power supply. When the reader and tag are within proximity, data can be

relayed between them according to the system in use. RFID is used for identification and does not host significant two-way communication.

- NFC (Near-field communication) is designed for short-range communication where devices are touched together, or brought into very close range to initiate data transfer. Unlike the RFID system, the data in an NFC tag can be rewritten. NFC can operate in card emulation mode, which is passive, reader/writer mode, which is active, and peer-to-peer mode.[18]
- EnOcean was primarily designed for automation but can be used in the IoT. It is designed to provide wireless connection by converting energy from the environment to energy used for communication. It is primarily used in HVAC IoT applications.[19]
- Thread is a protocol created by Nest. It started as a closed source product, but that changed with the implementation of OpenThread. It is built on IEEE 802.15.4, 6LoWPAN and IPv6.
- Wi-Fi is more familiar to the reader as the means by which we connect our computers to the Internet through a wireless router. The power requirements for Wi-Fi are greater and require larger batteries, or a wired connection to avoid power storage requirements. Additionally, Wi-Fi needs to stay connected to its access point, such as the wireless router. If connection is lost, Wi-Fi will need to reconnect, which can take time and consume power. Embedded Wi-Fi is a solution designed to run on low power, but at the cost of throughput.[20]

The protocols mentioned thus far offer transmission distances ranging from touch and centimeters to meters and tens of meters. However, some IoT applications require much longer ranges to achieve intended functionality. The following communication protocols are examples of solutions for long-range data transmission.

- SigFox is designed to transmit data using ultra narrow ban technology up to 50 kilometers with low power consumption. This protocol exists within a category of wireless communications called low-power wide-area network, or LPWAN.
- Cellular, like Wi-Fi, is another familiar solution for IoT networks. It is able to provide high speed connection and data transmission, but at the cost of higher power usage.[21]

Intelligence

What makes a thing intelligent? What makes a smart home smart? New technology is created to fill a perceived need or desire, such as monitoring health statistics with an activity tracker. Old technology can be given smart characteristics, such as with the Coke machine and toaster examples.

Intelligence, or smartness, should not be confused with automation. Automation has been around for some time. A dish washer is automated, as is a vacuum cleaner, but neither of these things communicates with other items in a home or business to report sensor data, coordinate services, or anticipate the needs of occupants. Automation is not connected to a larger context of interrelated sensors that can work together to detect and react to the surrounding environment. Intelligence is, and can adapt and possibly predict future needs based on current state and previously gathered information.[22]

Characteristics of intelligence cannot be defined in one chapter, or one book. However, intelligence as it relates to ubiquitous computing and smart design can better delineate automated from intelligent. There are some characteristics to consider when measuring the difference.

Devices taken as a whole, working together as a system, can extrapolate meaning from sensor data produced within the area of coverage. My smart home may know that I am in my garage based on sensors detecting sound and vibration in that part of the house.

The system can assume an existing state based on multiple data points. If there is more than one person in my garage the system might assume that we are preparing to leave the home. However, if I am the only person detected in the garage, the system may assume I plan on working there.

The system may try to predict what I am doing based on its understanding of the context. If other people enter the garage the system may assume that I will want the garage door to open for us to leave.

The system may, preemptively, open the garage door based on its prediction that a group of people in a garage wish to leave.[23]

The application of this kind of intelligence could also be found in business settings in which an IoT system within a building is monitoring, tracking, extrapolating data, and taking actions based on the numbers, locations, and actions of employees within the building. For example, in automated systems the building may be set to control lighting and HVAC on a timer that could include a seven-day week schedule to account for reduced usage on the weekend. In a 'smart' building, control of lighting and HVAC may be based on perceived activity.

During a week day there are more employees entering the building as the morning proceeds. The system may assume that it is a work day and adjust the lighting and HVAC accordingly. In the evening the building may, at some threshold of occupancy, assume the work day is concluding and again adjust the lighting and HVAC accordingly.

In an automated system, should the building not be occupied during a work day for any reason, the system will still adjust lighting and environmental controls according to schedule. The smart building will not, saving the cost of additional energy consumption.

However, as the activity of the system moves from processing data to prediction to action, errors can occur.[24] Extrapolating the intent of a number of people in a garage could result in the incorrect action of opening the garage door on an extremely cold evening, when the true intent was moving a heavy appliance from the garage to the interior of the home.

The descriptions, design options and functionality of IoT devices leans on the ability of systems to interpret data received from sensors embedded within the devices. Actions are then taken based on those interpretations. This ability requires storage and processing power to make the IoT work.

Distributed data storage and processing

The data generated by devices usually has to be transferred to another location to be processed as the devices themselves have limited ability. Once processed, the data is used to make decisions or take actions based on the intended function of the system within which the devices operate. Cloud computing, and fog/edge computing have become the means by which this part of the IoT ecosystem executes its intended purpose.

The amount of data to be transferred, stored, processed and presented has grown substantially over time. Cisco, in 2011, calculated that the number of Internet-connected devices surpassed the population of the planet.[25] The volume of data has continued to grow exponentially as billions of additional objects and devices have been connected to the Internet since that time.

Cloud computing

The National Institute of Standards and Technology offers the following definition.

> Cloud computing is a model for enabling ubiquitous, convenient, on-demand network access to a shared pool of configurable computing resources (e.g., networks, servers,

storage, applications, and services) that can be rapidly provisioned and released with minimal management effort or service provider interaction.[26]

This paradigm, of providing cloud computing to IoT devices, aids in the creation of a ubiquitous computing environment that operates unnoticed by humans. The cloud receives the data, analyzes and interprets it and provides web-based results to users. The advantage of the cloud is scalability and reliability, to meet the needs of the devices that are constantly generating and communicating data.[27]

However, as with the devices themselves, several potential weaknesses have been identified that may impact the device-cloud interrelationship adversely. A lack of standardization of cloud computer services creates interoperability problems when working with multiple providers, or transferring to new providers.[28] Additionally, the growing volume of data transferred to data centers that make up the cloud consumes energy and can result in data reception and processing latency.[29]

As seen earlier, devices may use any number of different protocols to communicate data, but data centers don't communicate using many of these protocols resulting in the need for data translation somewhere between the device and cloud.

These problems, left unaddressed, would frustrate efforts to grow the technology.

Fog/edge computing

Moving data processing closer to the edge, where the devices operate and generate data, has several potential benefits. Processing at the edge, or edge computing, can reduce the amount of data sent to the cloud, increase the speed of data processing, save energy, and generally reduce the load on the network.

Contending that the cloud was not designed to handle the volume, variety, and velocity of IoT-generated data, Cisco in 2015 defined fog computing as moving the cloud closer to the devices that produce and act on data. Of the benefits outlined in the white paper, latency minimization was listed as the first. In a manufacturing setting, a delay in information analysis as a result of communication translation, transfer to the cloud, analysis and return communication could result in a delay or system failure effecting output. The fog is designed to use nodes closer to the edge to process time sensitive data and return results within seconds or even milliseconds. Other benefits outlined included network bandwidth conservation, data security and reliability.[30]

Fog computing has been described as an architecture of edge computing wherein edge devices are used to process, store data and communicate locally.[31] The devices could then determine what data to send to the cloud. This, again, would reduce network traffic and speed up data processing needed for time sensitive tasks and decision-making.

THE PROBLEM OF HETEROGENEITY

Older technology such as computers, and now to a degree, mobile phones enjoy a level of homogeneity not as common in the IoT ecosystem. There are a few vendors that develop operating systems for personal computers such as Microsoft, Apple and the creators of the various flavors of Linux. The same is true for mobile devices with operating systems developed by Apple, Android, Research in Motion and Microsoft. Even in this case, we see the Apple iOS and Android assuming ever larger control of the operating system market for these devices.

The same is true of hardware. Though there are different chip, motherboard, graphics card and hard drive manufacturers, these products are built to a standard that allows some level of interchangeability and common support for operating systems and communication protocols. Hobbyists' can purchase all the parts needed to build a computer with some attention paid to compatibility requirements and end up with a system that works as well and with the same capabilities, operating system, and interoperability as a complete computer purchased off the shelf, or off the Internet.

The same is not true for the IoT. The myriad of devices features a diversity of hardware architecture and can use a wide variety of operating systems, or even have proprietary hardware and software.[32] They can utilize a number of different communication protocols, or again, utilize proprietary protocols.

This is a complicating issue for security design that research is attempting to address.[33] The ramifications for security and forensics will be address more fully later in this chapter.

CURRENT STATUS AND FUTURE TRENDS

While it is important for business, industry, academia, government and even private consumers to understand how the IoT may reap benefits for society, it is also important for these same entities to understand possible issues concerning data confidentiality, integrity and availability. The CIA of information security is no less important, or impacted, by the presence of an ecosystem that constantly collects and communicates data outside of the awareness of people than it is within the context of multiple computers in a network that we are accustomed to analyzing and protecting from intrusion. Conversely, outside of awareness, criminal enterprise may find it easier to harvest data, including personally identifiable information, corrupt it, prevent transmission, or control the devices themselves.

Familiarity with some current trends and statistics may help individuals within the various enterprises and institutions come to a better global understanding of potential impacts of the IoT. From this beginning the section will move to the sectors that researchers have described as utilizing IoT capabilities.

Statistics

The IoT market is projected to grow to 75.4 billion devices by 2025, from 15.4 billion in 2015.[34]

By 2021, 1 million new IoT devices will be purchased every hour.[35]

Those are impressive numbers given the nascent nature of the technology not so many years ago. Now by focusing in on two sectors that are seeing an expansion in the deployment of IoT devices, we gain a better understanding of the impact they can have on our lives.

In 2020, 75% of new cars will be able to connect to the Internet.[36]

The body sensor market, driven primarily by healthcare and sports, includes devices like heart monitors and activity trackers. Shipments of these devices are projected to increase from 2.4 million units in 2016 to 92.1 million in 2022.[37]

Overall, the amount of money invested into the ecosystem continues to grow, as would be expected from the statistics offered above.

Globally, the IoT market is projected to grow from about 170 billion USD in 2017 to 561 billion USD in 2022.[38]

As with any technology, there are attendant problems that can negatively impact the adoption and use of the IoT. The introduction and expanding use of person computers drove new opportunities for criminal conduct utilizing the technology against the users. While activities such as theft and fraud had long been classified as crimes, the introduction of personal computers provided a new avenue to commit those crimes. New criminal activity has also come with the growth of such technology, including cyber stalking and denial of service attacks. The IoT potentially offers the same criminal opportunities through the use and exploitation of the devices, communication networks and data storage and processing capabilities at the cloud and fog/edge.

When surveyed, 90% of responding developers did not believe that IoT devices in use had proper security in place, while 85% said they had felt pressure to rush devices to market despite security concerns.[39]

Finally, 70% of IoT devices were found to be vulnerable to attack with each device averaging 25 vulnerabilities.[40]

Trends

IoT devices can be tools used to commit crime, recorders of crime, or the target of crime. As a tool used to commit a crime, malware can compromise a device for use in a botnet for DDoS, or other attacks. They can also become a gateway to internal protected data in those areas where IoT devices have not been included in an overall information security strategy. While these attacks are not new to computers, IoT devices offer an expanded attack surface that must be addressed.

NEW TARGETS AND TOOLS OF CRIME

The Mirai botnet is one of the more well known, but not only, examples of the compromise of IoT devices for use in a botnet to launch DDoS attacks. In a way, this makes the devices both the target and tool of a crime. Discovered in August 2016, Mirai propagates by first infiltrating routers, DVR's and webcams. It then uses a dictionary of potential username and password pairs to gain administrative access to other IoT devices.

While they don't have the processing power of regular computers, there is a far greater quantity of IoT devices that can be compromised and used. The fact that they are always on and connected, combined with poor security, make them easy targets for subversion into a botnet.[41]

A Mirai 1.1Tbps DDoS attack using 148000 IoT devices broke records. The botnet grew from 213000 to 483000 devices within two weeks.[42] This example clearly shows that while individually weak in comparison to computers and servers, the sheer number of IoT devices has the ability to overwhelm resources to an extent not realized with their larger more powerful predecessors.

Devices used in botnets raise the issue of how devices are compromised to begin with. As with any other computer, mobile device, server or router, researchers are discovering a multitude of ways that an IoT device can be attacked.

The Mirai example of using a dictionary attack on default or weak passwords to gain control of the device brings up the relevance of changing passwords to stronger ones with greater complexity. So, in this one example we see a common, well known problem and solution that is no different from any other aspect of technology in use by us. How often do you change your passwords from the default? How complex to you make your passwords? How many different accounts do you have that use the same password?

If you are using the same password for your home computer user account, bank, email, auto loan and work, you have created a single key enabling multiple attacks. If you use slight variations on the same password for each account, you are not anticipating the ability of cryptographic tools to permutate and concatenate a dictionary to possibly find those variations.

The attack surface grows with every IoT device introduced to the ecosystem. There are several points of entry to be concerned with including the devices themselves, the network they communicate through, fog computing services in those systems that use them, and the cloud.

There is, as of yet, no commonly accepted model of the IoT. Several have been proposed by academia, government and industry that attempt to clarify the various levels of the overall infrastructure, with none having gained universal acceptance. There are, however, commonalities within them that can help build a general understanding of the separate parts that work to form the whole ecosystem.

THREATS AT THE EDGE/PERCEPTION/SENSING LAYER

At this layer exist the devices themselves. The sensor data created at this layer is the primary driver of IoT utility and is susceptible to a number of threats. Broadly, they can be classified as environmental threats and human threats.[43]

Environmental threats are described as those to the hardware from humidity, temperature, water damage, and infestation of insects or small animals. Natural disasters such as floods, tornadoes, and earthquakes are also included at this threat level. Often, protection is built into the devices to mitigate the impact of many environmental threats.

Human threats are a larger concern though. Device destruction is the most basic of the threats from this source. Beyond that, insecure interfaces can allow a device to be compromised by a malicious device on the network as can insecure initialization after reboot, during which time an actor can gain control of a device. Once access has been gained, devices can be subjected to jamming attacks, spoofing, and deprivation attacks. Deprivation takes advantage of devices that run on battery power. This attack causes a device to stay awake, causing power to drain from its battery faster than normal.[44]

Other attacks at this level can include node replication attacks, which allow an actor to add new, malicious, nodes to the system of existing ones, allowing access to traffic within the system, and malicious code attacks. Malicious code attacks occur before or during device fabrication. Code placed in devices during this process can be keyed to activate for a particular trigger.[45]

THREATS AT THE NETWORK/COMMUNICATION LAYER

Data transmission between the other layers occurs here. As to be expected with any network layer in any model where technology communicates, there are a large number of threats arising from a large number of weaknesses that need to be addressed through security planning and best practice. As demonstrated by the powerful DDoS attacks utilizing IoT devices, network security in this ecosystem is vital to the proper functioning of the IoT and protection of the devices, cloud and fog that make it work.

DoS attacks that concentrate on denial of service consist of jamming attacks used to reduce the performance of the system or completely prevent communications. Replay attacks occur as a result of the replication of captured packets exchanged between devices. These duplicate packets can then be sent again by malicious devices. Eavesdropping, or man-in-the-middle attacks, allows actors to access data passing between nodes, which if unencrypted has the potential to expose confidential information outside of the knowledge of system administrators and users. This attack could also enable additional attack types when the information is analyzed for configuration, identification information, passwords, etc.[46]

Other potential attacks familiar to security specialists and IT professionals within this layer include insecure nearest node discovery, buffer overflow, routing attacks, sybil, session hijacking, selective forwarding attacks, sinkhole and wormhole attacks, HELLO flood attacks, and traffic analysis attacks.[47] [48] [49]

This potential for network breaches is, again, complicated by the fact that the IoT is designed to operate in the background, outside of human awareness. In addition, failure to account for potential breaches within this part of the network could propagate attacks to the more traditional networks that organizations are familiar with and accustomed to protecting.

THREATS AT THE CLOUD/FOG LAYER

Cloud and Fog are treated as separate in some models while combined in others. Some research focuses on the applications that run on these services. Generally speaking though, the threats can be summarized with respect to each without fear of minimizing or neglecting one or the other. Generally speaking, and as will be seen shortly, there are a larger potential number of threats to the cloud then have thus far been identified.

Threats identified at this level include some that are very difficult to identify and mitigate. Malicious insider attacks can be among the most destructive and difficult to avoid. Along with these threats are users, or insiders, that mean no harm but are still successfully phished in email, releasing malware on the systems running the cloud service. Closely related, or possibly as a result of these activities, unauthorized access is another threat in which someone has illegally acquired a legitimate account to gain access to data.[50]

The cloud, or cloud computing, presents additional complications to the ones already mentioned. Cloud infrastructure utilizes virtualization, or virtual machines, to accomplish the goal of providing a seamless and scalable solution for data storage and processing. Resource provisioning can be accomplished within the environment to meet the growing demand for resources to serve a customer's IoT network. Additionally, through the virtual environment, many users share the same physical equipment, but are logically separate from one another.

Attacks have been developed for use against virtual machines and the hypervisors that create and run the virtual environments on physical hardware. There are several examples starting with attacks using virtual machines.

VM poaching is a DoS attack using a malicious virtual machine to consume more resources than allocated, starving other virtual machines within a hypervisor. VM sprawl is accomplished when unused virtual machines continue to use system resources. Computing resources cannot be reused during this attack. VM migration, while not intended as an attack as it allows a virtual machine to move from one host to another, becomes an attack when a malicious actor intercepts the VM and alters it during migration. In addition to the potential loss of data, if the virtual machine is infected with malware, it can spread to other host machines and virtual machines.

The attacks against the hypervisor represent a danger to multiple virtual machines as can be seen in the following examples. VM rollback uses a malicious hypervisor to revert to an older version of a virtual machine. This allows the attacker to delete data and history. Returning to an older version will also undo patches, making the virtual machine vulnerable. Hyperjacking takes control of the hypervisor to gain access to the virtual environment. Doing this allows attackers access to all the virtual machines running in the hypervisor. This effects the logical separation between virtual machines, as well as the host machine.

In addition to attacks on virtual machines and hypervisors, malicious actors can attack the hardware itself. Complex side-channel attacks attempt to gain information about the physical implementation of hardware in order to locate weaknesses for exploitation.[51]

Research geared toward applications in the IoT have identified several weaknesses that can lead to exploitation. Insecure interfaces, in which web interfaces to the IoT ecosystem are targeted for vulnerabilities represent potentially exploitable weaknesses. Other weaknesses include insecure software and operating system misconfigurations. Middle-ware provides communication between different kinds of devices and interfaces since there are a multitude of protocols. Lack of middle-ware security could provide access at this level and represent another weakness.[52]

RECORDERS OF CRIME

As the recorder of crime, a device could contain, or communicate, data that helps to solve a crime. Recognizing the potential value of data recorded by devices, investigators and examiners have increasingly turned toward them as a source of investigative information.

On September 19, 2016, a Middleton Ohio man stated he awoke to a fire in his home. He said that he packed some property in a suitcase and bags, broke out his bedroom window with a cane, and threw the property out of the window before climbing out himself and carrying the property to his car.

The police, who said that his statements were inconsistent with evidence located at the scene, had found gasoline on the man's clothing and indications that the fire started in multiple places. At some point that man had also told police that he had a pacemaker.

Investigators obtained a search warrant to obtain data from the pacemaker which recorded heart rate, cardiac rhythms, and pacer maker demand. Data requested was for the time before, during, and after the fire. Upon reviewing the data, a cardiologist determined that it was unlikely the suspect actually performed the actions he claimed based on his medical conditions.[53]

The story continues, revealing the potential legal gray areas that investigators and examiners find themselves exploring. Attorneys for the suspect argued that presenting the pacemaker evidence at trial would be a violation of his physician-patient privilege as well as a violation of his constitutional rights. As of this writing the 12th District Court of Appeals was scheduled to hear oral arguments concerning the admissibility of the pacemaker data.[54]

In December of 2015, police arrived to a homicide at an Ellington Connecticut home. The husband described a violent encounter with a masked assailant that had tied him to a chair and assaulted him with a knife. The attacker then shot his wife in the basement.

Among digital evidence collected from the scene, including door movements and alarm settings, was the murder victim's fitness tracker record that she had walked 1,217 feet around the house during the time of the alleged attack. This was well beyond the 125 feet she should have traveled from the garage to the basement according to her husband's statement. He was subsequently charged with murder.[55]

In another homicide case a fitness tracker provided exculpatory evidence. On May 21, 2016, in Wisconsin, a woman's body was found three miles from her home. The night the victim was murdered, she and her boyfriend, with whom she had a child in common, had been out with friends drinking. As the evening wore on, they ended up at different locations. Eventually, the boyfriend went home to bed. The next day, when she had not arrived home, he called friends and family to begin searching for her. He also reported her as a missing person.

After her body was found, the boyfriend was questioned, arrested, but eventually released without charges. Another suspect was identified, arrested, and charged with the murder. Defense attorneys for the suspect offered the theory that after finding his girlfriend and the suspect together in consensual sex, he murdered her, forcing the suspect to help move her body afterward.

Data from the boyfriend's fitness tracker indicated that he was asleep at the time of the incident and had not walked the three miles between his home and the location of her body. The defense tried, without success, to have the evidence from the fitness tracker suppressed stating that the data was unreliable. The suspect would eventually be convicted of murder.[56 57]

These are a few older examples of the growing recognition of evidence gathered from IoT devices that has encouraged research into, and development of, forensic techniques for obtaining that evidence. However, unlike well-developed tools and techniques used on computers, external storage and mobile devices such as cell phones, IoT devices represent a new and complex challenge for researchers, examiners and investigators, as will be explored shortly.

It is also important to recognize that IoT devices, as well as networks, cloud, and fog/edge may play more than one role as alluded to above. They can be any combination of tool, target and repository of information.

FOCUS FOR EXECUTIVES, DIRECTORS, AND MANAGERS

Business, government and academic leaders are now presented with a new technology to fold into long-range planning, development, deployment and monitoring. Additionally, security may be more challenging than simply applying standard models or practices to the new technology. A clear understanding of the IoT may assist in the deployment of proper security practices.

It is important to understand several key pieces of information detailed to this point. A substantial number of heterogeneous IoT devices are created and deployed on a daily basis. Security is not always built into the devices. There are no common operating systems or communication protocols used by the devices. They are ubiquitous and may fall 'under the radar' when deploying and monitoring security systems.

Physical security is just a starting point. A device that can be physically accessed is one that can be potentially compromised. Network security, application security, fog and cloud security follow up with additional requirements to protect the entire ecosystem.

While it may be extremely difficult to protect every device, communication point, application, and service, a well-rounded strategy of defense in depth, monitoring, logging, access control, server hardening and penetration testing, along with other practices such as white and black box testing when available, will provide a basis to build a security plan.

It is also important, for the purpose of having a clear view of the IoT, to understand the different contexts in which IoT technology can be found.

IoT devices today

As IoT devices can be purpose built, they find their way into a number of different domains as specialist objects with a specific purpose. As would be expected, the domains are diverse and are intentionally, and sometimes unintentionally, interlinked by the devices. Information for the domains provided below address technology currently in use as well as speculative assessments on potential technologies based on the domain in which they are proposed for use. In either instance, the focus should be on understanding the impacts, and potential impacts, that so many interconnected devices may have on security and forensics. Including speculative technology as a forecast allows the planning necessary to develop strategies for approaching those instances where forensics may be necessary.

Home and wearable devices

The home domain, or smart home, is probably the most familiar concept to readers and a well-researched area of interest. Items such as the Amazon Echo, Google Home, and Nest products are a few examples of the increasing intelligence applied to homes. These products, along with others such as 'smart' refrigerators, televisions, and other appliances add convenience, provide expanded options for entertainment and education, and can monitor energy consumption.

Within the home can also be found monitoring systems for intrusion, fire, carbon monoxide, and moisture. Home management systems may also control heating, air conditioning, and other utilities while relaying data to the utility companies on energy and water usage. This is one example of the crossover from domain to domain.

Wearable devices can play a part in this domain as very often the home network becomes the backbone source of communications for these devices. An activity tracker may relay information via Bluetooth to the user's smart phone, which could then communicate that data to the cloud via the Wi-Fi connection in the home.

Other wearable devices cross over into the domain of healthcare and wellness along with the activity tracker. Pacemakers are becoming smart devices that include the ability to connect to Wi-Fi for the purpose of transmitting data for review by health care providers. Security, clearly, is of deep importance to any individual who uses one of these devices to regulate heartbeat.[58]

As the home becomes more intelligent with the addition of sensors, controllers, and networked communication, the attack surface grows. If an actor can compromise part of this network, access can be gained to the entire network and the devices connected to it.[59]

Utilities/energy

Utility companies increasingly deploy IoT in the form of smart metering. Electricity usage in the home is monitored as information is shared out to the utility. Monitoring assists in the efficient use and modification of the way energy is used in the home. This is expanded out to a wider context with the smart grid. Data communicated in an IoT grid can assist in maintaining a proper load balance to ensure effective service in a wider geographic area.[60]

Water, included in the utilities category, may also utilize IoT. As with electric use measurement, smart metering applied in water delivery services will provide a closer to real-time measure of usage. Other aspects include the ability to constantly measure water quality, and more effectively detect issues such as water main breaks. Waste water networks can also be monitored for utilization and treatment.[61][62]

Energy is a more generalized term that encompasses the domain of utility in its functionality. Included is the concept of utilizing devices in an energy management system that will maintain balance between renewable energy and fossil fuel sources. Theoretically, the IoT domain of energy would constantly measure the supply of renewable energy for the purpose of detecting excess supply or shortage. In the first case action could be taken to reduce the amount of renewable energy fed into the grid, and in the second, power from traditional fossil fuel sources could be increased during the shortfall.[63]

Health/wellness

With an expected increase from 10.5 billion to 52 billion connected medical devices over a 10-year period, the health domain is seeing substantial growth.[64]

The patient is the direct beneficiary in this domain as there are a growing number of devices, such as the connected pacemaker, which can provide both lifesaving services and report medical conditions through the IoT network to healthcare workers.

In addition to walking or running distance, or number of steps taken, activity trackers can also monitor heart rate and sleep patterns among other capabilities. This is another example of the crossover between domains as activity trackers are often utilized by employers to incentivize healthy employee behavior by lowering out of pocket insurance costs for meeting certain benchmarks within a particular period such as a week or month. Data can also be delivered to the health domain for analysis by medical professionals.

Other devices can provide the same functionality of devices found in physicians' offices and hospitals such as blood pressure and oxygen saturation level sensors. With the ability to gain data from remote body worn sensors, health care decision may be made without the need for office or hospital visits.

At a global level, Bluetooth-connected devices can track trends based on numerous data points collected from the network. One example is a company, Kinsa, which produces Bluetooth-connected thermometers that can transmit temperature data to smart phone apps. The data is then relayed to the company for aggregation and trend mapping of the spread of illness in a particular area or region.[65]

Business/industrial

The industrial IoT concept has been expanded by research to include flexible definitions of exactly what it is and does. It has been described as the backbone of the IoT; the infrastructure that needs to be built in order to enable other IoT applications.[66]

In this view it is the underlying architecture of the entire ecosystem, regardless of domain. The same research also points to the concept that the industrial IoT serves the vital function of connecting critical services to each other. In this view, the services connected are so vital that failure could lead to catastrophic results such as threats to life or other emergencies. Examples include possible failures in healthcare, transportation energy, and industrial control systems (ICS). This is separate from what are considered consumer level devices such as activity monitors and smart home applications.

In this view of the industrial IoT, such potential for catastrophic failure would demand a security by design approach that encompasses all phases of design and implementation, from software to hardware. The project management for these applications would find it beneficial to include security at every stage of development from the very beginning.

Another view of the industrial domain limits IoT applications to use in industrial control systems, supervisory control and data acquisition (SCADA), and programmable logic controllers (PLCs).[67]

IoT as a supplement to, or replacement for, these controllers and functions is, as with other domains, nascent. While possible approaches to IoT forensics are addressed in this chapter, operations technology, SCADA and ICS forensics are addressed in greater detail in Chapter 6 of this book.

The same research also places agriculture under the industrial domain. From backyard gardeners to farmers, IoT devices can provide data on soil moisture, nutrient levels, and sunlight exposure. Intelligent systems can also be used for watering and feeding functions.

The Internet of Cows exists. Livestock management has also begun the deployment of IoT technology. Demand for animal products will clearly increase with the world population. The IoT can be used to improve the health of livestock and bring efficiency to the industry to meet this growing demand.[68] Motion and temperature sensors can be used with other sensors to capture data about livestock for analysis to determine if there is a risk or presence of disease.[69]

Fujitsu Kyushu Systems offers a service that monitors livestock for breading purposes. This service collects and analyzes data from device sensors to provide farmers insight into when their livestock are ready to breed.[70]

Business, commerce, and finance are interested in streamlining services, improving customer experience, reducing costs, and increasing prophet. This part of the domain includes devices such as portable credit card readers that can be attached to mobile phones, point of sale NFC, and inventory tracking. Business is also interested in gleaning buyer habits from data points provided by IoT devices.

Data generated by the IoT combined with big data analysis creates another opportunity to serve all of those interests. However, the challenge for big data is the substantial and ever-growing data yield. With the Fog/Edge filtering data to reduce traffic and provide better real-time decision-making, there may be a need to strike a balance between data that should be sent to the cloud for analysis, and data that does not meet the threshold of relevance for analysis.

The immediate recognition of a production line problem may be important for remedial action to prevent delayed delivery, but that information may not be as important to analytics as knowing where that product is sold most, and to whom. Whereas each can have an effect on the other, they are treated differently based on context. Production is concerned with system design and execution where sales is concerned with prediction and strategy.

Transportation

As stated earlier, it is projected that by 2020, 75% of all vehicles produced will be able to connect to the Internet. These connections provide for navigation, entertainment, and communication. Additionally, software and firmware updates can be uploaded to vehicles that have systems and sensors that can help drivers stay in a lane or break at the appropriate time. These systems can also be expanded to the enterprise level to help monitor and properly deploy company or government fleet vehicles.[71]

The ability to, in real time, sense the motion of a vehicle relative to other vehicles, traffic directions in the form of lane markers, traffic signs and signals, and react to weather conditions such as snow and rain has ramifications for autonomous vehicle research and development. More importantly is the speed with which vehicle intelligence would need to observe, process, and react to unforeseen events such as poor driving on the part of another vehicle, accidents, and obstacles such as debris or animals.

The technology has not matured to the level of true autonomy, but there are numerous examples of drivers who have tested it with poor results. At this point the ability to instantaneously receive, interpret, and act on data is done better by the human brain. In a few years, this may not be the case.

Expanding on this domain, data can be obtained from vehicles and a variety of other sensors to provide real-time traffic pattern information to assist drivers, or their vehicles, in determining the most efficient route to their destination.[72] Eventually, autonomous vehicles may communicate with each other, and devices within the public infrastructure, to determine the most efficient path to the desired location.

The designs, deployed technology, and ideas from many of the domains addressed may well be encompassed in whole, or in part, in the next domain.

Smart cities

The idea of a smart, or IoT-connected, city is one in which much of the automation is replaced with IoT-driven intelligent decisions. Public lighting would be more efficient in use when done in concert with sensors detecting levels of vehicle or foot traffic. The transportation domain would work within the smart city to provide efficient and effective traffic flow, while the utilities domain would provide the service of bringing efficiency to electric and water distribution. Safety, parking control, public building energy efficiency, public transportation, and air-quality control are listed as those functions of a government that may be positively impacted by the IoT.[73]

The societal interest in efficient and effective government will rely on coordination between the domains. The example below shows a hypothetical instance where elements of the domains mentioned above coordinate efforts to ensure a positive outcome for the consumer.

A house or building fire would be detected by IoT devices created and sold by private industry. The alert send out by these devices would prompt action by first responders that rely on efficient traffic pattern analysis and control to expedite their arrival to the fire. Smart utilities, detecting an emergency stemming from the fire, could turn off electric and gas service to the structure, while ensuring the prompt delivery of a sufficient amount of water for the fire department. Body worn health monitoring devices could detail the effects of the fire on occupants, allowing for the preparation and staging of medical personal both at the scene and in the hospitals most likely to receive patients.

The local and global nature of data creation and transfer may speed up the process of controlling outcomes, detecting and diagnosing issues, streamlining services, increase profit or savings, and aid in planning. However, the amount of data created and transmitted will continue to grow with the number of devices generating those data. As has been seen, there are many complicating factors that make it difficult to secure the devices and data they produce. Growth will only amplify the potential issues surrounding confidentiality, integrity, and availability.

VULNERABILITIES/RISKS/EXPOSURE

There is always the possibility that an action taken will lead to loss. That is risk. The difficulty rests in determining the level of risk associated with any technology, let alone the IoT. If an entity such as a company, government or individual decides the risks are not high enough to negate the potential reward, the risks will likely be taken. But what are the risks?

Throughout this chapter risks have been addressed in terms of the technology at work within the IoT. From the very lowest level of the device to the cloud, there are risks that must be properly evaluated and included in any assessment used to determine acceptable levels.

Determining the probability of an outcome in the IoT can be difficult because of a general lack of understanding. The remainder of this section summarizes the risks addressed to this point.

Devices

IoT devices themselves are resource constrained and heterogeneous. Constraints make deploying robust security difficult as it may demand a substantial amount of the device's resources, or more than the device can even provide. Heterogeneity works against security as there is no common framework or architecture in which standardized protocols and security designs are deployed.

Think of Microsoft Windows operating systems and their share of the market. Software patching is done on a regular basis and a large number of anti-virus programs are designed to work in these operating systems. Additionally, and as mentioned earlier, hardware that can work with these operating systems are fairly homogeneous, though firmware patches are often the responsibility of the system owner.

Security is also often neglected in device design and deployment due to pressure to bring the product to market. Refer to the statistics provided earlier that revealed the level to which developers believed security was an issue.

Physical damage to devices can occur from environmental sources such as water, tornados, and earthquakes. Physical damage may also occur as the result of malicious intent by a human actor. The risks assessment from these possibilities may determine that damage to a device, without additional risks to exposure, is acceptable. A device is usually small, inexpensive, and easily replaced. If the occasional isolated incident results in the loss of one, the cost of replacement may be acceptable when compared to the overall benefit provided.

Devices often have weak access passwords that are not changed by the consumer upon deployment. Additionally, communication among the devices and between the devices and consumer may not be encrypted, or encryption may be weak. Firmware updates that address exploitable weaknesses may not be automatic and may also be neglected by the consumer. Since these devices are designed to run without the need for human intervention, it is easy to imagine circumstances where consumers in business, government, and at home may neglect to perform the appropriate firmware updates.

The following examples of home user incidents provide insight into risks and exposure. While not comprehensive in covering all domains, it is easy to extrapolate outcomes within each as a result of the outcomes seen in these examples.

A homeowner outside of Chicago said that while standing outside the door of his young child's room, he heard a deep male voice. Initially he believed it to have come from a baby monitor, but when he was downstairs he heard the voice again. He discovered that it was coming out of a Nest camera, one of several in the home. He could tell by the comments made that the individual could see his family through the cameras.

He later noticed that the Nest thermostat in the upstairs part of the home had been raised to 90 degrees. He believed the individual who had gained access to the cameras was likely responsible for that as well.

Google, the parent company of Nest stated that the system was not breached, but that access had been gained through a compromised password exposed through breaches on other websites. They suggested that customers use two-factor authentication. Google also reset passwords that had been previously exposed.[74]

In another instance in Tennessee a mother installed a Ring camera in her daughters' room so that she could monitor them using her cell phone. Just a few days after the installation one of her daughters told her that she heard music and a voice through the camera. When the mother watched a recording of the incident she heard the intruder taunting the child. The parents disconnected the camera with plans to return it to the vendor.

A spokesperson for Ring stated that their security had not been breached and suggested that the owner's password had likely been used for several accounts, some of which may have been breached, resulting in the theft of the password. Ring also suggested two-factor authentication as well as the use of complex passwords, along with a periodic change of those passwords.[75]

Recently, a release from the FBI warned that a 'smart' TV could be used as a conduit to your home network. IoT-enabled television sets are network connected and can be used to browse the web and consume entertainment from streaming platforms. Additionally, these TVs include other features such as cameras, microphones, and facial recognition, allowing verbal commands to replace remote controls.

As with other IoT technology, however, these TVs can have poor security implementation. An intruder controlling the TV could be listening to, and watching, occupants of a home or business. The intruder could also control what is shown on the set. Additionally, the TV could provide access to the router and network that it manages.[76]

Once in the network an intruder would have access to the many other devices connected to the router. As stated, the potential attack surface has expanded with the introduction of IoT technology. Where in the past, there may have been one or two computers in a home or small office, there are now an array of 'smart' objects such as televisions, refrigerators, mobile phones, activity trackers, tablets, Nest technology, Ring technology, home alarm systems for burglary, smoke, and moisture detection, etc. Clearly the list is far larger than the few items mentioned here.

In addition to the danger of being monitored by an outside intruder, or having your devices slaved to a botnet, the processing power available in your home or business would be attractive for other uses. As mentioned earlier, the processing power of individual devices is relatively weak compared to those of a PC or laptop computer, but thousands or tens of thousands or more working together offer an effective resource for hackers to utilize.

Malicious software has been found on 'smart' devices such as refrigerators that allowed hackers to control the processing power of the device for cryptocurrency mining. The owner or user of the device may notice a little performance lag while CPU cycles are used for the hashing function necessary to obtain cryptocurrency such as Bitcoin or Monero. This may not be enough to attract attention or concern and go unaddressed.[77]

Researchers have also targeted the lack of security in design in 'smart' vehicles, showing the alarming impact it can have outside of the home. In a 2015 article written for Wired, Andy Greenberg shared his experience driving a Jeep Cherokee that had been hacked by researchers Charlie Miller and Chris Valasek.[78] In the case of Chrysler products, Miller and Valasek found that they were able to gain access to vehicles through the company's Uconnect interconnected computer. This feature controls entertainment, navigation, and provides phone service as well as a Wi-Fi hot spot. Once identified, the vehicle's IP address allowed the researchers access. More concerning, Miller and Valasek found that they could access the vehicle from anywhere in the country using a mobile phone running on the Sprint network and a laptop computer.

The researchers were able to gain control of the vehicle by sending commands through the entertainment system to another chip in the same head unit. Once there they were able to rewrite the chips firmware. When complete the firmware could send commands through the vehicle's computer network to the engine, wheels, and other components.

What were they able to do once they had access? In detailing his experience as the driver of the Jeep, Greenberg stated that Miller and Valasek started by controlling the air conditioning system, the radio, windshield wipers and wiper fluid. They also appeared on the digital display. As the demonstration went on, they disabled the transmission causing the vehicle to slow to a near stop while on the highway. During this time, Miller and Valasek were able to communicate messages to Greenberg over the radio. After restarting the car, Greenberg was able to leave the highway and drive to a parking lot for further demonstrations at lower speed.

The researchers were able to turn off the engine and control the breaks by either engaging or disabling them. At the time the article was written, Miller and Valasek stated they were working on better steering control as they were only able to control the system when the vehicle was in reverse. Additionally, access to the GPS system allowed tracking of vehicle location and speed.

Two years earlier, Miller and Valasek experimented with vehicle hacking, but had to be in the vehicle with their computers hardwired to the vehicle's diagnostic port. The ability to access vehicles remotely, from any location in the country, had clearly progressed at a rate faster than was addressed by auto manufacturers.

Using the phone and computer method for the Jeep hack, the researchers were able to demonstrate the ability to find vehicles anywhere in the country. They located vulnerable vehicles in California, Michigan, and Texas. Miller estimated that (at that time) there were 471,000 vehicles running the Uconnect system.

The article pointed out that Chrysler was not the only manufacturer deploying vulnerable systems. The researchers believed that to some degree, nearly every vehicle manufactured had some form of vulnerability.

These few examples are illustrative of the potential impact this technology could have in every sector. The same weakness and shortcoming that led to the events described above exist in IoT devices used in business, government, utilities, and every other place the IoT ecosystem exists. Vulnerability created by a lack of attention to security during device development can lead to use of the device for surveillance, privacy invasion, or data theft. It can also lead to device use as a control or malware proliferation agent, or worse.

Networks

The examples given above make it clear that most access to devices occurs through networks. Wireless networks used to identify and attack vulnerable vehicles and Wi-Fi networks in the home used to access cameras, televisions, refrigerators, and other devices are providing an easy access point to open, unencrypted communication between devices and between devices and people.

In the business setting, operation of IoT devices is no different and no less vulnerable. Many businesses may have plans in place to protect legacy networks for computers and servers including virtual private networks (VPNs), account controls, firewalls, IDS/IPS, demilitarized zones (DMZs) and air gapping, to name a few methods, but what of the IoT devices brought to work by employees, or even customers?

BYOD or bring your own device policies may help protect a company, government, or utility by establishing a set of procedures that address, among other concerns, security. These entities may establish policies concerning password usage and complexity, device use authorization, requirements for software download and usage, and data access by employee role.[79]

Mobile device management (MDM) attempts to separate company data from private information held in a device. Steps taken can include data lockdown and, upon separation of the employee or loss of the device, remote wipe.[80]

These policies and practices are common in the case of cell phones and computers brought to work by employees, but what does it do to address the small and unobtrusive IoT devices worn to work? The nature and operation of a device may not cause it to rise to any level of attention on the part of the employer, be it a business, government, academic institution, utility, or health care provider or any of the vast number of entities that rely on employees to operate.

Additionally, many retail businesses, governments, college campuses, and other service providers offer free Wi-Fi access to their consumers. In the context of laptops, tablets, and mobile phones, we are accustomed to seeing individuals utilize this service. As with the employer, employee relationship, how is this effected by the presence of customer-owned IoT devices?

Entities allowing BYOD or offering free network access would do well to understand that many consumer-level IoT devices, designed for convenience and ease of use with less attention paid to security, can significantly increase the chance of malware infiltrating the network. IoT devices can be a threat to the network they are connected to, as well as any other devices connected to the same network.[81]

Remember that many wireless networks are often served by a wired network backbone. Anything that infiltrates through a wireless connection will expose the wired network and devices connected to it.

In a network connecting numerous computer system, security response to an intrusion incident will likely include remediation processes that also address those systems. However, attention needs to be paid to devices specifically designed to be ignored as they operate autonomously, whether those devices belong to the entity attacked, or to an employee or customer who introduced them to the network. Vulnerability created by the lack of attention to IoT device security risks the unnecessary compromise of the network and all systems connected within.

Cloud

The amount of data in the cloud presents an enticing, large target for malicious actors. Confidentiality, integrity, and availability are all at risk when considering the amount and types of data stored in the cloud, as provided by IoT devices.

As mentioned earlier, IoT devices can utilize fog/edge computing to provide real-time, or near real-time, decision-making. Data, filtered at this level, is then sent to the cloud for higher level processing, storage, and data mining. Data can include any sensor-generated information from areas such as health monitoring, industrial processes, business, and government collection of personally identifiable information and the like.

The vulnerabilities of virtual environments and the hypervisors that enable them can lead to the risk of data manipulation, theft, and/or loss. This is compounded by attacks that potentially cross from one virtual machine to another, exposing even more data.

Often, these attacks stem from employee error or malicious insiders' intent on doing harm. Though a large risk, it is not the only one. Network attacks, exploits from unpatched systems and poor access controls, among other examples, can lead to cloud data storage exposure.

Privacy invasion, device control loss, data exposure, and other potentially harmful events at the consumer level represent risks to individuals. Exposure or loss of data at the cloud level risks the privacy and security of tens, or hundreds of thousands of people. To that end, many companies could, and do, examine methods to decouple personal information gained through IoT sensors from the identities of those from which it was gleaned.

THE ROLE OF CYBER FORENSICS

Cyber forensics, or digital forensics, is usually found within the response and recovery cycle of an organization's deployed security plan. However, the question of its utility as a preemptive measure arises when considering the heterogeneous and ubiquitous nature of the IoT. Can cyber forensics serve security prior to its traditional use in an incident response? What will performing cyber forensics on the different parts of the IoT, from edge device to cloud, reveal that will help develop a sound security framework?

THE FORENSIC PROCESS

Over the last few decades digital forensics has matured in its practices and procedures. The National Institute of Standards and Technology outlines four steps to digital forensics that will assist the layperson in understanding the practice. In order they are;

- Collection
- Examination
- Analysis
- Reporting[82]

Each phase, or step in the process, includes a number of sub steps or other considerations that outline the best practices of digital forensic examiners. These steps are shown in Figure 2.1.

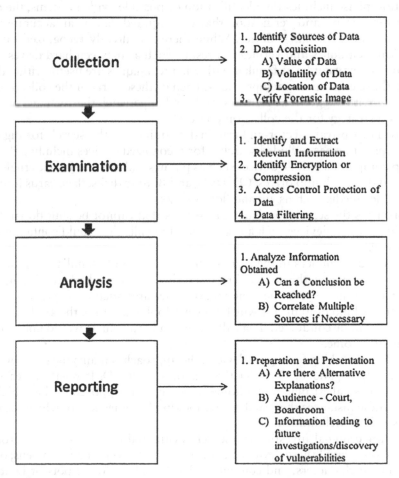

Figure 2.1 Steps of the forensic process[83]

It is important to note at this point that within digital forensics there are, as with other forensic disciplines, private sector examiners and law enforcement examiners. While there are private sector laboratories that do work for law enforcement agencies, this text will treat them separately for the purpose of illustrating the differing opportunities both may have when working toward the end goal of analyzing IoT devices. This 'two lanes' approach may offer perspective on the relative advantages and drawbacks examiners face in the context of both private sector and law-enforcement investigations.

The collection phase

Crime, obviously, predates the digital world of today. In the past the traditional approach to investigations included a search and analysis of physical evidence, the crime scene, interviews of victims, witnesses, and suspects. With the introduction of computers, and mobile devices after that, there were new tools for use by criminals. This naturally resulted in the expansion of searches to include data that may contain evidence of the crime. As a result, computers, mobile phones, other digital storage devices, and even networks became a target for digital evidence extraction. These devices became a new digital crime scene to be investigated.[84]

The collection phase includes the identification of possible evidence items, the acquisition of data from those items and verification that the acquired data is an accurate representation of the data on the original device. While there are already recognized differences in data acquisition techniques and verification results such as between hard drives and mobile phones, examiners may not be as familiar with what techniques are useful within the IoT and how to verify those extraction attempt results. Each of these parts of the collection phase are discussed below with suggestions from research and experience that may act as a guideline when building a strategy for the collection phase.

Common to both private sector and criminal examiners is the search for digital storage devices. In the past this has included searches for recognized devices including desktop computers, laptop computers, mobile devices such as phones and tablets, and external media such as external hard drives, thumb drives, CD, DVD, and Blu-ray disks, flash cards found in many cameras and older media such as zip and floppy disks.

Also important is the search for data from servers that cannot be shut down during data extraction and network devices such a routers and firewalls that could contain a wide array of logged activity.

Investigators and examiners in the criminal field are seldomly familiar with the area to be searched before arriving at the scene. Very often they will have had training in the proper way to conduct searches of a home, building, or other area such as a field, keeping in mind the specific item or kind of items for which they are looking. Often, the kind of item dictates where the search can be conducted. Logically, a search for a vehicle would not include checking the closets of a home.

However, when searching for digital devices, the approach remains the same but the places to look are greatly expanded. Consider the size of a MicroSD. It would be easy to justify searching even the smallest places of a home or business as opposed to the vehicle example above. Still, in these instances the searchers are looking for objects with which they are familiar, however small or well hidden.

A challenge for criminal investigation searches can, and often does, come from 'camouflage' devices. Examples of these devices include thumb drives embedded in pens, or that look like popular movie characters, and computers that look like flower pots or boats, or other objects.

A private sector examiner may have a better idea what to expect when searching for devices in the environment in which he or she already works. A contract examiner may also have the advantage of working with corporate security or IT when trying to locate items. In either case, camouflaged devices brought in by employees may present a challenge in this context as well.

In the case of the IoT, both corporate and criminal examiners may be challenged by the search for devices containing data relevant to investigations. In these instances, it could be useful to gain as much information about the crime, or violation, as possible prior to conducting the search. Understanding what happened may help the examiner gain a rough idea of what IoT devices may have recorded, been the target of, or used to commit the offence.

For example, in the case of a burglary, a criminal examiner may expand the search to IoT devices on or in doors and routers that record connections of devices at about the same time as the crime, indicating the suspect had been there, and connected, in the past. For the corporate examiner the process may be similar when investigating an unauthorized access event. Sensors on doors, or motion sensors in rooms, may contain data useful to the investigation. These examples are simplistic but illustrative of the need to consider this new ecosystem when conducting investigations, as was the case with the initial move from physical to digital evidence crime scenes.

Further complicating this issue is the possibility of creating new data during the process of searching for devices at a scene. Recall that, unless disabled, IoT devices contain sensors that are constantly active and constantly recording the activity they were designed to collect. They can act as witness to a crime and also to the subsequent investigation of that crime as investigators and examiners approach and analyze the scene. This can complicate the determination of what evidence is relevant, and what represents 'contamination' subsequent to the original event. This interaction makes it essential to document all activities at a scene, and with the devices, so that follow up analysis may be able to differentiate between evidence relevant to the investigation and data created during the investigatory phase.[85]

During the process of identifying items to be examined, it is important to recall that while a particular IoT device may produce data, it may not store that data. Remember that real-time operating systems are often deployed with these devices for the purpose of transmitting data almost immediately after creation.

Data scientist Usama Salama presented three evidence categories that will help guide the examiner to locate sources of evidence produced in the IoT. These categories are:

1. Evidence from the IoT devices themselves.
2. Evidence from the infrastructure that enables network communications such as servers, mobile devices, routers, firewalls, etc.
3. Evidence from the infrastructure outside of the network such as Internet service providers, mobile network providers, and the cloud.[86]

In fact, researchers have recognized that mobile and cloud forensics, already more established in the digital forensics field, are complimentary to IoT forensics due to the use of fog, and cloud storage, and the transmission of data to mobile devices for use by the consumer.[87] In the case of a home monitoring system, much of the data generated by the various sensors in a house are transmitted to the cloud to be processed and then subsequently sent to the home owner's mobile phone for review, or even for remote action to be taken by the owner.

Figure 2.2 shows a relatively simple IoT implementation in the home that can be used to demonstrate where data may be found.

In this instance we can see the potential for all three sources of information to come into play. Using, once again, the door sensor as an example, the sensor may detect the movement

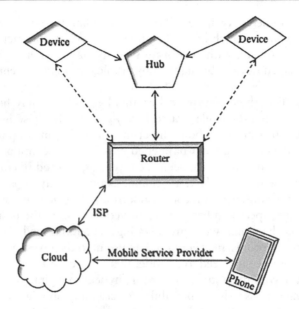

Figure 2.2 Possible areas of interest for data collection[88]

of a door and immediately transmit that data via Wi-Fi through the local network to the Internet service provider (ISP), then to the cloud. The data is then sent to the owner's mobile device, all in near real time.

The question for the examiner or investigator is where is the data stored? Does the door sensor retain any data such as logs of opening and closing events with an associated date and time? Does the network, specifically the home router, show activity at a particular date and time from the sensor? The sensor may be wirelessly connected to a hub via a proprietary communication protocol. In this instance, the hub is receiving data from the sensor and moving it from the proprietary protocol to Wi-Fi for transmission through the local network, to the ISP, then to the cloud. Does the hub retain any data? Did the router log the activity from the hub? Will the ISP log specific activity from the home?

Considering the use of network address translation by the router in which there is a single external facing IP address, and multiple internal router assigned IP addresses, will it be possible to differentiate the data transmissions of all the potential connections within the home when looking at ISP records? What of the cellular network that transmits data from the cloud to the mobile phone and back?

The research has suggested that answers to those questions are highly dependent on the specific IoT devices being used. Recall the large number of proprietary hardware implementations, software packages, and communications protocols designed into devices. Due to this heterogeneity there have been no standard or set of standard tools and techniques developed either for data source identification or retrieval.

Once a potential source of data is located and identified the process of data acquisition begins. While mobile, network and cloud forensics are more familiar to examiners, IoT devices present a challenge as research into forensic procedures on them is relatively sparse. As a result, the examiner will need an opportunity to identify and research the specific IoT device in question to know what, if any, possibility exists for data extraction. Recall that the heterogeneity identified throughout this chapter includes a lack of common interfaces, storage, or standard protocols between different kinds of devices. This is a major challenge to examiners attempting to extract data from devices.[89]

Assuming there is data in a non-RTOS device to be extracted the challenge then becomes how exactly to perform the extraction. In some cases, the examiner may find an interface that allows for data extraction that utilizes purpose-built tools. In their study on IoT forensics, Servida and Casey found that different IoT devices offered different opportunities for data extraction from serial connections, network traffic, smart phone companion applications, and cloud.

They would later develop plugins for use in the open source Autopsy program to parse the data. This stage will be discussed shortly. In their particular study, they found that smart phone application analysis was the most fruitful in obtaining and examining data associated with IoT devices.[90] This positive development is useful to the examiner proficient in mobile device analysis, but not all IoT devices, including those geared toward the consumer, have smart phone application associations.

Each potential data source is addressed separately. Whether from the device, the network or the cloud, each has its own unique opportunities and challenges. This section will cover each, with emphasis on the device as the other sources of evidence have been better developed within digital forensics.

The steps taken to extract data from the device can be broken down into several potential solutions starting with manual. This procedure simply uses the devices own system to display what is in its memory.[91] This process, while not necessarily a forensic procedure, may be the only way that information can be extracted.

A second choice for extraction, if available, would be to locate a port on the device that allows connection to a computer for the purpose of reading and extracting data. However, serial connections on devices are not universal and should not be assumed to exist prior to disconnecting and removing a device for analysis. Doing so may cause a loss of data that cannot be retrieved.

A device should be researched to any extent possible to determine if extraction should happen in place, or if the device can be removed and brought to a controlled location such as a lab. As with any of the methods for data retrieval listed below, it is beneficial for the examiner to communicate with other members of the community through message boards and direct contact.

Professional organizations such as the International Association of Computer Investigative Specialists (IACIS) and the International Society of Forensic Computer Examiners (ISFCE), among others, offer opportunities for examiners to communicate with one another on specific issues related to the extraction of data from a range of devices.

Some devices come with the ability to connect through standards such as Bluetooth. Tools such as Cellebrite Universal Forensic Extraction Device (UFED) Touch offer the ability to extract data from some phones using Bluetooth, and this may hold true for extracting data from IoT devices. However, as has been addressed earlier, many different standards and protocols are used within the IoT. Bluetooth, as has been seen, is just one of many communication methods.

Though the following techniques for data extraction are addressed in other chapters in this work, they bear repeating here for their potential utility with IoT devices. As with mobile devices, more intrusive methods of data acquisition may be required. These techniques include In-System Programming (ISP), Joint Test Action Group (JTAG) and chip-off. While ISP and JTAG may not result in the destruction of the device upon which they are used, the chip-off procedure will.

JTAG, implemented for circuit board verification and testing found usefulness in the forensic community for its ability to allow connection to points on the board, or Test Access Ports (TAPS) that subsequently enabled the extraction of data.[92] Very often this is accomplished with specialized tools connected to wires that are soldered, or connected by some other means, to the TAPS on the board. Many examiners have gone through the somewhat arduous process of determining which TAPS perform the necessary functions to allow for extraction, and then shared that information with the community. In the case of IoT devices, if JTAG is available, examiners will need to familiarize themselves with the process of testing possible connection points to determine which ones, if any, will allow for data extraction. This method may not be available with many IoT devices. Soldering skills are a plus.

ISP allows for the programming of chips while in circuit, or on the board. It eliminates the need for the chip to be programmed before placement in a system. This only works where the system supports it. This technique also became useful to forensic examiners for the same reason that JTAG did. It was found that particular points on the circuit board could be used to access and extract data from a chip.[93] Again, soldering skills are a plus.

Finally, chip-off should be used as a last resort. In this procedure, the memory ship is removed from the circuit board so that data can be extracted from it. This usually requires the use of an adapter in which to place the chip and connect to a computer for extraction by specialized software on the computer.

There are dangers associated with this procedure. First, the chip needs to be disconnected from the board. This is done by applying an amount of heat necessary to melt the solder, but not so high as to destroy the chip. In many cases a chip can be too thin to survive this method. In these cases, a lathe, or similar tool, is used to shave the circuit board off of the chip while taking care not to damage the connections on the bottom of the chip. Second, this will destroy the device as few examiners have the equipment necessary to reconnect the chip and place the device back into working order.

JTAG, ISP and chip-off are also affected by the presence of encryption. On newer mobile phones, data in the memory chip is encrypted. This second hurdle may make the processes described useless without the ability to decrypt the extracted data. IoT devices, to this point, may not offer that additional hurdle as security design has lagged behind development and deployment.

Network examinations, as with the cloud, are more familiar territory for the forensic examiner. Network components such as routers, firewalls, IDS/IPS, and some switches offer logging capabilities allowing the examiner to track activity within the network. At this level, such as with the home sensor example, the fog may come into play as a potential source of data for extraction. As the goal of the fog is to reduce data transferred to the cloud, and speed up decision-making, there may be data that is not found in the cloud.

The network also presents potential legal issues that are covered further in the discussion of the cloud. Often, network communications traverse through portals that do not belong to the individual or company conducting the investigation. Networks covering large regions often belong to telephone and cable companies and other similar service providers. Each will have their own requirements for providing legal documentation necessary to obtain data.

With cloud associated examinations, just as with many network examinations, there are legal requirements that often must be met to obtain data. The law enforcement examiner will often need legal process such as a subpoena or search warrant to obtain the data associated with an IoT device. This is further complicated by the fact that cloud services are often not within the same state, region or even country of the examiner's jurisdiction. The same is true for Internet and mobile service providers. In either case, when the request for information crosses international boundaries, laws governing legal requests, privacy and other related issues become a factor that may delay or even negate legal process. In these instances, time sensitive information may be lost before a resolution can be reached. Many service providers may only maintain data for a specified period such as thirty or sixty days for example.

In the private sector, the cloud may be maintained by the corporation for which the examiner works. Otherwise, service level agreements and other contractual obligations may need to spell out circumstances under which an examiner may request data and in what manner the request should be issued. Again, data maintenance may be time sensitive. Often, violations such as intrusions are not detected for many days, weeks, or even months, if at all. If the data retention period is shorter than the amount of time it took to identify and begin the remediation process for an intrusion, that data may be lost.

When first approaching the task of conducting an IoT investigation, with consideration given to where data may be located, it is important to categorize the data for the purpose of determining the order in which it should be collected. This idea comes from the more traditional processes involving computers, from personal to servers.

When conducting the initial evaluation of a scene from which evidence may be seized, the investigator or examiner is trained to evaluate the current condition of the devices present. If a computer is on, the examiner will attempt to determine if there is encryption present on the system. Turning the system off, without addressing encryption first, may result in the extraction of an encrypted and unbreakable forensic image.

The second consideration is volatile data from RAM, or random-access memory. If the computer is powered down, data in RAM will be lost, and this can include passwords and data changed but net yet saved to storage. The person responsible for the seizure may decide to extract the RAM data prior to powering off the device. This action will result in changes made to data on the computer, but is unavoidable. Actions taken with the computer should be well documented for the purpose of explaining changes made to data.

The potential sources of data within the IoT should be triaged with the same system of prioritization. The guidelines from NIST again provide useful to the investigator and examiner. Factors effecting prioritization include determining the likely value of the data, as mentioned, the volatility of the data and the effort needed to obtain the data.[94]

For example, when determining value, data from temperature sensors throughout a home or building may not contain information useful for investigating a burglary, but may have information useful for investigating an arson when attempting to determine where a fire started. Time spent extracting data of little or no potential value is time wasted.

The effort required may demand more resources and time than an organization or government agency is willing or able to provide. In the section covering the potential for encountering international boundaries and legal requirements for obtaining data, the entity seeking the information may conclude that the effort required does not justify the potential value.

The last part of this phase is the verification of the data. Using computers as an example once again, the standard procedure for extracting data is to connect hard drives found within a computer to a write blocking device and from there to a laboratory computer. The purpose of the write blocking device is to ensure that no changes are made to data on the evidence drive by the examiner's actions or computer operating system. Once the connections are made a forensic image is created that is essentially a bit for bit copy of all the data on the evidence drive. This includes data that still logically exist and can be seen by the computer user, hidden data, operating system protected data, deleted data and unallocated space which may still contain remnants of data.

Hashing algorithms, or hash values, are then used to verify that the data in the forensic image exactly matches the data on the evidence drive. Best practice is to perform the hashing function on the target drive after connecting it through a write blocking device. Create and then obtain a hash value for the forensic image, and perform the hashing function again on the evidence drive. In all cases the value should be the same. A match of the first two values would indicate that the image is an exact, bit for bit, copy of the evidence drive. The third value would verify that the imaging function did not change data on the evidence drive. If any of these values does not match the others, there was a problem.

Mobile devices, with which IoT devices share characteristics more than they do with computers, present a new dynamic that makes the verification process more difficult, if not impossible. Generally, extractions of data from a mobile device require that the device be powered on during the process. This action, though required for most extractions, changes data.

Tools used to extract data from phones often do not provide for a hashing function of the device before and extraction is attempted. As a result, many examiners will extract data, obtain a hash value of the data, process the data through software to enable examination, and then obtain a hash value of the extraction again. The two values are then compared to confirm that data was not changed during the processing step.

In either case, the goal is to produce an image or data extraction that, when analyzed, will consistently produce the same results when performed using the same techniques, procedures and software packages. In the case of IoT devices and the creation of custom scripts or other software packages, best practice would be to verify the functionality of the tools through third party testing and/or testing on reference devices with known data. Untested, or unverified tools, may face challenges in court proceedings that result in evidence suppression. In addition, availability of the tool may be required during the discovery process so that opposing counsel can verify functionality.

The examination phase

This is described by NIST as the phase during which the examiner assesses the data and extracts the artifacts that may be relevant to the investigation. A large data set can present a challenge during this process as the amount of data associated with a violation may be minuscule compared to the total amount that exists. In addition, encryption, compression, access control and other software features that can obscure data further complicate the process of locating relevant artifacts. Many tools have been developed to address the issue of encryption. Vendors of forensic training, software and hardware solutions often provide training specific to cryptography, geared toward general knowledge and the use of specific tools and techniques for obtaining data from encrypted containers.

It is useful, during this stage, for the examiner to be as familiar as possible with the facts and circumstances of the case. Knowledge of specific information related to the violation will allow the examiner to conduct key word searches, separate and exclude file types that will likely not have useful information, filter system generated files that are unnecessary and also filter by dates and times for the purpose of focusing on the period the violation may have occurred.

It is very helpful for the examiner to have possession of any report material generated during the investigation process. From these documents, he or she will be able to translate information into filters that will greatly assist in gathering relevant artifacts. Without information, the examiner will not be able to efficiently reduce the material to a manageable data set. This will increase time and resources requirements to an extent that the organization may decide the cost is not worth the potential benefit. The danger of this, particularly for intrusion cases, is that exploited vulnerabilities may not be discovered, and the probability for future intrusions will remain.

Fortunately, in the case of IoT or fog devices with limited memory, the data set will not be as large as that normally found in computers, or even mobile devices. As stated earlier, many devices have no storage capacity and limited volatile memory storage. However, information from networks, service providers and cloud services will increase the amount of available data.

The analysis phase

This third step involves time spent evaluating data collected during the examination phase for its relationship to the investigation. The purpose of the analysis is to identify specific artifacts that will inform the investigator of specific details of a crime or other violation. Analysis may identify the tools, specific dates and times, locations and even individuals involved in a violation. As there are multiple potential sources of data such as from devices, networks, the fog and cloud and service providers, analysis is also concerned with data correlation.

For example, an examination of network logs along with data from fog or cloud storage may create a timeline of events in combination with reported sensor data that allows the examiner to draw a conclusion. When exactly did an employee enter a restricted area? If there was video surveillance, what did it show once the employee was in the area? Were there any logs indicating body worn IoT devices belonging to the employee connecting with the company network during that time and in that location? Did the employee user his or her own access credentials, or those of someone else?

With the proper application of data source correlation, multiple data points may be combined to create a complete, or near complete picture of the events in question. Individually assessed, the data may be meaningless as there would be no overriding context within which to place it. Additionally, where some data may be impossible to obtain, other sources of data may exist that are accessible and corroborative of data already analyzed.

The possibility of data contamination was addressed early on in this section during the discussion of how investigators and examiners on a scene may inadvertently add data when IoT devices sense and record their activity. In addition to this there is another potential complication presented by false sensor data. Rahman, Bishop, and Holt found during their research that motion sensors on doors sometimes reported false positives and negatives. In another test they found that motion sensors may assume, after a person has fallen asleep, they have left the residence. Finally, they found false negatives once again with the use of an activity monitor. The monitor reported that a test subject had not walked four days in seven.[95]

While some sensors could be adjusted to an optimal sensitivity setting the same capability cannot be assumed true for all IoT sensors. Additionally, the examiner and investigator have no control over sensor adjustments prior to responding to an investigation. At best, entities utilizing IoT technology would be well advised to test devices and set appropriate sensitivity parameters prior to deployment thus decreasing the creation of bad data that can clog subsequent investigations. This of course depends on whether those adjustments can be made to a device.

In the absence of adjusting sensitivity, other mitigating steps may include effective data correlation between sources, attention to timeline analysis and detailed investigator notes about the violation and subsequent steps taken during the investigation. Like many aspects of IoT forensics, there is no 'one size fits all' solution.

There are other possible outcomes. Analysis may support a conclusion not anticipated by an investigation. Ethical standards require reputable examiners to report facts and draw conclusions based on analyzed data and no other considerations. Impacts on the investigation, payments for contract service, employment status and other such considerations should have no bearing on the examination findings. Examiners will, or should, include any exculpatory information revealed during the analysis.

Another possible outcome is that no conclusion can be drawn. A common question in a trial setting is 'can you say that my client was sitting at the keyboard when the crime was committed?' That answer is almost always 'no,' for the simple reason that the examiner was not there. In this instance a jury, judge, or boardroom will be confined to evaluating the submitted data and other investigative findings to determine what, if anything, the preponderance of evidence indicates.

The reporting phase

Finally, after data has been extracted, verified, examined and analyzed, it is time for the examiner to put pen to paper, or fingers to keyboard. Considerations here include explaining the findings, understanding who will receive the report, and imparting any actionable information found during analysis.

The examiner, in the face of inconclusive results, should be prepared to offer alternative explanations for findings. This occurs when the analysis results are inconclusive. Two or more possible explanations may exist for an event, and it is advisable for the examiner to address each one in the report. In addition, the analysis should include efforts to prove or disprove each explanation.

Know your audience. A report provided for criminal investigation will probably look very different from one prepared for high level management review. A criminal investigation features reports that are extensive and highly detailed. There are usually multiple copies of reports provided; one for the investigator and one for the prosecutor. A third may be produced for the defense during the process of discovery. Additional copies of evidentiary data may also be required for third party analysis or review. In this case, the report generated by the examiner may be used to determine what steps were taken to obtain the data and what tools were used during analysis. A third-party examiner may test those procedures and findings by replicating the actions taken to confirm accuracy of the conclusions. Proper forensic practices and procedures will enable this step in the process.

It is important to be aware of any existing legal guidelines for releasing evidentiary data in your jurisdiction. Some data may be illegal to release during a process such as discovery and doing so could subject the releasing party to criminal charges. In these instances, accommodations such as providing a review room within the laboratory for defense are acceptable alternatives that usually satisfy discovery requirements. Doing this can help ensure that data does not leave the controlled environment.

High level management will probably want a report that is closer to a high-level view or synopsis than a detailed and extensive report. This audience may simply want a display of what happened, how it happened and suggestions for preventing it in the future. Supplements to this report may include the cost of prevention so that alternative strategies such as risk acceptance or risk transference can be considered.

It is advisable for the examiner to produce a detailed report, and then use that as a framework for the high-level review. When the executive staff decides what actions to take, the detailed report will assist those responsible for implementing the decision.

Actionable information is very often uncovered during the analysis. The report should include this information. The examiner may have found an exploitable vulnerability or system backdoor that needs to be addressed. Planned crimes can also be discovered, and new suspects identified for further investigation. In some instances, the information found may require the examiner to communicate with interested parties due to the freshness of the information, the extent of the vulnerability or the likelihood that any delay in notification may lead to negative or even catastrophic results.

As stated numerous times, the addition of IoT devices in the workplace, at home and on the body, have significantly expanded the attack surface. Actionable information discovered during analysis will likely expand with the new technology.

EXAMPLE RECOMMENDATIONS

Preparing for a future IoT interconnected world may seem daunting. This dynamic technology adds a burden to information security practices that must account for the increased utilization of always on, always communicating sensors.

These efforts will not necessarily be helped by some of the mainstays of information security due to device heterogeneity and resource limitations. Other methods common in traditional network and computer security will help add the hurdles necessary for defense in depth and other strategies.

Though separated into sections by place within the ecosystem, the parts of the IoT, such as devices at the edge and the network, will often benefit from the same mitigation strategies, just as vulnerabilities at one level can have an effect on another and addressed at both. As strategies are discussed for each area, it is useful to consider how each may find usefulness in the other sections.

The recommendations provided below are examples only and in no way a comprehensive guideline for security. There have been volumes written by academics, researchers, professional organizations and governments centered on information security planning, implementation, monitoring and incident response. Likewise, there are numerous organizations that offer classes, training, degrees and certifications geared toward information security.

RISK MITIGATION AND PREVENTATIVE STEPS

This part of the chapter will touch lightly on risk mitigation generally, as the topic already receives more comprehensive coverage in other chapters of this text and through other outlets as just mentioned. However, steps found in research that may assist the process with regards to the IoT specifically will be addressed more fully here, including those practices common to other areas of security.

An effective information security strategy will likely include a global view of all components that require protection. These systems are often designed to monitor in real time those components using a variety of designed and deployed software and hardware solutions. Policy concerning access control, permissions, passwords and training are also built into these strategies to protect the entire network, all of its attached components, and users. However, 71% of IoT security specialists do not monitor IoT devices in real time.[96]

This substantial difference between practices creates a clear risk that vulnerabilities in the IoT will lead to successful attacks on the entire information infrastructure within an organization.

Securing the devices

The particular order in which to start securing a network and its devices is determined by the security professionals tasked with the responsibility. The order of information provided here is not intended to suggest the order in which steps should be taken to implement security.

In 2016 the U.S. Department of Homeland Security released Alert TA16-288A which offered preventative steps for securing IoT devices in light of attacks by botnets such as Mirai.[97]

The advice to prevent malware infections included;

- Change default password to stronger ones. Many organizations have policies in place governing the creation and periodic changing of passwords.
- Apply patches as soon as available. Again, this practice is usually addressed by organizations in policy. Patches are often tested to assure that other vulnerabilities are not created after deployment.
- Purchase devices from reputable companies.
- Device users, both at home and in business, need to familiarize themselves with device capabilities. Also, with reference to the first step, users should determine of the device has a default password or open Wi-Fi connection. If so, passwords should be changed and connections secured.
- Monitor or disable ports that can be used for remote access and malware infiltration.
- Disable Universal Plug and Play in routers if possible.
- Examine the capabilities of home medical devices. If these devices can transmit data or be accessed remotely, there is a potential for malware infection.

Though addressed to the consumer, it is easy to see that these basic steps can be helpful in many of the domains covered earlier in this work. Access and password security are a significant part of the foundation of security planning.

As mentioned earlier, physical threats to the devices themselves may come from natural or environmental threats. Entities that deploy devices should be aware of the potential dangers posed by these threats and plan accordingly. For example, an IoT device designed for use outside, over a large geographic area should have some level of protection built in to mitigate this particular set of dangers.

Any business or government considering the use of such devices should determine the level to which physical protection is built in. How rugged is the device? How will the network of devices react to the loss of one or more during a natural disaster, or other environmental event? Additionally, physical threats from human actors may require the use of access controls and user authentication mechanisms that limit exposure of the device.[98]

Of the threats posed by human actors, it would be accurate to split that population into two groups; those who mean harm and those who unintentionally compromise devices. In the former case access control and authentication are likely the most effective mitigations. The latter problem may be mitigated by those actions, as well as education. Though many employees mean well, mistakes are made. As a result, many companies create policies that reward observation of information security practices and sanction repeated violations of standards. Sanctions can often lead to termination based on the frequency or egregiousness of violations.

Pre-testing is another method that can lead to effective practices both at the device level, as well as the network, fog and cloud level. Pre-testing can include penetration testing to evaluate the effectiveness of deployed security. It can also include testing of updates. This activity will also assist in determining what information should be logged and what information should not be stored in the system.[99] A form of pre-testing would also be useful for forensic examiners and will be discussed later in this section.

Data encryption can be demanding on resources and as a result, difficult to deploy at the device level due to resource constraints, but not impossible. Lightweight protocols have been developed that only use a few cryptographic operations along with smaller key and message sizes. Companies that develop and offer IoT devices are starting to use encryption methods including AES, AES-CBC, AES with SHA-1 and SHA-512, AES with 3DES, SSL/TLS, 0Auth 2.0 authentication, Elliptic curve Diffie-Hellman and RSA.

Verification of digital signatures also challenges resource-constrained devices. Individually validating signatures would have a negative impact on the real-time operations of IoT devices. A solution to this is batch validation of signatures using lightweight digital signature algorithms.[100]

These concerns are not as pronounced with IoT devices that have a constant source of wired power and abundant resources to call on. However, resource constrained or not, many devices may not offer encryption and signature algorithm options. It would serve an entity well to thoroughly examine any proposed IoT solution prior to purchase and extensively test prior to deployment.

Securing the network

Encryption carries over into the network as encrypted communications find their way from the edge to the fog, if present, and finally to the cloud. There are additional options for securing the network that are already found in security practices implemented by entities and individuals.

Internal networks are often protected by firewalls, intrusion detection and prevention systems, and demilitarized zones that separate external, less secure networks from internal networks. Security planning very often includes the creation of diagrams that map out the network and connected physical devices with included security features. As the size of the network and number of nodes increases, so does the complexity. IoT devices, easily overlooked, need to be included in these plans.

Other techniques include reliable routing to combat attacks against routing protocols and role-based authorization that protects against requests by both intruders and malicious nodes.[101]

Other account control techniques such as account timeout, lockout and two factor authentication, are useful as computers are often connected to the same network as IoT devices and can be used to gain access to data and the devices themselves. Account timeout will automatically log a user out of their account after a set period of inactivity. A second technique is to provide an authorized user with a token that communicates using one of the short-range protocols such as NFC. In this instance, when the user leaves without logging out the computer will recognize the absence of the token and log out automatically.

Account lockout occurs after a preset number of login attempt failures. The account may be locked out, requiring administrative reset or an emailed password reset link. An alternative would be the activation of a CAPTCHA. The reader will be familiar with the CAPTCHA as a requirement to type the characters from a picture, or choose pictures of a particular category from a larger set when creating or logging into an account. This practice will prevent brute force attacks against an account and help mitigate DoS attacks trying to overwhelm a system.

Two-factor authentication uses a second means of authentication to allow account user log in. Authentication methods are based on 'what you know,' 'what you have,' and 'what you are.' What you know would be a user name and password. What you have could be a phone that receives a pass code to be entered after logging a password. What you are would be a biometric measure of a unique physical characteristic such as a fingerprint or retina scan.

Two-factor authentication would require any two of the three authentication techniques. If, for instance, an attacker was able to obtain a password to log into an account, they would also need the user's phone to receive and subsequently input a sent code, or an identical fingerprint to scan. Without both factors, access to the account would be denied.[102]

Air-gapping devices in a network can be particularly beneficial for protecting the IoT from those parts of the network that provide access to employees or visitors. An organization may opt to maintaining a separate network for IoT devices that cannot be accessed via a standard network. Other devices would then be needed to provide the update files and software for those devices that have been air-gapped.[103]

Physical security does play a role at this level as well. Restricting access to areas such as server rooms and network router and switching locations has long been a practice of many private and government entities. Secure rooms, physical security controls, motion sensors and video surveillance all play a role in ensuring access control and mitigation should a successful physical intrusion take place.

Other physical security measures such as smoke detectors, fire suppression systems, seismic server racks and storm-resistant structures provide a measure of physical security against mechanical, environmental and natural threats to the network and its connected components. Many organizations will often evaluate these threats and build physical infrastructure in low threat regions to protect vital operations. They may also choose to divide operations into two or more regions that also provide an opportunity for redundancy in the form of hot, warm, or cold disaster recovery sites.

A hot site has the equipment necessary to match the infrastructure of an operational site. It runs concurrently as data is synced between the sites during operation. If there is an event that effects the main site to a degree that it cannot function, the hot site can immediately take over functions to protect business continuity. A cold site contains space and resources such as electricity and environmental control but not the servers, workstations and other infrastructure necessary to immediately take over operations.

A cold site would require significant work and support to ramp up operations for continuity, but the space is already there and waiting. Of course, the cost difference between hot and cold sites is substantial and may be the deciding factor in which to utilize. Warm sites are, as expected, somewhere in the middle. It is a place that may already have some of the infrastructure needed, such as servers and workstations, but it is not synced with the main site and will require software installation, configuration and data migration. It will take some time to resume operations, but not to the extent a cold site would.[104]

Operations continuity planning is important for companies that provide continuous 24/7 service to other companies and individuals. A small local retail outlet would not necessarily benefit from such extensive planning, but a cloud service provider obviously would. Such a provider without these plans in place would be negligent and quickly find that other security measures do not matter when the cloud is down and customers have lost access to, or simply lost, all their data.

Securing the cloud

Assuming that physical protections are in place, including effective disaster recovery plans, there are other measures that can be implemented to secure data in the cloud. Since the cloud is essentially a virtualized environment that runs on physical servers, many of the steps described above also work at this level.

Physical access control measures will help to protect the physical infrastructure upon which the virtual machines work. Along with this, access controls will help ensure that only those authorized are permitted access to the system. This will not, however, guarantee against a trusted insider attack which has perhaps the most potential for harm. This is a threat that will always vex human resources and information security departments.

Event logging can help the security professional track numerous activities such as failed account login attempts, breach attempts and other activities that he or she considers unusual or of potential interest. This preemptive activity may separate a mistake from a malicious act, but in both cases potential harm may be averted.[105]

Data encryption, using lightweight encryption protocols at the device level will provide protection up to and in the cloud. Encrypted data in motion, even if intercepted, will likely maintain confidentiality. Once in the cloud and no longer needed for unencrypted processing, data at rest should also be encrypted.

Securing virtual environments, or the virtual infrastructure, within the cloud includes securing the hypervisor and the virtual machines that it runs. An attack on the hypervisor can lead to exposure of virtual machines, allowing the attacker access across environments. There are numerous proposed solutions that include integrity checking and the use of Trusted Platform Modules to ensure the correct state of the hypervisor both at boot up and while in a running state.

Virtual machines within the hypervisor can benefit from isolation techniques that prevent an attacker within a virtual machine from accessing the host or other virtual machines. Another concern is vulnerability during the migration of a virtual machine. This process can benefit from the use of proxy servers to hide network information and provide encrypted tunneling from one location to the next. Finally, virtual machine introspection, VMI, allows the monitoring of virtual machines either from the hypervisor or another virtual machine. The VMI will monitor running processes and the operating systems within virtual machines to detect malicious behavior.[106]

The overlapping techniques discussed in each section benefit the others by providing protections the make the task of intrusion more challenging. However, each point in the ecosystem represents a vulnerability that must be accounted for and addressed through security planning and implementation. As stated earlier, this short list of mitigation and protection steps is not comprehensive, but rather, a sample.

Cyber forensics is normally included in incident response. It isn't usually considered one of the techniques used, or steps taken to secure a system prior to an incident. The question prompting this work was if forensics can be used as part of planning and design? Additionally, can forensics be used as a preemptive measure, and not just a response?

CYBER FORENSIC PROCESSES

The idea that forensics can be pro-active may be a somewhat new concept, but for the purpose of dealing with the IoT ecosystem, researchers have recognized its utility in helping the examiner prepare for future incidents. IoT device designs are extremely diverse and may require a variety of techniques for data acquisition. If the examiner is not familiar with a particular device prior to examination, data could be missed or even destroyed during the extraction process.

Examining devices prior to an incident for the purpose of determining what tools and methods are necessary to obtain data is a pro-active step that will help the examiner prepare the necessary procedures for any given device.[107] An add-on benefit would be the discovery of device vulnerabilities and shortcomings that may assist security designers in determining the best methods to protect devices.

Engineering examination solutions

Examiners in the private sector, particularly those who work for a company 'in-house' may have the greatest advantage when presented with IoT forensics. A private, or even public, entity that intends to deploy IoT would, in a perfect world, practice due diligence by thoroughly investigating devices that are being considered for deployment. Part of this process may include providing sample devices to the examiner for evaluation. The examiner will then have the opportunity to determine memory capacity, operating system type, data storage locations such as local to the device, or fog/cloud, physical access points, if any exist, JTAG, ISP or chip-off solutions, and so on.

Contract examiners who provide services to a variety of customers may enjoy some of the same opportunities as listed above. As with internal examiners, the entities may choose to provide IoT devices to them. These examiners would be able to build a knowledge base of numerous devices that may benefit them across contracts, providing greater value to their customers.

Law enforcement examiners may have the most difficult path to IoT forensics. Where the examiners in the previous examples may be able to build a reservoir of knowledge, tools, and techniques for IoT devices within their organizations, criminal examiners often do not know what they will encounter prior to receiving evidence for analysis. It may benefit these examiners to research which devices appear to be most commonly located during investigations.

For example, the devices used to control environmental systems at a business may not commonly find their way into a criminal investigation, where activity trackers and motion sensors can and do. In these instances, it may be possible to obtain example devices reflective of those likely to be seized for analysis. However, budget constraints may severely limit, or even prevent this practice.

In any case, communication and information sharing between examiners directly and through professional organizations can be a vital lifeline to success in IoT forensics. Engagement with the community can assist examiners in any context, from public to private, one-person laboratories to teams.

The opportunity to examine sample devices may provide answers to the following questions regarding hardware:

1. What sensors does the device deploy?
2. How is power supplied?
3. Does the device have memory storage?
4. If so, does data remain if the device is powered off, or is it lost?
5. Are there any ports from which data can be extracted?
6. If not, can the device be disassembled for access to the circuit board without damage?
7. Can the circuit board be mapped for connections allowing ISP or JTAG procedures?
8. If there is memory storage, what kind of module is utilized?
9. Is there a chip-reader for the module should no other alternatives exist?

Examining the device may also provide insight into software used and practical examination solutions, answering such questions as:

1. What operating system is deployed in the device?
2. What, if any, is the default password for device access?
3. Will changing the password provide effective data protection?
4. What is the default sensitivity setting, if any, for the device?
5. Can sensitivity be changed?

6. Will testing reveal accurate from erroneous data based on sensitivity settings?
7. Is the data encrypted if stored locally, and what encryption is used?
8. Will the vendor provide the means to open encrypted data?
9. If accessible, what is in the data? For example, communication logs, activation logs and/or sensor data.
10. In what format is the data stored or communicated?
11. Is data in local storage also transferred creating multiple copies of data?
12. Is there a difference between locally stored data and that which is transferred?
13. What communication protocol is used by the device?
14. If data is not stored locally, where is it sent?
15. Is the data transferred to an intermediate device or straight to the cloud?
16. Does the examiner have access to any fog devices utilized by the device?
17. How quickly can data sent to the cloud be acquired either through contract or legal process?
18. Can data transferred to the fog/cloud be captured by an intermediate device designed to collect information for future analysis?

Testing devices prior to deployment or real-world examinations will provide examiners with the opportunity to build necessary capabilities. Many examiners will find that, if they don't already possess the ability, they need to learn how to make custom scripts for the purpose of extracting and decoding data. Many forensic software packages are flexible and allow the use of custom scripts. However, these scripts should be tested to make sure that data extracted is done so accurately and consistently.

Understanding the storage format of data, no matter where it is stored is also of great importance. Testing a device by using it, noting date and time, may provide the insight necessary to correctly interpret output by comparing data points to the real events used to create them. Additionally, false positive or negative output may indicate a deficiency or needed adjustment. Either will be valuable to an organization evaluating device functionality prior to deployment.

These tests may also expose some of the vulnerabilities many IoT devices have, contributing to an organization's decision on device deployment. If the vulnerability is easily patched, the overall utility of the device may result in a decision to move forward. If there are multiple vulnerabilities and the notion of patching them all creates greater expense than benefit, the project may not move forward.

Discovering vulnerabilities during testing may be advantageous for the criminal examiner who can then utilizing them for data extraction. This is nothing new. Vulnerabilities have been exploited by forensic software providers who will include tool functionality aimed at taking advantage of those vulnerabilities. As patches are developed to close vulnerabilities, examiners, solution providers, hackers and hobbyist look for new ones. This game of 'cat and mouse' will likely always be played.

SUMMARY

In conclusion, a pro-active approach to forensics may provide the opportunity to strengthen an organization's security stance. Knowing vulnerabilities ahead of time is tantamount to having a crystal ball that tells us where the problems are and where we need to concentrate our efforts. Considering the lack of security in the IoT, decision makers may conclude that it is vital to institute a robust evaluation regimen that includes forensics somewhere at the beginning rather than just the end.

Likewise, the criminal examiner would benefit from building capability before an exigent case leads to rushed on the job learning, or worse, a data destroying mistake. Budget constraints may make government entities hesitant to purchase devices strictly for experimentation, but those considerations should be weighed against statutory responsibility for public safety. Technology and technological development will continue to accelerate unabated. Public safety can greatly benefit from the use of technology to detect and solve crimes, as criminals benefit from its use to commit them. There is no return to a non-technological world.

The following are a series of questions which the reader and investigator may wish to consider when evaluating the interrelationships between IoT and the cyber forensics.

QUESTIONS TO CONSIDER

Planning questions

1. Does your organization plan on utilizing IoT devices?
2. Has your organization determined that the benefits of the IoT outweigh the risks presented by a larger attack surface?
3. What are the benefits expected from the deployment?
4. What are the risks?
5. Do you have a plan to vet devices before implementation?
6. Will vetting include forensic analysis?
7. Does your organization already utilize the IoT?
8. If so, were the devices included in your security framework?
9. What is your BYOD policy?
10. Does your BYOD policy include IoT devices owned and utilized by employees?
11. To what degree are you willing to control or restrict the use of personally owned devices?
12. Will IoT devices communicate with cloud services belonging to your organization, or another organization under contract?
13. What is included in your service level agreement with cloud service vendors?
14. Does the SLA address data retention policies relative to that created by IoT devices?
15. What security assurances and practices are required by the SLA?
16. Is your organization capable of securely expanding its technology utilization?
17. Is your network capable of expanding to meet demand?
18. Will your information security professionals be able to protect the additional resources as they are added to the network over time?

Security framework questions

1. Did the security plan include IoT devices at inception, or was it expanded to account for those devices after the fact?
2. If the original plan did not include these devices, can it be modified or will a new plan need to be made and implemented?
3. Do IoT devices access your secure network?
4. Are all devices on the network known?
5. Are devices actively monitored?
6. What are the known vulnerabilities of deployed devices?

7. If unknown, what steps are being taken to identify vulnerabilities?
8. What parts of the organization are most likely to be affected by IoT vulnerabilities, and how does this effect the overall security plan?
9. Are devices patched on a regular basis?
10. Are patches and other updates tested for functionality and security prior to deployment?
11. Does penetration testing include deployed IoT devices?
12. Do you utilize strong password requirements?
13. What authentication methods are utilized?
14. What physical security measures are in place?
15. What network security measures are in place?
16. Is there sufficient protection between your internal network and any external networks?
17. Are your IoT devices on your internal network?
18. Are devices brought in by customers isolated from the internal network?
19. Are devices brought in by employees isolated from the internal network?
20. If employee devices are allowed on the internal network, what specific steps are taken to prevent externally captured malware from migrating into your systems?
21. Is a culture of security encouraged by leaders within your organization?
22. Are IoT devices recognized within that culture?
23. Is security training offered to all company employees on an annual basis?
24. Does this training include personally owned devices?
25. Do your security professionals receive training in new technology?
26. Does your organization utilize 'in house' forensic services or contract for them?
27. Do you provide the resources necessary for your examiners to build knowledge and capability with the specific technologies used in your organization?
28. Do your examiners receive ongoing training?
29. Do contract examiners receive these resources from their employer?
30. Does your organization apprise contract security and forensic vendors of technology updates and utilization?

Legal and contract considerations

1. What are the regulatory requirements for your industry?
2. Is your security plan focused on preventing violations of those regulations?
3. What is your liability for the unintentional release of protected data violating those regulatory requirements?
4. Does your organization operate internationally?

   a. Are you aware of regulatory requirements within the nations you operate?

5. What resources do you have to learn and maintain an understanding of those regulations?
6. Do the IoT devices you utilize, or plan on using, gather and transmit any of the regulated data?
7. What amount of personally identifiable information is transmitted by your IoT devices?
8. What is your liability locally, nationally and possibly internationally for the unintended release of PII?

9. What is your organizational policy on self-reporting?
10. What legal requirements govern reporting within your jurisdiction, or in any jurisdiction you operate?
11. What is your liability for breaches originating from employee personally owned IoT devices allowed on your network?
12. What is your employee's liability?
13. What are the contract requirements for data retrieval from external vendors such as cloud service providers?

Law enforcement examiner questions

1. What legal processes are needed to retrieve data from cloud service providers?
2. Who in your organization is responsible for providing the legal process necessary to obtain data from external sources?
3. What legal authority do you have to conduct an examination on a device?
   a. Is the authority from a search warrant, consent or implied consent?
   b. What is the scope of the search warrant?
   c. Were any limitations placed on the examination by the individual providing consent?
   d. Did the individual providing consent have the authority to do so?
   e. What other information do you require from investigators prior to an examination?
   f. Do you receive a preliminary report or synopsis from investigators?

Non-law enforcement examiner questions

1. What authority is provided by your organization, or contract, that allows an examination to proceed?
2. Is the device to be examined property of the organization?
3. Is the device to be examined property of an employee or visitor?
4. What policies are in place that address any potential conflicts of interest arising from your findings?

General examiner questions

1. Do you maintain a consistent practice of tool and technique validation?
2. Do you have the tools and experience necessary to conduct an examination of a given device?
3. If not, what steps can you take to gain the resources necessary for the examination?
4. In the absence of these resources, will you be able to refuse the examination?
5. Do you have all the investigative information necessary to conduct a thorough examination?
6. Does your organization provide the budget needed to gain and maintain proficiency with new technology?
7. Do you have access to IoT devices for research prior to their deployment?
8. Will the methods you used to examine devices be repeatable by other examiners?
9. Will your examinations, either test or incident related, be used by your organization to improve its security practices?
10. Are there resources, such as professional organizations, that you have access to?

11. Are you bound by any ethical code of conduct stemming from certification or professional affiliation?
12. Does your organization expect you to be both an investigator and examiner?
13. If so, are you expected to provide both functions on the same incident?
14. Is your organization aware of the potential conflicts arising from this practice?
15. Are your reports written to assist the lay person in understanding your processes and findings?
16. Will you be able to adequately address IoT functionality and examination techniques within your report?
17. Will the methods you used to examine devices be repeatable by other examiners?

ACRONYMS

BLE	Bluetooth Low Energy
BYOD	Bring Your Own Device
CPU	Central Processing Unit
DHS	Department of Homeland Security
DMZ	Demilitarized Zone
DDoS	Distributed Denial of Service
DoS	Denial of Service
HVAC	Heating Ventilation and Air Conditioning
IACIS	International Association of Computer Investigative Specialists
ICS	Industrial Control Systems
IDS	Intrusion Detection System
IoT	Internet of Things
IPS	Intrusion Prevention System
ISFCE	International Society of Forensic Computer Examiners
ISP	In-System Programming
ISP	Internet Service Provider
JTAG	Joint Test Action Group
LPWAN	Low-Power Wide-Area Network
LoWPAN	Low-Power Wireless Personal Area Network
MDM	Mobile Device management
NIST	National Institute of Standards and Technology
NFC	Near-Field Communication
PC	Personal Computer
PII	Personally Identifiable Information
PLC	Programmable Logic Controllers
RF	Radio Frequency
RFID	Radio Frequency Identification
RAM	Random Access Memory
RTOS	Real-Time Operating Systems
ROM	Read Only Memory
SCADA	Supervisory Control and Data Acquisition
SLA	Service Level Agreement
TAP	Test Access Port or Test Access Point
VM	Virtual Machine
VPN	Virtual Private Network

NOTES

1 The inclusion of IoT product, forensic tool provider, training provider, and product names is not an endorsement of those providers or products by the author and have been included herein as reference, research, or as an example, of which there may be many within the specified marketplace.

2 Federal Trade Commission. 2015. "Internet of Things: Privacy & Security in a Connected World." Ftc.Gov. https://www.ftc.gov/system/files/documents/reports/federal-trade-commission-staff-report-november-2013-workshop-entitled-internet-things-privacy/150127iotrpt.pdf, retrieved March 19, 2020

3 Mocrii, Dragos, Yuxiang Chen, and Petr Musilek. 2018. "IoT-Based Smart Homes: A Review of System Architecture, Software, Communications, Privacy and Security." Internet of Things 1–2 (September): 81–98. https://doi.org/10.1016/j.iot.2018.08.009, retrieved March 22, 2020. Used with permission.

4 Radoglou Grammatikis, Panagiotis I., Panagiotis G. Sarigiannidis, and Ioannis D. Moscholios. 2019. "Securing the Internet of Things: Challenges, Threats and Solutions." Internet of Things 5 (March): 41–70. https://doi.org/10.1016/j.iot.2018.11.003, retrieved March 22, 2020. Used with permission.

5 Ashton, Kevin. 2009. "That 'Internet of Things' Thing | RFID JOURNAL." Www.Rfidjournal. Com. June 22, 2009. https://www.rfidjournal.com/that-internet-of-things-thing, retrieved April 10, 2020.

6 Teicher, Jordan. 2018. "The Little-Known Story of the First IoT Device." IBM Industries. February 7, 2018. https://www.ibm.com/blogs/industries/little-known-story-first-iot-device/, retrieved June 18, 2020.

7 "The Internet Toaster." n.d. BroadbandNow. https://broadbandnow.com/internet/i/ia_myths_toast.htm, retrieved June 22, 2020.

8 Mocrii, Dragos, Yuxiang Chen, and Petr Musilek. 2018. "IoT-Based Smart Homes: A Review of System Architecture, Software, Communications, Privacy and Security." Internet of Things 1–2 (September): 81–98. https://doi.org/10.1016/j.iot.2018.08.009, retrieved March 22, 2020

9 Holdowsky, Jonathan, Monika Mahto, Michael Raynor, and Mark Cotteleer. 2015. "Inside the Internet of Things (IoT): A Primer on the Technologies Building the IoT." Deloitte University Press. https://www2.deloitte.com/content/dam/Deloitte/pe/Documents/technology/Inside%20 The%20Internet%20Of%20Things.pdf, retrieved March 16, 2020. Email correspondence with Mark Cotteleer, used with permission. Table recreated by author.

10 Mountrouidou, Xenia, Blaine Billings, and Luis Mejia-Ricart. 2019. "Not Just Another Internet of Things Taxonomy: A Method for Validation of Taxonomies." Internet of Things 6 (June): 100049. https://doi.org/10.1016/j.iot.2019.03.003, retrieved March 8, 2020.

11 Mocrii, Dragos, Yuxiang Chen, and Petr Musilek. 2018. "IoT-Based Smart Homes: A Review of System Architecture, Software, Communications, Privacy and Security." Internet of Things 1–2 (September): 81–98. https://doi.org/10.1016/j.iot.2018.08.009, retrieved March 22, 2020.

12 Aly, Mohab, Foutse Khomh, Mohamed Haoues, Alejandro Quintero, and Soumaya Yacout. 2019. "Enforcing Security in Internet of Things Frameworks: A Systematic Literature Review." Internet of Things 6 (June): 100050. https://doi.org/10.1016/j.iot.2019.100050, retrieved April 22, 2020.

13 Meffert, Christopher, Devon Clark, Ibrahim Baggili, and Frank Breitinger. 2017. "Forensic State Acquisition from Internet of Things (FSAIoT)." Proceedings of the 12th International Conference on Availability, Reliability and Security, August. https://doi.org/10.1145/3098954.3104053, retrieved March 25, 2020.

14 Milinković, Aleksandar, Stevan Milinković, and Ljubomir Lazić. 2015. "Choosing the Right RTOS for IoT Platform." INFOTEH-JAHORINA 14 (March): 504–9. https://infoteh.etf.ues. rs.ba/zbornik/2015/radovi/RSS-2/RSS-2-2.pdf, retrieved March 15, 2020.

15 Vikas G, and Arvindpdmn. 2017. "IoT Operating Systems." Devopedia. March 9, 2017. https:// devopedia.org/iot-operating-systems, retrieved May 20, 2020.

16 Mocrii, Dragos, Yuxiang Chen, and Petr Musilek. 2018. "IoT-Based Smart Homes: A Review of System Architecture, Software, Communications, Privacy and Security." Internet of Things 1–2 (September): 81–98. https://doi.org/10.1016/j.iot.2018.08.009, retrieved March 22, 2020. Used with permission. Table redrawn by author.

17 MacDermott, Aine, Thar Baker, and Qi Shi. 2018. "IoT Forensics: Challenges for the IoA Era." 2018 9th IFIP International Conference on New Technologies, Mobility and Security (NTMS), February. https://doi.org/10.1109/ntms.2018.8328748, retrieved March 23, 2020.

18 Al-Sarawi, Shadi, Mohammed Anbar, Kamal Alieyan, and Mahmood Alzubaidi. 2017. "Internet of Things (IoT) Communication Protocols: Review." IEEE Xplore. May 1, 2017. https://doi.org/10.1109/ICITECH.2017.8079928, retrieved March 12, 2020.

19 Salman, Tara, and Raj Jain. 2017. "A Survey of Protocols and Standards for Internet of Things." Advanced Computing and Communications 1 (1). https://doi.org/10.34048/2017.1.f3, retrieved March 15, 2020.

20 "Ultimate Guide to Internet of Things (IoT) Connectivity." n.d. Argenox. Accessed March 19, 2020. https://www.argenox.com/library/iot/ultimate-guide-iot-connectivity/, retrieved March 15, 2020.

21 Al-Sarawi, Shadi, Mohammed Anbar, Kamal Alieyan, and Mahmood Alzubaidi. 2017. "Internet of Things (IoT) Communication Protocols: Review." IEEE Xplore. May 1, 2017. https://doi.org/10.1109/ICITECH.2017.8079928, retrieved March 12, 2020.

22 Mocrii, Dragos, Yuxiang Chen, and Petr Musilek. 2018. "IoT-Based Smart Homes: A Review of System Architecture, Software, Communications, Privacy and Security." Internet of Things 1–2 (September): 81–98. https://doi.org/10.1016/j.iot.2018.08.009, retrieved March 22, 2020.

23 Edwards W.K., Grinter R.E. 2001. At Home with Ubiquitous Computing: Seven Challenges. In: Abowd G.D., Brumitt B., Shafer S. (eds) Ubicomp 2001: Ubiquitous Computing. UbiComp 2001. Lecture Notes in Computer Science, vol 2201. Springer, Berlin, Heidelberg. https://doi.org/10.1007/3-540-45427-6_22, retrieved May 30, 2020.

24 Ibid.

25 Evans, D., 2011. "Cisco Internet Business Solutions Group (IBSG) The Internet of Things How the Next Evolution of the Internet Is Changing Everything." https://www.cisco.com/c/dam/en_us/about/ac79/docs/innov/IoT_IBSG_0411FINAL.pdf, retrieved June 2, 2020.

26 Mell, Peter, and Timothy Grance. 2011. "The NIST Definition of Cloud Computing," September. https://doi.org/10.6028/nist.sp.800-145, retrieved July 3, 2020

27 Gubbi, Jayavardhana, Rajkumar Buyya, Slaven Marusic, and Marimuthu Palaniswami. 2013. "Internet of Things (IoT): A Vision, Architectural Elements, and Future Directions." Future Generation Computer Systems 29 (7): 1645–60. https://doi.org/10.1016/j.future.2013.01.010, retrieved March 23, 2020.

28 Hoefer, C. N., and G. Karagiannis. 2010. "Taxonomy of Cloud Computing Services." 2010 IEEE Globecom Workshops, December. https://doi.org/10.1109/glocomw.2010.5700157, retrieved April 10, 2020

29 Risteska Stojkoska, Biljana L., and Kire V. Trivodaliev. 2017. "A Review of Internet of Things for Smart Home: Challenges and Solutions." Journal of Cleaner Production 140 (January): 1454–64. https://doi.org/10.1016/j.jclepro.2016.10.006, retrieved July 6, 2020.

30 "Fog Computing and the Internet of Things: Extend the Cloud to Where the Things Are." 2015. https://www.cisco.com/c/dam/en_us/solutions/trends/iot/docs/computing-overview.pdf, retrieved May 15, 2020.

31 Mountrouidou, Xenia, Blaine Billings, and Luis Mejia-Ricart. 2019. "Not Just Another Internet of Things Taxonomy: A Method for Validation of Taxonomies." Internet of Things 6 (June): 100049. https://doi.org/10.1016/j.iot.2019.03.003, retrieved April 9, 2020.

32 Yaqoob, Ibrar, Ibrahim Abaker Targio Hashem, Arif Ahmed, S.M. Ahsan Kazmi, and Choong Seon Hong. 2019. "Internet of Things Forensics: Recent Advances, Taxonomy, Requirements, and Open Challenges." Future Generation Computer Systems 92 (March): 265–75. https://doi.org/10.1016/j.future.2018.09.058, retrieved April 11, 2020.

33 Casola, Valentina, Alessandra De Benedictis, Massimiliano Rak, and Umberto Villano. 2019. "Toward the Automation of Threat Modeling and Risk Assessment in IoT Systems." Internet of Things 7 (September): 100056. https://doi.org/10.1016/j.iot.2019.100056, retrieved March 17, 2020.

34 Lucero, Sam. 2016. "IoT Platforms: Enabling the Internet of Things." IHS Technology. March 2016. http://cdn.ihs.com/www/pdf/enabling-IOT.pdf, retrieved April 17, 2020.

35 Levy, Heather. 2015. "Gartner Predicts Our Digital Future." Gartner. October 6, 2015. https://www.gartner.com/smarterwithgartner/gartner-predicts-our-digital-future, retrieved July 8, 2020.

36 Greenough, John. 2015. "The 'connected Car' Is Creating a Massive New Business Opportunity for Auto, Tech, and Telecom Companies." Business Insider. February 19, 2015. https://www.businessinsider.com/connected-car-statistics-manufacturers-2015-2?IR=T, retrieved July 8, 2020.

37 "Smart Clothing and Body Sensors." 2017. London: Informa PLC. https://tractica.omdia.com/research/smart-clothing-and-body-sensors/, retrieved July 9, 2020.

38 "Internet of Things Market Size, Share and Global Market Forecast to 2022." MarketsandMarkets. June 2017. https://www.marketsandmarkets.com/Market-Reports/internet-of-things-market-573.html, retrieved July 5, 2020.

39 Flittner, Kelsey. 2015. "Surprised? Turns out, Consumers Don't Trust IoT Security." Auth0 - Blog. Auth0. November 6, 2015. https://auth0.com/blog/surprised-turns-out-consumers-dont-trust-iot-security/, retrieved July 12, 2020.

40 "HP News - HP Study Reveals 70 Percent of Internet of Things Devices Vulnerable to Attack." 2014. Hp.Com. July 29, 2014. https://www8.hp.com/us/en/hp-news/press-release.html?id=1744676, retrieved July 13, 2020.

41 Kolias, Constantinos, Georgios Kambourakis, Angelos Stavrou, and Jeffrey Voas. 2017. "DDoS in the IoT: Mirai and Other Botnets." Computer 50 (7): 80–84. https://doi.org/10.1109/mc.2017.201, retrieved July 10, 2020.

42 Angrishi, Kishore. 2017. "Turning Internet of Things (IoT) into Internet of Vulnerabilities (IoV): IoT Botnets." ArXiv:1702.03681 [Cs], February. https://arxiv.org/abs/1702.03681v1, retrieved July 12, 2020.

43 Radoglou Grammatikis, Panagiotis I., Panagiotis G. Sarigiannidis, and Ioannis D. Moscholios. 2019. "Securing the Internet of Things: Challenges, Threats and Solutions." Internet of Things 5 (March): 41–70. https://doi.org/10.1016/j.iot.2018.11.003, retrieved March 22, 2020.

44 HaddadPajouh, Hamed, and Reza Parizi. 2019. "A Survey on Internet of Things Security: Requirements, Challenges, and Solutions." Internet of Things, November. https://doi.org/10.1016/j.iot.2019.100129, retrieved March 13, 2020.

45 Aly, Mohab, Foutse Khomh, Mohamed Haoues, Alejandro Quintero, and Soumaya Yacout. 2019. "Enforcing Security in Internet of Things Frameworks: A Systematic Literature Review." Internet of Things 6 (June): 100050. https://doi.org/10.1016/j.iot.2019.100050, retrieved April 22, 2020.

46 Ibid.

47 Ibid.

48 HaddadPajouh, Hamed, and Reza Parizi. 2019. "A Survey on Internet of Things Security: Requirements, Challenges, and Solutions." Internet of Things, November. https://doi.org/10.1016/j.iot.2019.100129, retrieved March 13, 2020.

49 Radoglou Grammatikis, Panagiotis I., Panagiotis G. Sarigiannidis, and Ioannis D. Moscholios. 2019. "Securing the Internet of Things: Challenges, Threats and Solutions." Internet of Things 5 (March): 41–70. https://doi.org/10.1016/j.iot.2018.11.003, retrieved March 22, 2020.

50 Ibid.

51 Aly, Mohab, Foutse Khomh, Mohamed Haoues, Alejandro Quintero, and Soumaya Yacout. 2019. "Enforcing Security in Internet of Things Frameworks: A Systematic Literature Review." Internet of Things 6 (June): 100050. https://doi.org/10.1016/j.iot.2019.100050, retrieved April 22, 2020.

52 HaddadPajouh, Hamed, and Reza Parizi. 2019. "A Survey on Internet of Things Security: Requirements, Challenges, and Solutions." Internet of Things, November. https://doi.org/10.1016/j.iot.2019.100129, retrieved March 13, 2020.

53 "Telltale Heart: Pacemaker Data Leads to Arson, Fraud Charges." 2017. Fox News. February 8, 2017. https://www.foxnews.com/us/telltale-heart-pacemaker-data-leads-to-arson-fraud-charges, retrieved July 9, 2020.

54 Pack, Lauren. 2020. "Middletown Arson Case Involving Pacemaker Data Delayed by Coronavirus Precautions." Journal-News.Com. Journal-News. April 23, 2020. https://www.journal-news.com/news/crime--law/middletown-arson-case-involving-pacemaker-data-delayed-coronavirus-precautions/tCWbseOfqtf5AtelMlESCO/, retrieved July 9, 2020.

55 Hauser, Christine. 2017. "In Connecticut Murder Case, a Fitbit Is a Silent Witness (Published 2017)." The New York Times, April 27, 2017, sec. New York. https://nyti.ms/2qbhr2M, retrieved July 10, 2020.

56 Hay, Andrea, and Brittany Schmidt. 2018. "Douglass Detrie Testifies in George Burch Murder Trial." WBAY. February 21, 2018. https://www.wbay.com/content/news/WATCH-LIVE-Day-three-of-testimony-in-George-Burch-trial--474710733.html, retrieved July 25, 2020.

57 Hay, Andrea, and Brittany Schmidt. 2018. "Jury Finds George Burch Guilty of Murdering Nicole VanderHeyden." WBAY. March 1, 2018. https://www.wbay.com/content/news/WATCH-LIVE-Final-witnesses-and-closing-arguments-at-Burch-murder-trial-475535703.html, retrieved July 27, 2020.

58 Rizvi, Syed, RJ Orr, Austin Cox, Prithvee Ashokkumar, and Mohammad R. Rizvi. 2020. "Identifying the Attack Surface for IoT Network." Internet of Things 9 (March). https://doi.org/10.1016/j.iot.2020.100162, retrieved April 2, 2020.

59 Ibid.

60 Gubbi, Jayavardhana, Rajkumar Buyya, Slaven Marusic, and Marimuthu Palaniswami. 2013. "Internet of Things (IoT): A Vision, Architectural Elements, and Future Directions." Future Generation Computer Systems 29 (7): 1645–60. https://doi.org/10.1016/j.future.2013.01.010, retrieved March 23, 2020.

61 Ibid.

62 Perwej, Yusuf, Kashiful Haq, Firoj Parwej, and Mumdouh M. 2019. "The Internet of Things (IoT) and Its Application Domains." International Journal of Computer Applications 182 (49): 36–49. https://doi.org/10.5120/ijca2019918763, retrieved August 18, 2020.

63 Ibid.

64 Ibid.

65 Kinsa Data Team. 2020. "As Missouri Schools Consider Reopening, Kinsa Data Signals Trouble Ahead." Kinsa Inc. August 13, 2020. https://www.kinsahealth.co/blog/health-weather/as-missouri-schools-consider-reopening-kinsa-data-signals-trouble-ahead/, retrieved August 21, 2020.

66 Perwej, Yusuf, Kashiful Haq, Firoj Parwej, and Mumdouh M. 2019. "The Internet of Things (IoT) and Its Application Domains." International Journal of Computer Applications 182 (49): 36–49. https://doi.org/10.5120/ijca2019918763, retrieved August 18, 2020.

67 Rizvi, Syed, RJ Orr, Austin Cox, Prithvee Ashokkumar, and Mohammad R. Rizvi. 2020. "Identifying the Attack Surface for IoT Network." Internet of Things 9 (March). https://doi.org/10.1016/j.iot.2020.100162.\, retrieved April 2, 2020.

68 Egger-Danner, C, B Fuerst-Waltl, P Klimek, O Saukh, and T Wittek. 2019. "Internet of Cows -Opportunities and Challenges for Improving Health, Welfare and Efficiency in Dairying." ICAR Technical Series, no. 24. https://www.icar.org/Documents/technical_series/ICAR-Technical-Series-no-24-Prague/Egger-Danner.pdf, retrieved August 14, 2020.

69 Vyas, Shivank, Vipin Shukla, and Nishant Doshi. 2019. "FMD and Mastitis Disease Detection in Cows Using Internet of Things (IOT)." Procedia Computer Science 160: 728–33. https://doi.org/10.1016/j.procs.2019.11.019, retrieved August 14, 2020

70 "Fujitsu to Rollout Global Sales of 'GYUHO' SaaS." 2013. Fujitsu. October 15, 2013. https://www.fujitsu.com/global/about/resources/news/press-releases/2013/1015-01.html, retrieved August 22, 2020.

71 Rizvi, Syed, RJ Orr, Austin Cox, Prithvee Ashokkumar, and Mohammad R. Rizvi. 2020. "Identifying the Attack Surface for IoT Network." Internet of Things 9 (March). https://doi.org/10.1016/j.iot.2020.100162.\, retrieved April 2, 2020.

72 Gubbi, Jayavardhana, Rajkumar Buyya, Slaven Marusic, and Marimuthu Palaniswami. 2013. "Internet of Things (IoT): A Vision, Architectural Elements, and Future Directions." Future Generation Computer Systems 29 (7): 1645–60. https://doi.org/10.1016/j.future.2013.01.010, retrieved March 23, 2020.

73 Perwej, Yusuf, Kashiful Haq, Firoj Parwej, and Mumdouh M. 2019. "The Internet of Things (IoT) and Its Application Domains." International Journal of Computer Applications 182 (49): 36–49. https://doi.org/10.5120/ijca2019918763, retrieved August 18, 2020.

74 Sundby, Alex. 2019. "Hacker Spoke to Baby, Hurled Obscenities at Couple Using Nest Camera, Dad Says." Www.Cbsnews.Com. January 31, 2019. https://www.cbsnews.com/news/nest-camera-hacked-hacker-spoke-to-baby-hurled-obscenities-at-couple-using-nest-camera-dad-says/, retrieved August 3, 2020.

75 Holley, Jessica. 2019. "Family Says Hackers Accessed a Ring Camera in Their 8-Year-Old Daughter's Room." WMC Action News 5. December 10, 2019. https://www.wmcactionnews5.com/2019/12/11/family-says-hackers-accessed-ring-camera-their-year-old-daughters-room/, retrieved August 3, 2020.

76 Min, Sarah. 2019. "Your Smart TV Might Be Spying on You, FBI Warns." CBS News. December 3, 2019. https://www.cbsnews.com/news/smart-tv-spying-fbi-says-the-device-may-be-spying-on-you-today-2019-12-03/, retrieved August 3, 2020.

77 "How Your Smart Fridge Might Be Mining Bitcoin for Criminals." 2018. CBS News. June 29, 2018. https://www.cbsnews.com/news/how-your-smart-fridge-might-be-mining-bitcoin-for-criminals/, retrieved August 3, 2020.

78 Greenberg, Andy. 2015. "Hackers Remotely Kill a Jeep on the Highway—With Me in It." WIRED. July 21, 2015. https://www.wired.com/2015/07/hackers-remotely-kill-jeep-highway/, retrieved July 8, 2020.

79 Vignesh, U., and S. Asha. 2015. "Modifying Security Policies Towards BYOD." Procedia Computer Science 50: 511–16. https://doi.org/10.1016/j.procs.2015.04.023, retrieved October 3, 2020.

80 Ibid.

81 Zeadally, Sherali, Ashok Kumar Das, and Nicolas Sklavos. 2019. "Cryptographic Technologies and Protocol Standards for Internet of Things." Internet of Things, June. https://doi.org/10.1016/j.iot.2019.100075, retrieved March 27, 2020.

82 Kent, Karen, Suzanne Chevalier, Tim Grance, and Hung Dang. 2006. "Guide to Integrating Forensic Techniques into Incident Response," August. https://doi.org/10.6028/nist.sp.800-86, retrieved June 28, 2020.

83 Figure 1, Steps of The Forensic Process, original artwork by author. Material for Figure 1 is based upon NIST Special Publication 800-86, Guide to Integrating Forensic Techniques into Incident Response, which is available at https://doi.org/10.6028/nist.sp.800-86, retrieved June 28, 2020.

84 MacDermott, Aine, Thar Baker, and Qi Shi. 2018. "IoT Forensics: Challenges for the IoA Era." 2018 9th IFIP International Conference on New Technologies, Mobility and Security (NTMS), February. https://doi.org/10.1109/ntms.2018.8328748, retrieved March 23, 2020.

85 Servida, Francesco, and Eoghan Casey. 2019. "IoT Forensic Challenges and Opportunities for Digital Traces." Digital Investigation 28 (April): S22–29. https://doi.org/10.1016/j.diin.2019.01.012, retrieved March 17, 2020.

86 Usama Salama. 2017. "Smart Forensics for the Internet of Things (IoT)." Security Intelligence. March 22, 2017. https://securityintelligence.com/smart-forensics-for-the-internet-of-things-iot/.

87 Servida, Francesco, and Eoghan Casey. 2019. "IoT Forensic Challenges and Opportunities for Digital Traces." Digital Investigation 28 (April): S22–29. https://doi.org/10.1016/j.diin.2019.01.012, retrieved March 17, 2020.

88 Figure 2, original artwork created by author.

89 Meffert, Christopher, Devon Clark, Ibrahim Baggili, and Frank Breitinger. 2017. "Forensic State Acquisition from Internet of Things (FSAIoT)." Proceedings of the 12th International Conference on Availability, Reliability and Security, August. https://doi.org/10.1145/3098954.3104053, retrieved March 25, 2020.

90 Servida, Francesco, and Eoghan Casey. 2019. "IoT Forensic Challenges and Opportunities for Digital Traces." Digital Investigation 28 (April): S22–29. https://doi.org/10.1016/j.diin.2019.01.012, retrieved March 17, 2020.

91 Zia, Tanveer, Peng Liu, and Weili Han. 2017. "Application-Specific Digital Forensics Investigative Model in Internet of Things (IoT)." Proceedings of the 12th International Conference on Availability, Reliability and Security, August. https://doi.org/10.1145/3098954.3104052, retrieved March 19, 2020.

92 Willassen, Svein. 2006. "Forensic Analysis of Mobile Phone Internal Memory." In Advances in Digital Forensics, edited by M. Pollitt and S. Shenoi. Springer, Boston, MA: IFIP — The International Federation for Information Processing, Vol 194. https://doi.org/10.1007/0-387-31163-7_16, retrieved October 16, 2020.

93 Silveira, Claudinei Morin da, Rafael T. de Sousa Jr, Robson de Oliveira Albuquerque, Georges D. Amvame Nze, Gildásio Antonio de Oliveira Júnior, Ana Lucila Sandoval Orozco, and Luis Javier García Villalba. 2020. "Methodology for Forensics Data Reconstruction on Mobile Devices with Android Operating System Applying In-System Programming and Combination Firmware." Applied Sciences 10 (12): 4231. https://doi.org/10.3390/app10124231, retrieved October 18, 2020.

94 Kent, Karen, Suzanne Chevalier, Tim Grance, and Hung Dang. 2006. "Guide to Integrating Forensic Techniques into Incident Response," August. https://doi.org/10.6028/nist.sp.800-86, retrieved June 28, 2020.

95 Rahman, K M Sabidur, Matt Bishop, and Albert Holt. 2017. "Internet of Things Mobility Forensics." Edited by Scar de Courcier. Forensic Focus. https://articles.forensicfocus.com/2017/05/17/internet-of-things-mobility-forensics, retrieved March 23, 2020.

96 Casola, Valentina, Alessandra De Benedictis, Antonio Riccio, Diego Rivera, Wissam Mallouli, and Edgardo Montes de Oca. 2019. "A Security Monitoring System for Internet of Things." Internet of Things 7 (September): 100080. https://doi.org/10.1016/j.iot.2019.100080, retrieved March 26, 2020.

97 "Heightened DDoS Threat Posed by Mirai and Other Botnets." 2016. Us-Cert.Cisa.Gov. Cybersecurity and Infrastructure Security Agency. October 14, 2016. https://us-cert.cisa.gov/ncas/alerts/TA16-288A, retrieved September 20, 2020.

98 Radoglou Grammatikis, Panagiotis I., Panagiotis G. Sarigiannidis, and Ioannis D. Moscholios. 2019. "Securing the Internet of Things: Challenges, Threats and Solutions." Internet of Things 5 (March): 41–70. https://doi.org/10.1016/j.iot.2018.11.003, retrieved March 22, 2020.

99 Aly, Mohab, Foutse Khomh, Mohamed Haoues, Alejandro Quintero, and Soumaya Yacout. 2019. "Enforcing Security in Internet of Things Frameworks: A Systematic Literature Review." Internet of Things 6 (June): 100050. https://doi.org/10.1016/j.iot.2019.100050, retrieved April 22, 2020.

100 Zeadally, Sherali, Ashok Kumar Das, and Nicolas Sklavos. 2019. "Cryptographic Technologies and Protocol Standards for Internet of Things." Internet of Things, June. https://doi.org/10.1016/j.iot.2019.100075, retrieved March 27, 2020.

101 Aly, Mohab, Foutse Khomh, Mohamed Haoues, Alejandro Quintero, and Soumaya Yacout. 2019. "Enforcing Security in Internet of Things Frameworks: A Systematic Literature Review." Internet of Things 6 (June): 100050. https://doi.org/10.1016/j.iot.2019.100050, retrieved April 22, 2020.

102 Rizvi, Syed, RJ Orr, Austin Cox, Prithvee Ashokkumar, and Mohammad R. Rizvi. 2020. "Identifying the Attack Surface for IoT Network." Internet of Things 9 (March). https://doi.org/10.1016/j.iot.2020.100162, retrieved April 2, 2020.

103 Ibid.

104 Spacey, John. 2016. "Cold Site vs Hot Site." Simplicable. November 16, 2016. https://simplicable.com/new/cold-site-vs-hot-site, retrieved September 20, 2020.

105 Rizvi, Syed, RJ Orr, Austin Cox, Prithvee Ashokkumar, and Mohammad R. Rizvi. 2020. "Identifying the Attack Surface for IoT Network." Internet of Things 9 (March). https://doi.org/10.1016/j.iot.2020.100162, retrieved April 2, 2020.

106 Aly, Mohab, Foutse Khomh, Mohamed Haoues, Alejandro Quintero, and Soumaya Yacout. 2019. "Enforcing Security in Internet of Things Frameworks: A Systematic Literature Review." Internet of Things 6 (June): 100050. https://doi.org/10.1016/j.iot.2019.100050, retrieved April 22, 2020.

107 Servida, Francesco, and Eoghan Casey. 2019. "IoT Forensic Challenges and Opportunities for Digital Traces." Digital Investigation 28 (April): S22–29. https://doi.org/10.1016/j.diin.2019.01.012, retrieved March 17, 2020.

Chapter 3

Cyber forensics

Examining commercial Unmanned Aircraft Systems (UASs) and Unmanned Aerial Vehicles (UAVs)[1]

Albert J. Marcella

CONTENTS

INTRODUCTION

Digital forensics is a branch within the field of forensic science that is concerned with retrieving, storing and analyzing electronic data that can be useful in digital criminal investigations. This includes information from computers, hard drives, mobile phones, and other data storage devices. In recent years, more varied sources of data have become important, including motor vehicles, drones, and the cloud. Digital forensic investigators face challenges such as extracting data from damaged or destroyed devices, locating individual items of evidence among vast quantities of data, and ensuring that their methods capture data reliably without altering it in any way. Advancements in technology have resulted in following a non-traditional digital forensic approach. Non-traditional digital forensics is when examiners face challenges related to data flow (e.g., cloud storage).

There are many different definitions of electronic or digital evidence. The Council of Europe Convention on Cybercrime, also called 'Budapest Convention on Cybercrime' or simply 'Budapest Convention' refers to electronic evidence as evidence that can be collected in electronic form of a criminal offence. The United States Department of Justice defines digital evidence as 'Information stored or transmitted in binary form that may be relied on in court,' as mentioned in the Forensic Examination of Digital Evidence: A Guide for Law Enforcement. In general, though, most definitions seem to summarize that digital evidence is digital data that can be used to help establish (or refute) whether a crime has been committed.[2]

In this chapter we are going to discuss the digital forensic examination of commercial Unmanned Aircraft Systems (UASs), which embodies an Unmanned Aerial Vehicle (UAV) colloquially known as a drone, a ground control station (GCS) and a controller (tablet, mobile device, etc.). The use of UAVs for commercial purposes (legal and illegal) is growing exponentially. In March 2020, there were 1,563,263 registered drones in the United States according to the Federal Aviation Administration (FAA). Yes, registered, so this does not account for those flying illegally, which have not been registered.

Of those 441,709 are registered as commercial drones and 1,117,900 as recreational small hobbyist drones. The FAA also reported that as of March 2020, 171,744 remote pilots had been certified.[3]

In its report 'FAA Aerospace Forecast Fiscal Years 2020–2040 Full Forecast Document and Tables,' the FAA states the professional grade commercial UAS sub-sector stands to expand rapidly over time, especially as newer and more sophisticated uses are identified, designed, and operationally planned and flown.

Should the professional grade UAS meet feasibility criteria of operations, safety, regulations, and satisfy economics and business principles and enters into the logistics chain via small package delivery, the growth in this sector will likely be phenomenal.[4]

Business Insider Intelligence predicts total UAV global shipments to reach 2.4 million in 2023 – increasing at a 66.8% compound annual growth rate (CAGR). Drone growth will occur across the four main segments of the enterprise industry: Agriculture, construction and mining, insurance, and media and telecommunications.[5]

As technology advances, cost of ownership declines and access to better and more sophisticated UAVs becomes easier, not only legitimate business applications of UAVs but also illegal uses will continue to expand. Oversight and enforcement of UAV registration will continue to be an issue. Those seeking to circumvent the law will continue to discover new ways to employ UAV technology and platforms to conduct illegal activities.

After a brief introduction on the broader field of UAS and UAVs this chapter will examine the growing urgency for organizations, departments, and agencies to be able to forensically examine a UAV, which is suspected to have been used in engaging in illegal, unauthorized activities.

In this chapter we will answer important questions: What type of information can be found on a UAV? Where is this information stored? How is this information accessed? What are some of the examination concerns when considering extracting data from a UAV?

We will also review proposed UAV forensic examination frameworks, in an effort to provide an insight into this new and emerging field. of UAV forensics Without a current standard or standardized, accredited, accepted method for performing a digital, forensic examination on a UAV, these frameworks provide the best guidance currently available for cyber forensic examiners.

When reviewing these frameworks, it is to be understood that the authors do not endorse or recommend any one of the frameworks presented. As a cyber examiner deciding to implement any of these frameworks, it is strongly advised that you should first review each framework's focus, objectives, strengths, and weaknesses, prior to applying said framework to an active case investigation.

WHAT IS AN UNMANNED AIRCRAFT SYSTEM (UAS)?

Technically speaking, a drone refers to an unpiloted aircraft or spacecraft. For purposes of this chapter and analysis, our focus will be on those aircraft which remain in Earth's atmosphere. However, drone refers mainly to an 'unmanned aircraft which is mostly used in a military context.' In broader conversation, drone in the common language is more commonly used to designate any type of aerial unmanned vehicle.

As to categories of unmanned aircraft (UA), we have consumer (mini, hobby, professional, selfie and racing), commercial, and military drones. Our focus again, will be on UA used for commercial purposes.

The term Unmanned Aerial Vehicle (UAV) popular and, in vogue, may be slowly fading into the sunset. This designation is typically used to define the flying objects employed for recreational and professional civilian applications. It appears that the international community is opting for different, more refined definitions.

An Unmanned Aircraft System (UAS) by definition includes the airframe, ground control station, command and control links, and crewmembers – all of the equipment necessary for the safe and efficient operation of that aircraft. See Figure 3.1, for an example of the components of a UAS.

Organizations such as the Federal Aviation Administration (FAA), the European Aviation Safety Agency (EASA), and the Unmanned Aerial Vehicle Systems Association (UAVSA) favor the use of UAS.

An Unmanned Aircraft (UA) is a component of a UAS. It is defined by statute as an aircraft that is operated without the possibility of direct human intervention from within or on the aircraft (Public Law 112-95, Section 331(8)).

UA includes a broad spectrum (or subsets) of aircraft, from drones (generally weighing less than 25 kg), unmanned free balloons, and model aircraft to highly complex remotely piloted aircraft (RPA) operated by licensed aviation professionals. See Figure 3.2, for a representative presentation of the topology of unmanned aircraft.

The International Civil Aviation Organization (ICAO) is trending use of the term, Remotely Piloted Aircraft System (RPAS). To better reflect the status of these aircraft as being piloted, the term 'remotely-piloted aircraft' (RPA) is being introduced into the lexicon.

An RPA is an aircraft piloted by a licensed 'remote pilot' situated at a 'remote pilot station' located external to the aircraft (i.e., ground, ship, another aircraft, space) who monitors the aircraft at all times and can respond to instructions issued by air traffic control (ATC), communicates via voice or data-link as appropriate to the airspace or operation, and has direct responsibility for the safe conduct of the aircraft throughout its flight.

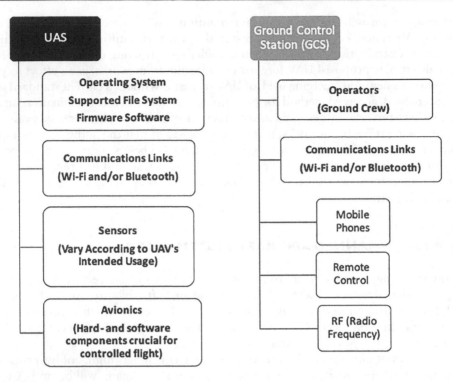

Figure 3.1 UAS components

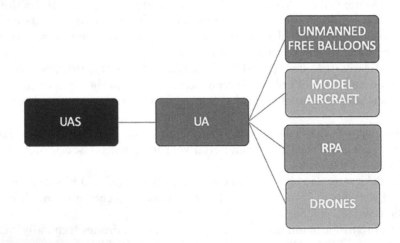

Figure 3.2 Topology of unmanned aircraft

An RPA may possess various types of autopilot technology but at any time the remote pilot can intervene in the management of the flight. This equates to the ability of the pilot of a manned aircraft being flown by its auto flight system to take prompt control of the aircraft. RPA is a subset of unmanned aircraft.[6]

Globally, The European Organisation for the Safety of Air Navigation (Eurocontrol), the European Aviation Safety Agency (EASA), the Australian, Civil Aviation Safety Authority (CASA), the Civil Aviation Authority (CAA) in New Zealand, and the Belgian Unmanned Aircraft Systems Association (BeUAS) are following suit in employing RPAS when issuing formal communique, discussing, and referring to unmanned aircraft.

As the unmanned aircraft field/market continues to grow and expand globally, the usage of UAS and RPAS is destined to become the dominant or preferred usage of form for referring to unmanned aircraft.

USES AND MISUSES OF UNMANNED AERIAL VEHICLES (UAVs)

The commercial uses for UAS/RPAS as an organizational competitive tool are continuing to be discovered, with new, tactical uses emerging daily.

As reported in the Federal Aviation Administration (FAA) Aerospace Forecast for Fiscal Years 2018 to 2038, non-model UAS are primarily used for aerial imaging and data collection, including real estate photography, industrial and utility inspection, and agricultural applications, including crop inspection.

Increasingly, state and local governments are using UAS for emergency services, including search and rescue operations. As the sector grows, there will be many more non-model UAS in use.[7] See Figure 3.3 for UAS utilization (as of this writing) by industry sector.

Some examples of how organizations are implementing UAVs as enterprise-wide, strategic tools include, but are certainly not limited to:

- Film making
- Forest fire fighting assistance
- General security surveillance (private residences, corporate offices, public spaces)
- High rise commercial building maintenance and safety inspection
- Land survey, management
- Livestock/range management
- Monitor road races, crowd control/management
- Package delivery
- Photography of previously inaccessible places/spaces/perspectives
- Pinpoint pesticide delivery
- Pipeline security, management, maintenance, survey
- Power line maintenance and safety inspection
- Real estate sales
- Real-time geo mapping

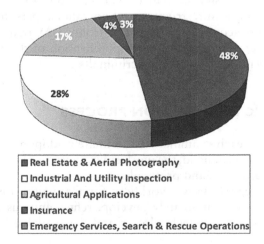

Real Estate & Aerial Photography
Industrial And Utility Inspection
Agricultural Applications
Insurance
Emergency Services, Search & Rescue Operations

Figure 3.3 UAS usage by industry sector

- Traffic monitoring
- Underground sewer, power, utility, maintenance and safety inspection
- Usage at sporting events (e.g., the Olympics!)
- Wildlife conservation
- Wind turbine maintenance and safety inspection

and the list goes on, and on....

The FAA's regulations (14 C.F.R. § 1.1) similarly define an 'aircraft' as 'a device that is used or intended to be used for flight in the air.' Because an unmanned aircraft is a contrivance/device that is invented, used, and designed to fly in the air, it meets the definition of 'aircraft.'

In addition, on December 16, 2015, the FAA the FAA promulgated an Interim Final Rule (80 Fed. Reg. 78594) that defined Unmanned Aircraft, Model Aircraft, Small Unmanned Aircraft, and Small Unmanned Aircraft System in 14 C.F.R. § 1.1. The FAA has promulgated regulations that apply to the operation of all aircraft, whether manned or unmanned, and irrespective of the altitude at which the aircraft is operating. For example, 14 C.F.R. § 91.13 prohibits any person from operating an aircraft in a careless or reckless manner so as to endanger the life or property of another.[8]

Unfortunately, UAV technology is also being used for illegal activities such as:

- Terrorism
- Privacy violations
- Smuggling and delivery of drugs into prisons
- Illegal surveillance
- Risk to emergency services
- Interfering with first responders
- Flying a UAV near an airport or airfield

The list will only continue to grow as threat actors, hackers, criminals discover new ways to employ this evolving technological platform and delivery system.

Reports of illegal UAS operation and sightings from pilots, citizens and law enforcement have increased dramatically since 2016. The FAA continues to receive increasing numbers of reports involving the illegal use of UAVs[9] (see Figure 3.4).

UAVs used for missions, which violate laws (municipal. state, Federal, international), are considered illegal and constitute a crime. When investigating such crimes, the tools used to commit said crimes will also be subject to investigation, and this includes the UAV itself and any associated communications infrastructure and operational personnel.

The ability to forensically acquire evidential data from a UAV that has been used in criminal activity may contribute significantly to the apprehension of the alleged criminal(s) and lead to successfully prosecution of the alleged criminal(s).

UAV CYBER FORENSIC EXAMINATION PROCESS

The authors recognize that each examiner will ultimately adopt an investigation workflow best suited for the examination at hand, the work environment and internal company, department, agency policies, procedures, and protocols.

Figure 3.5, presents a representative workflow for the digital, forensic examination of a UAV. Forensic examiners may ultimately develop, refine, and customize their own UAV examination workflow, as both the technologies and the available UAV forensic processes mature.

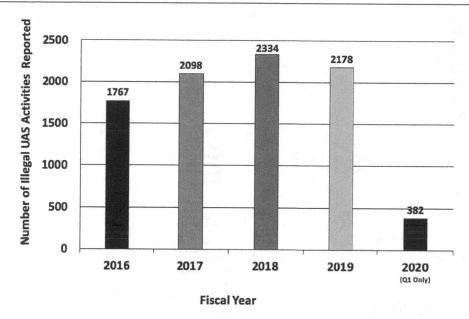

Figure 3.4 UAS illegal activities reported to FAA

The remainder of this chapter will examine the technologies, issues, challenges, and developments in performing a digital cyber forensic examination of a UAV. The chapter ends with a list of questions that should be raised, by the examiner or by management, before, during, and after the forensic examination of a UAV.

The authors recognize that UAV technologies are advancing exponentially with new applications for use introduced on what seems to be a daily basis and at times, in surprisingly new and previously unthought of ways.

The cyber forensic investigator will always be pushed and challenged to keep abreast of these technological changes and innovative uses for UAVs, be they legal or illegal. The authors hope that this chapter provides the reader with some additional information and insight in keeping pace if not a step ahead of this rapidly changing field.

CYBER FORENSIC EXAMINATION CHALLENGES

> No action taken by law enforcement agencies, persons employed within those agencies or their agents should change data which may subsequently be relied upon in court.[11]

> Wherever possible no actions taken during the seizing of any evidential material should cause that material to be changed and this is of particular importance when dealing with digital evidence which could be seen as prone to accidental 'tampering.'[12]

NIST Special Publication 800-101, Revision 1, Guidelines on Mobile Device Forensics, published in May 2014, isn't up-to-date with today's mobile technology and does not address the mobile components of a UAV, vital to a cyber forensic examination.

The Association of Chief Police Officers (APCO) Good Practice Guide for Digital Evidence, published in March 2012, does not provide UAV forensics guidance, nor does the European Network of Forensic Science Institutes (ENFSI), Best Practice Manual (BPM) for the Forensic Examination of Digital Technology, published in 2015.

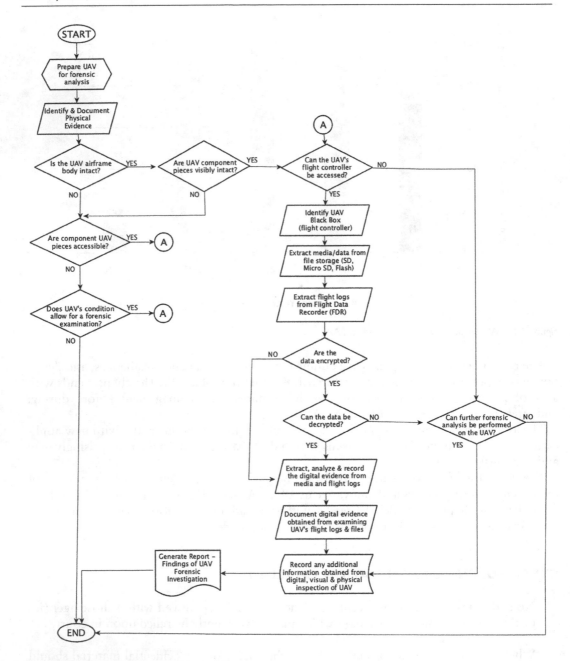

Figure 3.5 UAV forensic exam flowchart[10]

However, while these guidelines address the forensic examination of mobile devices, the underlining principles can be directly related to the examination procedures to be used in the examination of the UAV platform.

UAV and mobile technology

Like the many challenges that come with evidence protection when seizing and examining mobile devices, UAV forensic examinations also entail evidence protection challenges. These challenges are due to the UAV employing similar mobile device communications to control/ pilot and communicate with the UAV and ground control station (GCS), which manages flight operations. The challenge similarity comes from the technology used. If the UAV was operated and flown using a mobile phone as a remote control, this will then require the examiner to execute and follow mobile forensic techniques and procedures to acquire and then analyze digital evidence related to the UAV incident.

Data or digital evidence held in onboard ROM devices may be lost if power is not maintained to the UAV. Lost communications between the UAV and GCS or remote controller (e.g., mobile phone) represent possible digital evidence. Software logs maintained by the UAS should have a record of any lost communications with the UAV. These logs may potentially prove valuable to an investigation.

Control of the UAV is managed in some fashion via a ground control station (GCS). The GCS may consist of a dedicated GCS, a mobile device (e.g., mobile phone, tablet, etc.), or a computer or laptop.

Ground Control Stations are sets of ground-based hardware and software that allow UAV operators to communicate with and control a drone and its payloads, either by setting parameters for autonomous operation or by allowing direct control of the UAV. A drone ground control station will be based around a processing unit, which may be an off-the-shelf laptop with an Intel i5 or other common high-performance processor, or a bespoke system based on an embedded computing platform.

A wireless data-link subsystem will provide remote communication with the UAV. Telemetry data, commands, and sensor data such as video, images and measurements may all need to be transferred between the UAV and the GCS. These data can be stored in web cloud storage (e.g., DJI GO App), providing the examiner with an additional source of possible evidential data. Communication methods include analog and digital radio and cellular communications, with operational ranges extending to the hundreds of kilometers.

Portable UAV GCS are usually single or double screen units, and the dual-screen systems can often be set up so that two operators can work simultaneously – one pilot and one payload operator. The control system may be twin-stick, like common radio-controlled aircraft and small quadcopter controllers, or a HOTAS (Hands on Throttle and Stick) layout, which is an intuitive set-up originating from manned aviation that enables a high degree of flight control and versatility.[13]

An example of a professional dual screen UAV Ground Control Station, manufactured by Desert Rotor, is shown in Figure 3.6.

A GCS may be a complex piece of stationary equipment which provides various flight control output data or a basic mobile device (iPad, mobile phone, etc.), with an appropriate UAV-flight control application (e.g., DJI Fly, DJI Go 4, DJI GS Pro, etc.) installed.

Ground control station software, currently available, and the operating system platform on which it runs are shown in Table 3.1.

Mobile technology, GCS, and UAV forensic examination

In a more traditional seizure of digitally processing equipment (e.g., mobile phones, laptops, etc.), the decision to switch off the device or keep the device on and running, may be clearer based upon more knowledge of the device's specific operating system, device type, previous device experience or case specifics. The seizure and examination of a UAV, with its onboard communications link capabilities, present similar risks associated to examining mobile devices.

Many mobile devices offer the user the ability to perform either a remote lock or remote wipe by simply sending a command (e.g., text message) to the mobile device. The mobile

Figure 3.6 Dual screen UAV ground control station[14]

Table 3.1 Ground control station software and the operating system platform

Ground control station software	
Desktop	*Operating system*
APM Planner 2	Windows, Mac OS X, Linux
Micro Air Vehicle (MAV) Proxy	Linux
Mission Planner	Windows, Mac OS X (Using Mono)
QGroundControl	Windows, Mac OS X, Linux, Android and iOS
UgCS	Windows, Mac OS X, Ubuntu
Tablet/Smartphone	
AndroPilot	Android Phones and Tablets
MAVPilot	iPhone, iPad
SidePilot	iPhone, iPad
Tower (DroidPlanner 3)	Android Phones and Tablets

device should be seized along with associated hardware. Media cards, UICCs, and other hardware residing in the mobile device should not be removed.

Additional reasons for disabling network connectivity include incoming data (e.g., calls or text messages) that may modify the current state of the data stored on the mobile device. Outgoing data may also be undesirable as the current GPS location may be delivered to an advisory providing the geographic location of the forensic examiner.

The benefits of turning off the phone include:

- Preserving call logs and last cell tower location information (LOCI).
- Preventing overwriting deleted data.
- Preventing data destruction signals from reaching the mobile phone.
- Preventing improper mobile phone handling (i.e., placing calls, sending messages, taking photos or deleting files).[15]

Isolating a mobile device from all radio networks (e.g., Wi-Fi, Cellular, and Bluetooth) is important to keep new traffic, such as SMS messages, from overwriting existing data. Besides the risk of overwriting potential evidence, the question may arise whether data received on the mobile device after seizure is within the scope of the original authority granted.

The risks of turning off the mobile phone include possibly engaging authentication mechanisms (e.g., passwords, PINs, etc.). Exigency may dictate that the mobile phone remains on for immediate processing. If the mobile phone must be left on, isolate it from its network while maintaining power.

The cyber forensic examiner should be aware of any potential to disengage the network connectivity between the UAV and the GCS. Isolating the UAV to mitigate this disruption and potentially impact evidence collection should be given serious consideration. While engaged, a mobile phone, used as a remote controller is still capable of sending and receiving commands from the UAV. Thus, a potential to change, corrupt or delete potential evidential digital data.

Vulnerabilities may exist that may exploit a weakness related to software vulnerabilities from the web browser and OS, SMS, MMS, third-party applications, and Wi-Fi networks. The possibility of such vulnerabilities being exploited may permit the argument that data may have been modified during the forensic examination.[16]

Therefore, forensic examiners need to be aware and take precautions when securing mobile devices mitigating the chance of data modification.

Challenges facing UAV forensic investigations

One significant challenge facing the UAV forensic examiner is that for the moment, there are no generally accepted forensic examination processes, *specifically approved or vetted for UAV* forensic examinations and analysis. The field is that new. Now, that will change as such investigations and examinations become more common, more routine, and more cases are both tried and adjudicated throughout the legal system.

Roder et al. identified several significant challenges facing the UAV forensic examiner:

- There is no standard location or format for UAV flight data, and research is necessary to prevent missing evidence or misinterpreting extracted data. UAV manufacturers may store data in different formats, and currently there is no standardization.
- Freeware tools such as DatCon, designed to interpret .DAT files specifically from DJI UAVs, are unlikely to have been validated according to forensic requirements. These tools are unlikely to be forensically sound and artifacts obtained from using such tools may be inadmissible in a court of law.
- Data can be stored in several locations, such as the UAV, GCS, network routers, and so on. Storage locations can also be overt or covert, and one also needs to note that in some instances, there are inbuilt persistent storage media such as Micro SD cards.
- The recovery of artifacts from flash storage will typically require some form of direct connection.

- There is a very high likelihood that a UAV used in a criminal activity has been modified to either hinder forensic investigation or enhance certain features such as increased load carrying capacity (e.g., in drug smuggling activities across borders, or act as an improvised explosive device). Also, disabling software functionality such as the No Drone Zone (NDZ) function.
- Similar to a mobile device, a modern or advanced GCS is likely to have Wi-Fi, Bluetooth, or Internet connection. Therefore, there is a possibility that the device could be remotely wiped or modified.
- UAV forensics can also involve conventional storage media forensics (e.g., memory cards are copied) and live forensics (e.g., real-time access to a live UAV to view data stored on flash memory). Since most UAVs do not have a graphical user interface (GUI) or inbuilt interface, there is a real-risk that data may have been changed without the knowledge of the forensic examiner/investigator.[17]
- In certain circumstances, it may not be possible to remove storage devices, such as embedded multimedia card (eMMC) storage. Prior to conducting destructive examination techniques (chip-off, etc.), consideration should be given to performing live examination of the device.[18]

Additional challenges to a UAV forensic examination are noted by M. A. Hannan Bin Azhar et al., specifically:

- To re-create the actions taken by the drone, interpretation of the recorded flight data is essential, which is not a likely skillset of forensic investigator. At a minimum, the understanding of time stamped latitude, longitude, and altitude measurements is required, as well as speed, battery level and other data from a host of possible onboard sensors.[19]

Because UAV technologies and platforms are evolving constantly, there is yet no industry-wide standard for operating systems (OS). Therefore, the examiner should identify how a specific operating system would address data loss, destruction, or contamination when the device is powered off.

Depending on the circumstances and case at hand, if the examiner is in the field and will potentially examine a UAV seized on scene, the examiner will need to determine whether to separate (or isolate) the UAV from telecoms connectivity with the control unit/ground control station (GCS) or to maintain that link.

The examiner should be aware of the potential for data loss, contamination or corruption, if the UAV is powered down intentionally, loses power due to insufficient battery life, or is disconnected from or loses the uplink connection with the GCS.

If the UAV is in a powered down state when acquired, Clark, D., et al. indicated that their analysis of the DJI Phantom III that the UAV should not be turned on as turning it on changes data on the drone by creating a new DAT file and may also delete stored data if the drone's internal storage is full.[20]

While the DJI Phantom III represents a single class and model of UAV, care should be taken by the examiner when either powering up or down a UAV to avoid the loss or modification of resident data. Care should also be exercised due to the potential of data modification, which in turn may make data which are retrieved, inadmissible. Prudent forensic practices should be followed at all times when attempting to acquire data from the UAV.

UAV – Owner/registrant

According to FAA regulations, failure to register an unmanned aircraft, which meets registration requirements, may result in regulatory and criminal penalties. The FAA may assess civil penalties up to $27,500. Criminal penalties include fines of up to $250,000 and/or imprisonment for up to three years.[21]

An owner must register the UAV according to the intended operational use of the UAV when it is flown.

If the UAS is flown by Certified Remote Pilots including Commercial Operators (flying under Part 107), the FAA requires the drone owner to:

- Register the drone when flying under Part 107.
- Label the drone with the assigned registration number.

If the UAV is flown by recreational flyers and are flying for hobby or recreation only, the FAA requires the owner to:

- Register as a 'modeler'
- Label the model aircraft with the FAA issued registration number[22]

If the forensic examiner is in possession of a UAV, it is very likely that UAV was used in some type (possibly yet to be determined) criminal activity. Thus, requiring its examination. Persons who may utilize a UAV for unlawful activities may not be so law abiding as to follow the FAA requirements for registering the drone. Identification via registration number may then be moot.

If the UAV is seized, recovered, or simply found without the presence of a pilot in command (PIC) or ground crew, it may be difficult to determine (a) who was actually piloting the UAV; (b) the individual who actually registered the UAV (if registration was required), which may not be the same individual piloting the UAV; and (c) the owner of the UAV, who may be altogether different than the individual who registered or piloted the UAV.

If the required external registration number is not clearly visible, determining exactly who owns the UAV may require additional analysis of data stored on internal UAV system files. Who was actually piloting the UAV (if pilot is not on scene when the UAV is acquired) and who registered it may also need to wait until further analysis is performed by the forensic examiner.

WHAT TYPE OF DATA MAY BE FOUND ON A UAV?

The examiner should always follow proper safety practices when examining a UAV. It is advisable to disengage and remove all rotors/blades, prior to a further examination of the UAV.

Basic UAV forensic artifacts

Forensic artifacts are objects that have forensic value, i.e., a piece of data that may or may not be relevant to the investigation. These may be in the form of logs, files, time stamps, way points, photos, and videos.

In general, the cyber forensic examiner will find a copious number of artifacts during the examination of a UAV. These artifacts may be categorized as either physical, digital, or tangible documents.

An overview of the UAV forensic artifact categories are presented in Figure 3.7. The artifacts shown within each category below are representative and not a comprehensive inventory of artifacts within each category, which the examiner may encounter.

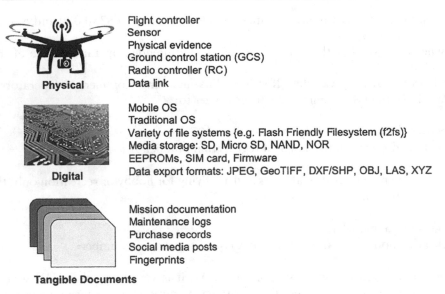

Figure 3.7 UAV forensic artifacts categories

What data are stored on the UAV?

There are two specific log sources, which will be most important to the UAV-forensic examiner. The UAV itself and the device used as the ground control station (GCS).

The basic unmanned aircraft system (UAS) interface between the UAV and the GCS is represented in Figure 3.8.

The UAV will typically be configured to allow for the use of flash storage or media storage devices (MSD). These devices can store flight logs and diagnostic data of the UAV every time it was powered on.

Depending on the make and model, there may be multiple MSDs. A camera's memory card usually stores media (images and videos), the detailed flight logs are typically stored on the motherboard's memory card.

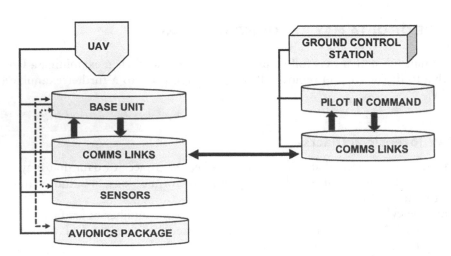

Figure 3.8 Interface between unmanned aircraft system (UAS) components: UAV and ground control station[23]

Table 3.2 Open sources and commercial UAV operating systems

Operating systems	
Open source	Commercial
Arducopter	Linux
Ardupilo	Windows
DroneDeploy	Mac OS X
MAVSDK	Android
FlytOS	iOS
NuttX OS	LynxOS 7.0 Real-Time Operating System (RTOS)
Open Robot Control Software (Orocos)	
PX4 Drone Autopilot	
Robot Operating System (ROS)	
Xenomai	
FreeRTOS	

To analyze the component by component of the various makes and models of UAVs on the market today would be a tremendous and lengthy exercise. There are many variants and manufactures are changing models, payloads, configurations, and technologies in an ever-evolving cycle. First to keep pace with the rapid deployment of technologies that make UAVs a go-to strategy for many commercial organizations (and criminals as well). Secondly, to remain competitive in a rapidly growing, very competitive market.

The examiner will encounter differences between UAV models, with respect to operating system (OS), data storage conventions (many of which are proprietary), and communication protocols.

Depending upon the age, make, model, and configuration of the UAV, data that are stored on the UAV, and associated GCS, *may include*:

- Flight controller:
  - It is the UAV's CPU. The 'brains' of the UAV. Refer to the next section in this chapter; Unmanned Aerial Vehicles: Where Data Can Be Found, for a closer look at the flight controller and its important role in UAV forensic examinations.
  - As with most CPUs you will also find an operating system.
    Operating systems in UAVs vary. Knowing the operational specifics of the particular operating system that is installed, on the UAV under examination, will provide the examiner with additional insights into the UAV's operational dynamics, capabilities, data storage capacity, and overall operating protocols.
    The most common operating systems, both open source and commercial, which are installed on most popular UAVs at the time of this writing are listed in Table 3.2.

- UAV-specific information
  - Model name
  - Model ID
  - Camera (if so equipped)
  - Primary circuit board inside the remote control that is used to control the drone
  - Battery type, manufacture
  - Launch point

- Maximum speed
- Acceleration and velocity values
- Individual motor rotational speed
- Pitch and Yaw
- Time stamp
- Version numbers for critical firmware
- Battery usage/levels (overall battery level and more detailed information about the battery capacity, temperature, current, and voltage for the individual cells)

     Battery levels can reveal the amount of time the UAV has been operational. The battery level may then be correlated with time. Using these data, the forensic examiner can re-construct the flight path of the UAV. Combining this information with other forensic artifacts from the UAV examination could provide valuable information to authorities in identifying the operator, especially when the UAV has been used in a suspected, flight-related crime.

- Coordinates of the UAV's home point
- Serial number(s) of firmware
- Software used to control the UAV
- Remote control device (mobile device or tablet)
- Tablet location
- Remote control status (such as throttle, rudder and elevator)
- UAV's MAC address
- The MAC address of the remote-control device listed as ground control station (GCS), which is often a mobile device combined with a radio controller, controlling the UAV
- The type of encryption (WEP/WPA/WPA2/OPN) used by the wireless communication network
- The network IP address
- System logs, containing details of software and hardware events from the UAV's internal operating system (OS) (version information, configuration data, mount information, file creation logs)
- Serial numbers (SN) of key UAV components. This serial number ties the physical airframe to the logs to the mobile device

Drone serial numbers are unique for every drone manufactured. With the serial number (and the proper warrants) a drone could be traced from the manufacturer down the supply chain eventually to the end purchaser. For this purpose, serial numbers on drones are unique. Table 3.3 identifies the serial number location, for several models of more popular UAVs.

The forensic examiner may be in possession of the complete UAV, making ownership identification easer. However, in some cases only pieces of the UAV may be available for examination. Having the ability to trace specific serial numbers found on the body, battery and remote controller of the UAV may assist in identifying the UAV's owner.

It may be advantageous to the examiner to establish permanent file documentation, recoding the individual component serial numbers from each UAV which is examined.

Table 3.3 presents a template, an example and an initial starting point for further collecting and developing an important piece of investigative documentation.

The flight controller is inside the UAV (see Figures 3.23 and 3.24). In order to verify the flight controller serial number, via the Go app, the UAV must be turned on. In IOS it is referred to as the 'Flight Controller Serial Number.' In Android it is called 'Flight Controller SN.'

Table 3.3 Serial number location, for several models of more popular UAVs

UAV component serial number	UAV model				
	DJI Mavic models (Mavic Pro, Mavic Pro Platinum, Mavic 2, Mavic Air)	DJI Inspire 2	DJI Phantom 3 models (Phantom 3 Professional, Phantom 3 Advanced, Phantom 3 Standard, Phantom 3 4K, Phantom 3 SE)	DJI Phantom 4 models (Phantom 4, Phantom 4 Advanced, Phantom 4 Pro, Phantom 4 Pro V2)	MATRICE 600 PRO
Aircraft body	Located in the battery compartment	Located on the left-hand side just underneath the battery location	Located on the underside of the body (tail)	Located on the inside of the battery bay	Under the battery #1 terminal. Remove the plastic cover for installing gimbals. It Is the small white sticker with a QR code on It
Battery	Found on the back of the battery	Located on the back of the battery	Located on the bottom of the battery	Located near the connectors	Located on the bottom of the battery
Remote Controller (RC)	Located on the back of the controller	Found on the bottom of the RC near the USB port	Found in back near the USB port	Located in back near the USB port	Located on the lateral side of the A3 flight controller

Keep in mind the proper protocol for examining technology. Determine first, what data/information might be lost or altered by turning the UAV on?

Alternatively, the examiner may open up the UAV, locate the flight controller, and there should be a serial number sticker stuck to the flight control board. This procedure will nullify the manufacturer's warranty; however, this is most likely not a concern of the examiner.

- Flight Information
  - Total flight time
  - Distance flown
  - Number of flights
  - UAV home point and coordinates of waypoints (in a DJI Phantom II, these data are stored in 16-bit character strings with Unicode Transformation Format (UTF)-16 little endian encoding[24])
  - Countries where the UAV was flown
  - Flight path (geo-data)
  - Duration of the flight

The UAVs flight data recorder (FDR) holds a trove of valuable information for the forensic investigator. See the following section for a closer look at the role which the FDR plays in providing possible UAV forensic evidence.

There are two primary sources for flight data from the DJI Phantom III standard UAV, as noted by Clark et al. These include:

1. TXT files created by the DJI GO mobile application and stored on the mobile device for controlling the UAV.
2. DAT files created by the drone itself and located on the drone's nonvolatile internal storage.

Both files are encrypted and encoded using two different proprietary formats. After decrypting and decoding these files, data regarding the GPS, motors, remote control, flight status, and other information can be extracted. These files essentially serve as the electronic flight recorder for the drone.[25]

Kovar, et al. identify additional flight information is available from the UAV, most notably,

- State change information such as launch/land
- Manual/waypoint operation
- GPS available or unavailable

Also observed by Kovar et al. is that two common flight controllers, PixHawk and DJI's family, write log messages from each subsystem as individual records as they come in so the structure is more similar to a network packet capture than an event log. Viewing this data as a table rather than as a series of distinct but related messages obscures valuable nuances in the data.[26]

- GPS position
  - Latitude
  - Longitude
  - Altitude
  - Time stamp
  - GPS location of the controlling application (this allows for identifying the location of the operator at the time of flight). Inertial measurement unit (IMU)[27] is located inside the UAV.

GPS coordinates can reveal from where the drone took off, or in the event of a crash, battery levels can reveal the time when the drone failed as it can be correlated with time. These data can also be used to re-construct the flight, which is especially important when the drone has been used in smuggling or other flight-related crime.[28]

It should be noted that persons engaging in unlawful activities may attempt to use anti-forensic tools, for example, GPS spoofing (the generation of a GPS signal which conforms to the GPS standard but contains incorrect information on satellite position or current time), to interfere with recording of accurate GPS coordinates. GPS spoofing could hamper investigators in identifying the location of the UAV launch site, owner/operator.

GPS data represent critical evidence related to the UAV's flight activities, for example, longitude, latitude, waypoints, each of which may be used to not only identify the UAV's path of flight but, may be used to pinpoint the UAV operator's location.

- Account details
  - User/owner's name
  - Photo (it is interesting to note that frequently the very first photo a user takes is during the maiden flight of the UAV. This photo typically includes recognizable landmarks and/or a selfie of the owner/operator. Such information may be very useful in both identifying and locating the UAV owner/operator, if the individual is not on scene when the UAV is taken into possession by law enforcement personnel).
    - Bio
    - Website
    - Country
    - Date of account creation
    - Social networks linked to this account
- Sensor Data: The type of sensor installed/carried by the UAV will tell you a lot about the purpose of the flight and potentially information that will allow investigators to profile and identify the UAV operator and/or owner.
  - Light Detection and Ranging (LIDAR) is a remote sensing method that uses light in the form of a pulsed laser to measure ranges (variable distances) to the Earth.
  - Optical (e.g., temperature, velocity liquid level, pressure, displacement (position), vibrations, chemical species, force radiation, PH-value, strain, acoustic field, and electric field). Most common sensor package used with a UAV.
    Artifacts
    - The Image that was or images that were taken
    - The image's Exchangeable Image File (EXIF) data files
  - Near-visible Infrared (NVIR) used to perform remote aerial thermography, crop yield estimation, and plant disease detection, etc.
  - Thermal sensors measure the relative surface temperature of objects (e.g., surveillance and security, infrastructure inspections, water source identification, livestock detection, and heat signature detection).

As UAV technology and construction advances, there are literally thousands of uses for which a UAV's payload may be configured, for both constructive/legal and illegal missions. An examination of the UAV's payload will assist in developing a profile of the owner/operator. Payloads modified beyond the manufacture's recommend use may be indicative of a UAV used for illegal purposes/missions.

Examining the EXIF data files, which are stored in the UAV's internal memory for each picture taken, may reveal considerable evidence to assist identifying and locating the UAV's launch point, operator, and flight path. In addition to any photos retrieved from the onboard camera, the EXIF data would include such information as:

- Creation date the photo was taken
- GPS location where the photo was taken
- Shutter speed
- Aperture
- ISO speed
- White balance
- Focal length
- Lens type (dependent on camera type, i.e., DSLR).

Figure 3.9 Putnam, Connecticut, town-scape image[29]

See Figures 3.9 and 3.10 for an example of an image taken via an onboard UAV camera and the corresponding EXIF data recorded for that image. The EXIF data shown in Figure 3.10 has been culled from 14 pages of EXIF data produced from this single image.

Each of these data, in combination with photos retrieved, may be useful, supplemental, evidence for law enforcement, or company management seeking to prosecute the illegal use of a UAV.

- Mission Information
- Photographs (e.g., .DNG, .JPG., .PDS)
- Audio recordings (e.g., MP4, .mov)
- Video recordings
- Mapping data (.CSV)

The UAV-forensic examiner, following appropriate and controlled forensic procedures, may wish to convert the .CSV data file to KMZ (Keyhole Markup Language) file format. This will allow the examiner to visualize the GPS data through an Earth browser such as Google Earth.

An additional consideration, dependent upon the UAV's configuration and accessories, is that data may be streamed onto a storage device (e.g., mobile device or the cloud) during flight operations.

This would then require the examiner to identify: (a) if such streaming was performed during the UAV's mission and (b) what type of device was the data streamed to. If the data were streamed to a third-party cloud service provider, this may require additional warrants to secure and gain access to the data.

If the data were streamed to a mobile device (laptop, mobile phone, etc.), the forensic examiner should follow forensic protocols already established for the examination of electronic devices.[31]

```
<x:xmpmeta xmlns:x="adobe:ns:meta/" x:xmptk="Adobe XMP Core 6.0-c002 79.164460,
2020/05/12-16:04:17      ">
    <rdf:Description rdf:about="" xmlns:drone-dji="http://www.dji.com/drone-dji/1.0/"
    <xmp:ModifyDate>2020-07-30T14:46:35-04:00</xmp:ModifyDate>
    <xmp:CreateDate>2019-11-05T08:36:34</xmp:CreateDate>
    <xmp:CreatorTool>Adobe Photoshop Lightroom Classic 9.3
(Macintosh)</xmp:CreatorTool>
    <xmp:MetadataDate>2020-07-30T14:46:35-04:00</xmp:MetadataDate>
    <dc:format>image/jpeg</dc:format>
    <drone-dji:GpsLatitude>+41.9128932</drone-dji:GpsLatitude>
    <drone-dji:GpsLongitude>-71.9077982</drone-dji:GpsLongitude>
    <drone-dji:SelfData>DJI Self data</drone-dji:SelfData>
    <aux:SerialNumber>0K8TFAH0020831</aux:SerialNumber>
    <aux:Lens>28.0 mm f/2.8</aux:Lens>
    <xmpMM:PreservedFileName>DJI_0684.DNG</xmpMM:PreservedFileName>
<xmpMM:OriginalDocumentID>20BFD84451FBA7EBEBE4F4FC6A1A546B</xmpMM:OriginalDoc
umentID>
    <crs:WhiteBalance>Custom</crs:WhiteBalance>
    <exif:ExifVersion>0231</exif:ExifVersion>
    <exif:DateTimeOriginal>2019-11-05T08:36:34</exif:DateTimeOriginal>
    <exif:ISOSpeedRatings> <rdf:li>100</rdf:li>
    <exif:SerialNumber>0K8TFAH0020831</exif:SerialNumber>
    <exif:ShutterSpeedValue>6643856/1000000</exif:ShutterSpeedValue>
    <tiff:Make>Hasselblad</tiff:Make>
    <tiff:Model>L1D-20c</tiff:Model>
    <exif:ApertureValue>4339850/1000000</exif:ApertureValue>
    </exif:LensInfo> <exif:Lens>28.0 mm f/2.8</exif:Lens>
    <exif:FocalLength>10260/1000</exif:FocalLength>
```

Figure 3.10 EXIF data for Putnam, Connecticut, town-scape image[30]

Data from the UAV, in part or in entirety, will also provide essential evidence to investigators of UAV flight accident.

A UAV will use a data-link to communicate with the GCS or PIC. Data-link uses a radio-frequency (RF) transmission to transmit and receive information to and from the UAV. This data-link can also transmit live video from the UAV back to the GCS so the pilot and ground crew can observe what the UAV camera is seeing.

Data-link information, which may be available for review, by the examiner, includes:

- Location
- Remaining flight time
- Distance and location to target
- Distance to the pilot
- Location of the pilot
- Payload information
- Airspeed
- Altitude[32]

This information, combined with various other data identified above, when analyzed by the examiner, may provide essential insight into the UAV's mission, launch point, flight path, owner/operator, and the type of data that the UAV may have collected.

Controllerless flight operation

Technology, in some UAV models, has moved away from the need to have a controller to actually 'control' the UAV. In some types of UAVs the 'controlling' GPS transmitter is worn by the user. When the transmitter is activated, the UAV will follow the user outfitted with the transmitter.

Follow Me is an intelligent flight mode which uses the UAV's GPS signal to tether the UAV to the phone or tablet of the operator. Wherever the operator moves with the mobile device, the UAV will follow.[33]

Currently there are two main technologies that are used in the Follow Me feature, GPS transmitter, and vision recognition tech.

The earliest Follow Me drones were programmed to follow a GPS transmitter or Ground Station Controller (GSC) that users had to wear. This created a virtual tether between the user and the drone. This GPS transmitter is usually built into the remote controller, which then becomes required for the drone to follow the operator.

Quite a few Follow Me mode drones use a GPS-enabled device such as a mobile phone, tablet, or a Ground Station Controller (GSC), along with a transmitter (wearable transmitter or mobile phone). The drone is programmed to follow the transmitter and to keep the subject in the picture at all times.

The software to program Follow Me is generally built into the overall drone application. DJI, for example, has integrated this software into the company's mobile-based GO 4 App.

The UAV forensic investigator may find using established and validated mobile forensic techniques of benefit when analyzing the GPS-enabled device such as a mobile phone, tablet, or a Ground Station Controller (GSC) for forensic evidence.

The tracking accuracy of GPS tech is unrivaled, providing much better precision than other techs. Most Follow me UAVs equipped with Follow Me technology can also remain stationary and track the subject by rotating, or the UAV is able to move along with the subject being tracked.

Several limitations can be noted in the GPS technology, a tracker/transmitter is always required for the system to work and obstacle avoidance is not a feature.

Vision recognition tech detects objects, people, and obstacles through data captured by a camera.

In general, the vision recognition technology works as follows:

- The drone cameras and sensors collect image and sound data, which is then transmitted to the processor.
- With sufficient data, the processor identifies the background parts of the scene and locates any moving object.
- As programmed, the aircraft starts to automatically follow the moving object.[34]

Sensors and recognition technology, along with software algorithms, give UAVs the ability to recognize and follow a person or object. This deep learning following drone technology allows the UAV to track a moving subject without a separate GPS tracker.[35]

Besides speedy response time, vision recognition also has the advantages that it is able to track all types of moving objects, including cars, bikes, people, and animals and does not require an external GPS tracker, instead using compatible software/app on the controlling device.

Recognition accuracy, however, can be affected by lighting conditions and the contrast between areas in direct light and those in shadow.

To overcome the limitations of GPS and vision recognition systems, DJI combined the two technologies into the ActiveTrack system.[36]

The ActiveTrack is a product feature, which debuted with the Phantom 4 in March 2016. When enabled ActiveTrack follows a chosen subject, whether they are walking along a trail, driving a car, or even swimming in the ocean. While intelligently tracking the subject, the drone will use its vision and sensing systems to maintain safe flight.

ActiveTrack options include:

Trace
- The aircraft tracks the subject at a constant distance.

Profile
- The aircraft tracks the subject at constant angle and distance from the side.

Spotlight
- The aircraft will not trace a subject automatically, but it keeps the camera pointing at the subject during flight.

The way ActiveTrack is able to identify and follow its subject is by color contrast between the subject and background. This means that the greater the color difference, the better ActiveTrack performs.[37]

Figure 3.11 shows the interface between the unmanned aircraft system (UAS) components, the UA,V and the autonomous, controllerless flight operation.

Prastya et al. state that GPS data, which could potentially be used as digital evidence, is always stored in the system log, contained within either the UAV's onboard internal storage (e.g., the Black Box), the UAV's SD card. If a mobile device is used as the flight controller, digital evidence may also be found there. Prastya also noted that GPS data is always stored, even if the system uses UAV flight mode without engaging the GPS. Thus, reinforcing the importance of examining the UAV's GPS data contained in the flight logs.

Based upon an examination of the amount of evidential data available from the UAV itself, Prastya and his team found that if a mobile device is used to control the UAV, the mobile device contained 50% more data for examination than the amount of examinable data available from the UAV itself (only 16%) (see Table 3.4).[39]

Most mobile devices, including those used to control a UAV, will typically contain removable media. Mobile device forensic tools will often perform acquisitions for these types of

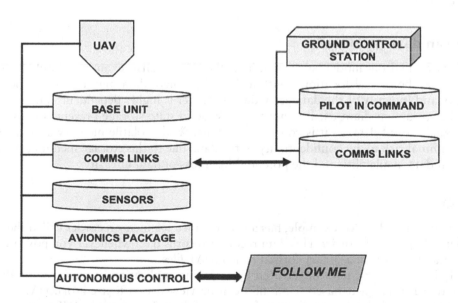

Figure 3.11 Interface between unmanned aircraft system (UAS) components: UAV and controllerless flight operation[38]

Table 3.4 Digital evidence artifacts available comparison – DJI Phantom 3

	Storage		
	UAV	Memory card	Smartphone
Acquisition Method	Live	Static	Live/Static
Type of Image	Physical	Physical	Logical
Image Format	.dd	.dd	.ad1
Acquisition Tool	FTK Imager	FTK Imager	FTK Imager
Digital Evidence Information Available			
GPS Location	X	X	X
Log coordinate flight path	X		X
UAV configuration information	X		X
Pictures/Videos		X	X
Flight Mode Information	X		X
UAV user information			X
UAV flight data information	X		X
Directions shooting			X
UAV signal strength information			X
Information UAV sensor condition			X
UAV power condition information	X		X
Information condition UAV controller			X

removable media. If a mobile device is powered on, the removable media should remain in the device during extraction. If a mobile device is powered off, acquire the removable media separately from the device. If the live device is required to interpret data on the removable media, a separate acquisition may be considered once the removable media has been returned to the device.[40]

Where can data be found?

Depending on the drone model being examined, the UAV forensic examiner should be aware that there may be several locations, within the UAV, where data may be stored. Both the availability and integrity of the data as evidential matter is highly dependent upon the volatility of the storage medium used within the UAV. Solid state storage presents minimal risk on the availability of data, as it is considered robust. While volatile memory such as RAM imposes minimal risk to the confidentiality of the data, due to this storage medium, there is a risk of both data availability and the potential for a loss of data integrity.

In the UAV

The DJI Phantom 3 (P3), for example, has a very detailed telemetry recorder called the Main Controller (MC) Data Recorder. This data recorder records data whenever you power on the P3. The data are stored in a binary format called .DAT files.

A .DAT file is a generic data file created by a specific application. It may contain data in binary or text format (text-based .DAT files can be viewed in a text editor). DAT files are typically accessed only by the application that created them. For example, DJI go app for DJI model UAVs.

Comma Separated Values (CSV) is a simple format for representing a rectangular array (matrix) of numeric and textual values. It is an example of a 'flat file' format. It is a delimited data format that has fields/columns separated by the comma character %x2C (Hex 2C) and records/rows/lines separated by characters indicating a line break.

There are several web sites that provide software, which can be used to read and translate .DAT files. The .DAT files are converted into CSV or other readable formats.

Such translation software may be found from these vendors at:

- DJI Log Converter (www.djilogs.com/#/)
- Flight Replay (www.flightreplay.com/)
- Litchi & Healthy Drones (now Airdata) (https://healthydrones.com/litchi)
- Maps Made Easy (www.mapsmadeeasy.com/drone_mapping)
- DatCon (https://datfile.net/DatCon/intro.html)
- CsvView (https://datfile.net/CsvView/intro.html)[41]

Removable memory card (SD, micro SD, etc.)

The internal, onboard SD card (not the SD card that may be found in the UAV's camera, if so equipped), contains flight logs and diagnostic data of the aircraft every time it was powered on. It serves as the flight controller's Black Box on the aircraft and records all important flight data. The SD card can also be a source for retrieving possible evidence for any flight accident.

DatCon reads the .DAT from a Phantom 3, Phantom 4, Phantom 4 Pro, Inspire 1, and Mavic Pro and produces output files that can then be used by CsvView, Excel, Dashware, and Google Earth. DatCon requires a 64-bit Java to run.

CsvView addresses some of the limitations of DatCon. CsvView provides the means to visualize log files data via graphs and a Google Earth like viewer. CsvView can accept these log file types

- .txt file produced by the DJI Go App.
- .DAT from the Phantom 3, Phantom 4, Phantom 4 Pro, Inspire 1, and Mavic Pro
- Litchi tablet app
- Autologic tablet app
- FPV tablet app

Note: The onboard .DAT for the Mavic Air, Mavic 2, and Mavic Mini is encrypted and cannot be processed by either DatCon or CsvView. However, the .DAT created on the mobile device by either the Go or Fly app is not encrypted and can be processed. Thereby allowing you to view the flight data in .TXT or .DAT flight log files.[42] Most .DAT files (other than the Mavic products noted above) can be read with DatCon.

Log file extensions differ depending if you look on the UAV itself to retrieve the file or on the mobile device used to operate and control the UAV. When retrieved directly from the UAV, the log file will have the file extension DAT and the file syntax will be FLYXXX.DAT (where XXX is the file sequence number).

Log files retrieved from a mobile device will have the file extension .TXT. For example, on a DJI Phantom, the file syntax for this type of log file would be DJIFlightRecord_2019-12-20_ [09-25-09].txt.

File formats vary among manufactures. Table 3.5 provides an example of several various flight file formats and their associated UAV manufacture and model types.

Table 3.5 Various flight file format by manufacturer

Manufacturer	Model	Flight file format
Aerialtronics	Zenith	0140_2020-04-07_15-22 -28 blue demo new fcc fixed props.bin "Lognumber""GPS Date".bin
Autel Robotics	EVO	autel_23030-05-25_(00-30-09-359)_750.txt
DJI	Phantom	DJIFlightRecord_2019-12-20_[09-25-09].txt
	Mavic Pro	DJIFlightRecord_2020-05-09_[15-33-28].txt
	Mavic Mini	field_flight.txt
Feeefly	Alta8 + Alta 6 (Synapse Flight Controller)	SYNLog-23-53-50_27-05-2020.csv
	Alta X + Alta Pro (PX4 Flight Controller)	03_55_28.ulg (can also produce .txt files)
Parrot	Bebop	droneFlight.txt
Yuneec	Typhoon H	Telemetry_00001.csv

DJI flight file formats will appear as DJIFlightRecord_2019-12-20_[09-25-09].txt, where flight logs should look like this for a Bepop model UA… droneFlight.txt.

Mobile device .DAT files, for Android will be found by searching /DJI/dji.go.v4/FlightRecord/MCDatFlightRecords. On OS devices, these files will be found will under File Sharing in iTunes – FlightRecords. The file should be first saved to disk. Next, open the sub-folder, MCDatFlightRecords for access to the data.

If the examiner does not have access to the mobile device used to control and communicate with the UAV under examination, the .DAT logs can be found on the UAV's internal micro SD card. It should be noted that the location and access to a UAV's SD card (if equipped) is not standardized. On some UAV models, the SD card may be hidden and require additional effort on the part of the examiner to locate and access the card.

Also, as noted previously, if the examination process will require dismantling the UAV, the examiner must first evaluate what, if any, impact such a dismantling process will have on the integrity of data and preservation of evidence. Additional consideration must also be given to the possible destruction of any physical evidence and the probably of not being able to reconnect or power up the UAV, if such action is required, at a later time.

Having determined the safest and forensically valid method of connecting to the UAV creates a disk image of the UAV's SD card and save it. If accessing the UAV's SD card is not possible via connection through the UAV's USB port (due possibly to damage to the UAV), following approved protocols, the examiner will need to safely removed the SD card from the UAV. Note this is a different SD card than the one found on the camera. It is a second SD card and not easily accessible to the user.

Working on a forensically validated copy of the original SD card, open the disk image and save the .DAT files to a folder on the examiner's computer.

An example of what the examiner should see when accessing the SD card for a DJI Phantom 3 UAV is shown in Table 3.6.

An examination of a DJI Phantom's .txt file using a product such as DJI Flight Log Viewer [https://www.phantomhelp.com/LogViewer/Upload/] produced the flight log shown in Figure 3.12.

Once the .txt file has been opened, several data retrieval options become available. The first option is to download KML data. The default file type for spatial data in Google Earth is KML (Keyhole Markup Language) or KMZ (a compressed or 'zipped' KML file). KML files are text-based and employ coding tags like to those used for XML or HTML programming.

Table 3.6 Example internal flight log (.DAT) files as found in the DJI Phantom 3 UAV

Name	Date modified	Type	Size
☐FLY000.DAT	4/3/2020 8:20 AM	DAT File	125,879 KB
☐FLY001.DAT	4/5/2020 10:20 AM	DAT File	78,902 KB
☐FLY002.DAT	4/6/2020 4:15 PM	DAT File	257,013 KB
☐FLY003.DAT	4/8/2020 10:00 AM	DAT File	56,980 KB
☐FLY004.DAT	4/12/2020 3:20 PM	DAT File	34,872 KB
☐FLY005.DAT	4/15/2020 4:30 PM	DAT File	200,533 KB
☐FLY006.DAT	4/21/2020 10:00 AM	DAT File	98,010 KB
☐FLY007.DAT	4/23/2020 3:00 PM	DAT File	67,340 KB
☐FLY008.DAT	4/25/2020 11:30 AM	DAT File	289,583 KB
☐FLY009.DAT	4/27/2020 4:15 PM	DAT File	88,763 KB

Time	Flight Mode	GPS	IMU Altitude	VPS Algtitude	Speed	Home Distance	Battery	Battery Voltage	Battery Cell 1 Voltage	Battery Cell 2 Voltage	Battery Cell 3 Voltage	Battery Cell 4 Voltage	Cell Deviation	Message
1m 43.7s	Starting Motors	10 satellites	0ft	0.3ft	0mph	14.0ft	94%	17.024v	4.21	4.272	4.264	4.278	0.068v	You are in a Warning Zone (Airport). Fly with caution
1m 44.7s	Manual Takeoff	10 satellites	0ft	0.7ft	0mph	14.1ft	94%	17.028v	4.211	4.273	4.265	4.279	0.068v	
1m 45s	Manual Takeoff	10 satellites	0ft	0.7ft	0mph	14.2ft	94%	17.028v	4.211	4.273	4.265	4.279	0.068v	
1m 47.1s	P-GPS	10 satellites	0ft	0.7ft	0.4mph	14.4ft	94%	16.996v	4.203	4.266	4.257	4.27	0.067v	Braking now. Return sticks to midpoints, then continue flying
1m 47.4s	P-GPS	11 satellites	0ft	1.0ft	0.2mph	14.5ft	94%	16.996v	4.203	4.266	4.257	4.27	0.067v	
1m 47.7s	P-GPS	11 satellites	0.3ft	1.3ft	0mph	0.3ft	94%	16.996v	4.203	4.266	4.257	4.27	0.067v	Home Point recorded. Return to Home Altitude: 98FT

P-GPS = GPS Positioning and Vision sensors are available
in this mode UAV is using GPS or the position

Figure 3.12 UAV-1 flight log.

To view the KMZ file on a PC, go to Google My Maps. Open the 'maps tab' [www.google.com/maps/d/]. Click 'create map.' Click 'import' and upload the .kmz file. Give the map a title and save it.

An example of the contents of the .kmz file retrieved from the flight record of the DJI Phantom UAV-1 is shown in Figures 3.13 and 3.14.

Information recorded in the .kmz file, such as shown in Figure 3.15, may indicate a pilot violation, in this case operating the UAV in a restricted area. Continued flight into a restricted area and disregard for warning would be document violation of flight operations in most every country.

Using the information provided by examining the .kmz file along with GPS information provided from the detailed flight log (see Figure 3.20), the investigator is able to identify the Home Point as recorded by the UAV. Further investigation is warranted; however, this Home Point information may lead to the address of the owner of the UAV. This is certainly a critical piece of information, which can be used in possibly identifying the UAV's owner or the potential target of an investigation.

The actual flight path of the UAV under investigation can be visualized by drilling down into a more detailed level within Google Maps. Figure 3.13 shows an example of the flight path information available from the .kmz file. Not much detail can be seen from this view. The investigator will need to expand and zoom in to obtain a more precise view of the UAV's flight path.

Once doing so, obtaining a visual of the flight area is available, again through Google Maps. Figures 3.16 and 3.17 are images of where DJI UAV-1 was flown.

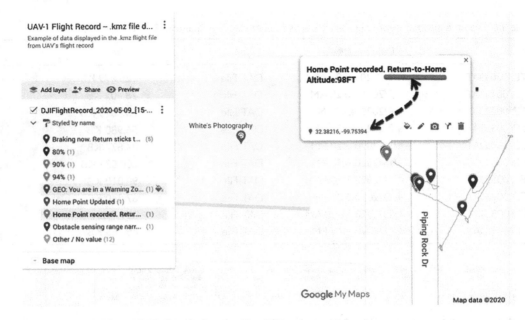

Figure 3.13 Flight record .KMZ showing home point GPS

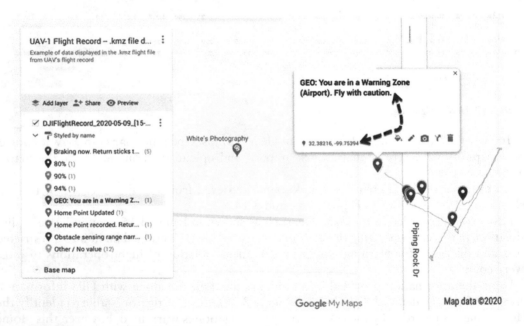

Figure 3.14 Flight record KMZ showing operation warning

The Investigator may wish a more granular view of this data and can obtain it my opening the .kmz file in Google Earth. Using Google Earth and the downloaded .kmz file, go to 'my location.' Opening the .KMZ file in Google Maps provides visual data on the flight location and path of the UAV.

Using a progressively tighter view and zooming in, Google Earth will display very detailed location information, recording the UAV's flight path. Figure 3.18 provides a three-step zoom into the map data centered on the UAV's initial GPS position.

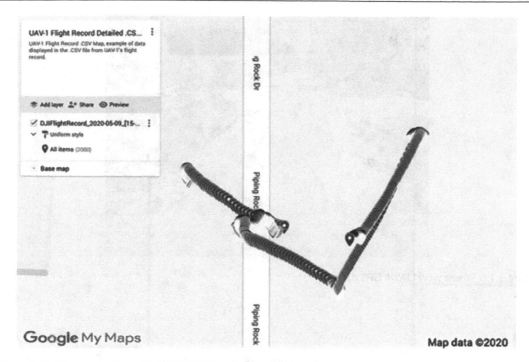

Figure 3.15 Flight record detailed UAV-1 flight path from the .CSV file

Figure 3.16 Street view UAV-1 GPS data

Figure 3.19 provides a map and graphical view of UAV-1's flight path. This data was obtained from the DJIFlightRecord_2020-05-09_[15-36-27].txt file.

Opening the DJIFlightRecord_2020-05-09_[15-36-27]csv.zip file gives you the file DJIFlightRecord_2020-05-09_[15-36-27].csv. Opening the .csv file will provide the detailed information shown in Figure 3.20.

Figure 3.17 Street view UAV-1 GPS data

Figure 3.18 Flight record – three-step zoom view of UAV flight path

The UAV's log file name is recorded as DJIFlightRecord_2020-05-09_[15-36-27].txt

This flight log contains a significant amount of data. Figure 3.20 displays partial log entries, however, item number two (2) indicates that there are 2,000 entries in the file. Each entry represents one second of flight time. The entire record logs 33:33 minutes and 33 seconds of flight time.

The detailed information from the .csv file (Figure 3.20) and the flight log (Figure 3.12) will enable the investigator to identify the UAV's launch point (starting motors).

The information contained in the .csv file is extensive. Examining a single log entry provides the investigator with an extensive data trove of information. For a single flight log entry, #1687 (of a total 2,000 log entries), Figure 3.21 illustrates the various types of data, about the UAV and the UAV's flight, which are recorded and retained in the .csv file.

Having discussed what type of data a forensic investigator may find on a UAV, the following section presents information and discussion on where those data may be found, within the UAV.

Figure 3.22 summarizes the various components that may be found on a UAS and UAV. Direct or easy access to these various components may be hampered by the UAV's internal design, location of media storage devices (e.g., SD cards), use of encryption on data by the UAV manufacture, and the overall condition of the UAV when received for processing.

Figure 3.19 UAV-1 flight path Google Earth

Id	Time(seconds)	Time(text)	Latitude	Longitude	FlightMode	Altitude(feet)	Altitude(meters)	VpsAltitude(feet)	VpsAltitude(meters)	HSpeed(mph)
1	1	103.7	1m 43.7s	32.38216647	-99.75394602	Starting Motors	0	0	0.3	0.1	0
2	2	103.8	1m 43.8s	32.38216646	-99.75394602	Starting Motors	0	0	0.3	0.1	0
3	3	103.9	1m 43.9s	32.38216647	-99.75394606	Starting Motors	0	0	0.3	0.1	0
4	4	104	1m 44s	32.38216651	-99.75394606	Starting Motors	0	0	0.3	0.1	0
5	5	104.1	1m 44.1s	32.38216646	-99.75394606	Starting Motors	0	0	0.3	0.1	0
6	6	104.2	1m 44.2s	32.38216642	-99.75394606	Starting Motors	0	0	0.7	0.2	0
7	7	104.3	1m 44.3s	32.38216641	-99.7539461	Starting Motors	0	0	0.7	0.2	0
8	8	104.4	1m 44.4s	32.38216636	-99.75394616	Starting Motors	0	0	0.3	0.1	0
9	9	104.5	1m 44.5s	32.38216631	-99.75394617	Starting Motors	0	0	0.7	0.2	0
10	10	104.6	1m 44.6s	32.38216626	-99.75394617	Starting Motors	0	0	0.7	0.2	0
11	11	104.7	1m 44.7s	32.38216619	-99.75394617	Manual Takeoff	0	0	0.7	0.2	0
12	12	104.9	1m 44.9s	32.38216604	-99.75394608	Manual Takeoff	0	0	0.7	0.2	0
13	13	105	1m 45s	32.38216603	-99.75394608	Manual Takeoff	0	0	0.7	0.2	0
14	14	105.1	1m 45.1s	32.38216598	-99.75394607	Manual Takeoff	0	0	0.7	0.2	0
15	15	105.2	1m 45.2s	32.38216595	-99.75394606	Manual Takeoff	0	0	0.7	0.2	0

DJIFlightRecord_2020-05-09_[15:   (1)

Find in table   (2)   1–200 of 2000

(1) The UAV's log file name recorded as DJIFlightRecord_2020-05-09_[15-36-27].txt

(2) Each entry represents one second of flight time. The entire record logs 2,000 entries (33.33 minutes of flight time)

Figure 3.20 Detailed flight log from UAV-1 .CSV file

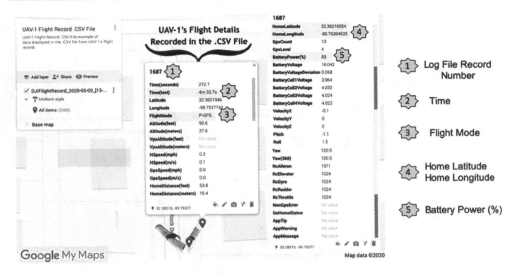

Figure 3.21 Detailed flight log entry 1687 from .CSV file

Hardware
WiFi Extender (on GCS)
Camera
CPU
GPS
Flight Controller
Flight Data Recorder (e.g., Black Box)
Micro SD Card
Ground Control Station (GCS)
Sensors
Electronic Speed Controllers
Autopilot
Batteries
Airframe (body, fuselage, propellers, brushless motors, multi-rotor frame and arms, landing gear
RC Receiver

Software
Operating System – Linux, MS Windows, Open Source, Real-Time Operating System (RTOS)
Firmware - operates from machine code to processor and afterward to memory access
Middleware manages flight control, navigation and telecommunications
Sensor Payload Software

Other
EXIF Data
.DAT files
.CSV files
.TXT files
UAV serial number (MC ID)

Figure 3.22 UAS-UAV components

UNMANNED AERIAL VEHICLES: WHERE DATA CAN BE FOUND

UAV anatomy

Before looking under the wing, so-to-speak, and collecting possible forensic evidence, it is good to know a bit more about the UAV construction, components, and parts. Having this knowledge will assist the examiner in narrowing down exactly where to look, where to spend time looking, and how to maximize the examination effort.

The soul of the UAV might be the pilot in command (PIC); however, the brains reside in the flight controller.

The flight controller reads all of the sensor data and calculates the best commands to send to the UAV in order for it to fly. The flight controller consists of flight stack software running on vehicle controller ('flight controller') hardware. Most flight controllers have 32-bit processors, while some older models may still be found with 8-bit processors. An example of flight controller hardware is the Holybro Kakute F7 All-In-One flight controller board shown in Figures 3.23 and 3.24.

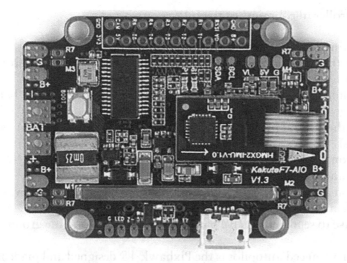

Figure 3.23 The Holybro Kakute F7 All-In-One flight controller[43]

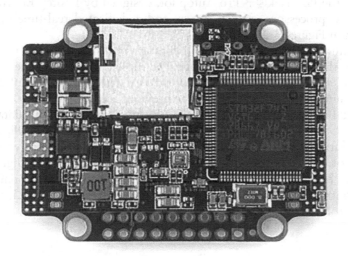

Figure 3.24 The Holybro Kakute F7 All-In-One flight controller (Reverse)[44]

All multirotor aircrafts require a flight control to make them fly. It takes the radio signals and mixes them to get each motor to do what it needs to do. It can be as basic as a gyro and accelerometer[45] with supporting hardware with no stabilization (used for acrobatic (acro) drone flight), add a few more sensors and a programmable chip and you can have basic stabilization (beginner UAV without a GPS functions), add a GPS function and a highly configurable ground station and you have an autopilot.[46]

An autopilot then is basically a flight controller with more features.

In general, most UAV autopilot systems include the following components:

- Accelerometer (measures acceleration forces)
- Barometer (measure the UAV's altitude)
- Compass/magnetometer (directional accuracy)
- Distance Sensors

- Flight Controller (the 'brains' of the UAV)
- GPS
- Gyroscope (measures rotational forces)
- Optic Flow (help to supplement and support GPS when flying in areas where GPS is difficult, for example, under tree cover, inside buildings)
- Power Module (used to convert the battery voltage from the UAV's battery down to a lower voltage that the autopilot uses)
- Processor (the central unit that runs the autopilot firmware and performs all the calculations)
- Pulse Position Modulation (PPM) Encoder
- R/C Inputs and Outputs
- Sensor fusion is software, which intelligently combines data from several different sensors because a UAV cannot operate on single sensor alone.
- Telemetry use to send and receive data between the UAV and the ground control station

An example of an advanced autopilot is the Pixhawk 4® designed and made in collaboration with Holybro® and the PX4 team and optimized to run the PX4 operating system, version 1.7 is shown in Figure 3.25. The Pixhawk 4's microcontroller comes configured with 2MB flash memory and 512KB RAM.

An example of the Pixhawk® 3 Pro Autopilot, designed by Drotek Electronics and PX4, featuring advanced processors, sensor arrays, and a dedicated real-time operating system (RTOS) is shown in Figures 3.26 and 3.27.

Several of the autopilot's components will capture and collect data that may be relevant and important to the UAV forensic examiner. The most critical being the UAV's Blackbox. The Blackbox maintains a log of all autopilot activity. The specific type of data and its relevance to an examination was discussed in the previous section.

A UAV's Black Box, depending on manufacture may contain useful data such as flight performance, flight paths, and drone ownership, a record of the entire flight for playback and analysis, operating practices and all of its maneuvers and interactions with other aircraft.

A central part of the operational UAV is the flight data recorder. The flight data recorder (FDR) is designed to be installed onboard a UAV in order to record all data from the aircraft's sensors, control commands (servos, ECU etc.), as well as control packets received from the Ground Control Station (GCS).

Figure 3.25 Pixhawk 4® Autopilot[47]

Figure 3.26 The Pixhawk® 3 Pro Autopilot (top view)[48]

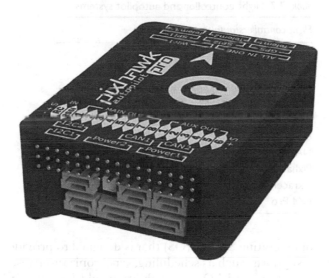

Figure 3.27 The Pixhawk® 3 Pro Autopilot (end view)[49]

An example of an FDR from UAV Navigation is shown in Figure 3.28. The UAV Navigation's FDR is equipped with 512 MB fast internal FLASH ECC memory for data circular buffer (up to 10 hours of data recording for standard UAV Navigation telemetry) and an external USB Mass Storage support (FAT32) with intelligent automatic data backups. Downloads of the latest flights occur only on each new connection.

The FDR provides advanced date and time stamping (date as a file name, 1ms precision time stamps, rechargeable Real-Time Clock (RTC) backup battery, built-in smart recharge circuit).

As technology advances, as UAVs evolve into more complex aircraft, the use of real-time operating systems (RTOS) will become more sophisticated, giving the autopilot systems installed on the next generation of UAVs even more capabilities, operability, and payload capacity.

Figure 3.28 Flight data recorder[50]

Table 3.7 Flight controller and autopilot systems

Flight controllers/autopilots
AMP
ArduPilot
BeagleBone Blue
CUAV V5
Cube
Emlid Navio2
NAZA-M V2
Pixhawk
Pixracer
PX4 Pro

An RTOS is a type of Operating System (OS) that is designed to provide real-time applications with several basic supports, such as scheduling, synchronization, resource management, precise timing, communication, and I/O. Its reliability would have direct impacts on safety operations of UAVs.[51]

Currently the most significant flight controller/autopilot systems, which the forensic examiner should be aware of, are shown in Table 3.7.

Flash memory (NAND, NOR, etc.)

Flasher tools

Some UAVs may store data in Flash memory, requiring the examiner to use specialized tools to capture and extract full memory copies of flash memory devices. Such extraction is similar to data capture performed on mobile devices.

Flasher boxes that are also known as flashers are a combination of software, hardware, and drivers. Flasher tools are the easiest and noninvasive way to read flash memory data.

Flasher boxes offer access to the phone memory unmatched by command-based methods. They also do not require the investigator to install any software on the target mobile phone and therefore do not disrupt the evidence in that way. This in turn means that they follow rules of evidence more closely than command based forensic software tools.[52] However, there is no guarantee that the flasher will preserve the evidence present in the device's memory intact.[53]

A lot of these flasher tools work in a similar way: They enter the bootstrap mode of a phone; upload dedicated flash loader software to RAM; execute this software; and then use it for low level access to the flash memory.

Pros and cons of using flasher tools

- Hardware connection is usually easy with a connector.
- Flash memory can be imaged without de-soldering of flash memory chips.
- Some tools do not make a full forensic image of flash memory (some do only parts of the memory space or skip spare area).
- It cannot be guaranteed that no data is written in flash memory.[54]

UAV FORENSIC EXAMINATION – FRAMEWORKS

Currently, there is no one universally accepted standard or standardized approach, methodology or framework to performing a digital, forensic examination of a UAV.

New makes, models, features, capabilities, and application of UAVs are entering the market almost weekly. The UAV forensic examiner will be required to consistently remain abreast of these changes. While established, certified forensic examination practices should always be followed, the examiner will need to be flexible in adopting new approaches and procedures in the forensic examination of commercial UAVs.

While there are several approaches to forensically examining a UAV, determining which approach is better than the next, without the benefit of any current singularly accepted standard or standardized approach for examination, makes this identification difficult.

Table 3.8 presents a summary of five, more extensive and detailed, proposed UAV examination frameworks. Each presents a slightly different approach. There is, however, a considerable overlap in the basic examination process of each framework; the examiner will ultimately select individual steps and processes, developing a customized examination process, based upon the prevailing circumstances of each case presented.

Until a universally standardized, forensically certified, legally accepted, UAV digital examination process is established and available, these and other frameworks will provide 'best available' guidance to examiners and others seeking to forensically exam a UAV.

The framework presented by Jain et al. focuses the examination on the UAV airframe. This proposed framework does not focus investigative efforts on retrieving data stored on the GCS. From this approach the examiner may be able to obtain data such as, potential latent fingerprints, registration numbers, weight, model/make/classification information, communications, and data storage capabilities. These data, collectively, may assist authorities in identifying the owner of the UAV. This is a hardware-focused framework.[55]

The examination framework proposed by Gülataş et al. focuses on the digital data stored on the UAV and extracting digital evidence to determine the flight path of the UAV. These data may provide evidence which proves the possible, illegal use of the UAV. This would be a more inherently data-focused framework.[56]

The INTERPOL framework, beyond a broad and necessary discussion of cyber forensic techniques, focuses on the digital forensic process for drones and drone controllers. If there are associated devices such as laptops, mobile phones, or tablets, this examination process is covered in the INTERPOL Global Guidelines for Digital Forensics Laboratories. Procedures addressed in the INTERPOL framework cover methods for conducting digital forensic analysis on drone-related electronic evidence in a digital forensic lab (DFL).

Table 3.8 Proposed UAV examination frameworks

Examination activities	Roder, A. Choo, K.K.R. Le-Khac, N.	INTERPOL framework for responding to a drone incident	Gülataş, I. Baktir, S.	Jain, U. Rogers, M. Matson, E.T.	Salamh, F. Karabiyik, U. Rogers, M.
Scene Control			✓		✓
Acquisition		✓			
Preparation	✓		✓	✓	✓
Identification/ Classification				✓	✓
Customization Detection			✓	✓	
Data Acquisition	✓		✓	✓	✓
Evidence Authentication	✓		✓	✓	✓
Examination	✓	✓	✓	✓	✓
Analysis	✓	✓	✓	✓	✓
Report Presentation	✓	✓	✓	✓	✓

The INTERPOL framework also features extracts from fundamentals of crime scene management and processing that are taken from the Crime Scene Responder Guide published by the United States National Institute of Justice.

According to Jürgen Stock, Secretary-General of INTERPOL, the INTERPOL framework document is designed as a reference tool for law enforcement worldwide.[57]

The forensic examiner must evaluate each situation independent of the last examination. When necessary, customizing the examination procedures to meet the current examination requirements, UAV make/model, physical condition, and any other sustaining factors. In the end, taking an approach that has the highest potential for the preservation of evidence (digital and physical).

The following is a discussion of one proposed examination framework, which provides a methodology for examining both the airframe and associated components as well as the data storage and communication capabilities of the UAV.

Writing in "Unmanned Aerial Vehicle Forensic Investigation Process: DJI Phantom 3 Drone as A Case Study," Roder et al. propose a three-stage, 20-step forensic examination process.[58] The forensic approach and methodology by Roder et al. is as follows:

Three-Stage Process; 20 steps

1. Preparation (Steps 1 to 6)
2. Examination (Steps 7 to 17)
3. Analysis/Report (Steps 18 to 20)

Preparation

Step 1 – Identify and determine the chain of custody
- Have all necessary procedures and documents, required to substantiate a legal and defensible chain of custody, been prepared and completed?

Step 2 – Have conventional forensic practices (e.g., DNA, fingerprints, and ballistic) already been implemented?
- The collection of traditional forensic evidence may provide additional useful information to authorities. This evidence may also corroborate evidence, conclusions, and report findings submitted by the forensic examiner.

Step 3 – Identify the role of the device in conducting the offence (Offence analysis)
- What were the circumstances that lead the UAV to be confiscated or obtained? Under what conditions was the UAV acquired? What role is the UAV suspected to have played in these circumstances? This information will assist the examiner in identifying evidence that may be specifically germane to the circumstances leading to the UAV's confiscation.

Step 4 – Photographs
- In a forensic examination, photographs will substantiate statements made by the examiner regarding the evidence obtained from the physical UAV. Its condition upon acquisition, the existence of any attached payloads, evidence of non-manufacture modifications. When it is opened, the condition of any surfaces, internal components, damage, etc. The UAV should be photographed from all possible angles and all possible surfaces.

Step 5 – Identify the make and model
- This may be a more challenging process than it seems. UAV manufactures are responding to a continuing increase in demand, newer, more sophisticated models seem to come to market monthly. This is only destined to continue.

  The UAV may have also been modified to conceal its owner's identity. In the United States, the FAA will issue a unique registration number beginning with either an 'N' or 'FA.' These numbers must be placed on the unmanned aircraft to be readily visible, or they may be inside a battery compartment or other place in the aircraft, provided no tools are needed to open the compartment. The registration number will be unique to the operator, if operating strictly as a Model Aircraft,[59] and unique to the aircraft, if operating other than as a Model Aircraft.

  The presence or lack of these identification numbers may be significant in an FAA investigation. For example, an operator may state that he or she is conducting an approved commercial activity, which usually requires registered aircraft. However, the absence of registration markings on the UAS may indicate that the aircraft is not registered, meaning the operation may not be authorized. Registration identification numbers may not be conspicuous from a distance because of the size and non-traditional configuration of some UAS.[60]

  The registered owners of UAS operating under an approved commercial or governmental authorization bearing identification numbers can be found by searching for the N-number on the FAA's website: www.faa.gov.

  Identifying the UAV make/model will also provide insight into (a) how data are stored within the UAV; (b) where those data are stored; (c) file syntax for data files; (d) the UAV's intended use (as designed by the manufacture) and operating specifications; and (e) other useful information valuable to the examiner.

Step 6 – Open source investigation to identify device characteristics, potential data storage locations, and available forensic/non-forensic tools

- This step, in combination with Step 5, will assist in determining if the UAV has been modified for use and purpose, not originally intended by the manufacturer. Once completing Step 5, the examiner may use information provided via the Internet to access the manufacture's website and obtain detailed and specific documentation regarding the UAV's original equipment manufacturer (OEM) design, components, storage capacity, configuration, etc. These data can then be compared to the UAV under examination to determine to what extent (if any) the UAV has been modified.

Examination

Step 7 – Identify capabilities (Video/Audio recording, carrying capacity, and technique)
- This step pulls together information gleaned from Steps 3, 5, and 6. This information should provide the examiner with a clearer indication as to the intended use of the UAV, any modified capabilities beyond those originally installed by the manufacture, what purpose these modifications may have served and what role the UAV may have played in any criminal offense.

Step 8 – Identify potential modifications
- The confiscated UAV should be examined for any obvious modifications, not in spec with the manufacturer's release for the same, commercially available, make and model. Information from Steps 3, 5, 6, and 7 should assist the examiner in accurately determining if the UAV would be capable of engaging in the suspected criminal offense.

Step 9 – Identify data storage locations
- Refer to the previous section in this chapter and the discussion of where data may be found and stored within the UAV and associated flight control devices (e.g., ground control station, flight controller/mobile device).

Step 10 – Identify ports
- Access to the UAV's internal components, including the flight controller (e.g., Black Box), data storage devices (e.g., SD cards) etc., is typically through an external port. That is if the UAV is intact and has not been damaged. If access to internal data storage via traditional ports is not feasible, this will require opening the UAV.

  The examiner must follow strict forensic procedures in all cases and importantly so here. As each case is different, there is no exact 'game plan.' Following forensic protocols is critically important.

  Opening the UAV may damage components and may make accessing any data more difficult/challenging. Processes used to access data, which are not preformed correctly, could 'brick' the device, potentially making access to these data impossible.

  Actions on the part of the owner/operator of the UAV could have intentionally corrupted any external ports in an attempt to make access to any onboard data more difficult. The UAV airframe may have been damaged during flight, upon landing or when confiscated. Any of these actions may also contribute to the external ports not functioning as designed or to be unavailable to use as an access point to the UAV's internal data.

Step 11 – Extract removable data storage mediums
- Roder et al. recommend the use of non-destructive removal methods to obtain data. As with any data collection process, which may result in the collection of evidential data, adherence strict identification, collection, processing, and documentation procedures is required.

Step 12 – Preserve evidence – Clone / forensic copy of storage medium
- This step is a common process within forensic examination protocols. The examiner should make forensically sound images of all obtained data. However, what is unique in the examination of UAV systems is the data read/write/storage process.

In their examination of the DJI Phantom 4, Roder et al. make an astute observation, essential to the forensic investigator...

When a new flight log is opened, it also has the secondary effect of closing the previous .DAT file. The last flight log is not viewable until the device is turned on. By cloning the removable storage device, the examiner is then able to replace the memory card with the cloned memory card, power on the device; thereby, closing the final .DAT file, and ultimately re-examining the memory card which now has the last recorded flight data (last recorded prior to seizure). Original data has not been changed, but new data has now become viewable.[61]

Step 13 – Traditional interrogation of storage medium – use certified forensic tools and
Step 14 – Extended interrogation of storage medium

- In traditional forensic examinations, the identification and use of proven, certified forensic tools is not difficult. However, in the emerging field of UAV platforms, data storage formatting is not regulated and there is no set standardization or agreed upon nomenclature for data representation. See Table 3.5 Various Types of Flight File Formats, earlier in this chapter for examples of various UAV data file formats. While there are many different tools available, designed to read UAV .DAT files, many have yet to be certified or proven in court. Thus, leaving the forensic investigator to tread lightly when considering the use of un-certified, third-party software to access and read UAV data files.

  As with every forensic examination, strict protocols for the collection of data should be followed and documented.

Step 15 – Interrogation of the UAV/drone – Potentially using a clone of any storage medium identified

- Depending on the make and model of the UAV and the condition of the airframe itself when confiscated, the removal of internal storage media may be problematic. Some UAV manufactures have configured their aircraft so that the location of any removable media is well concealed and, in some cases, designed not to be removed at all. Actually, gaining access to these devices, if at all possible, may require opening the UAV as was discussed previously.

  In such cases the examiner may elect to perform a live acquisition of the data via a direct USB cable connection to the UAV. Protocols and procedures that would be followed in the live forensic examination of mobile devices would be good to implement here.

  Another cautionary note, if performing a live acquisition on the UAV the examiner, should:

  - Photograph the UAV airframe from all angles (this can be in addition to photos already taken in Step 4).
  - Remove all blades, props and propellers
  - Remove any external payload (e.g., camera)
  - Secure the examination area
  - Wear protective eye wear, in case of accidental or intentional release of flying debris
  - Ground all working surfaces and the UAV to avoid any errant electrical discharge

Once again, the necessity to utilize examination tools that have yet to be certified may be the only recourse left for the examiner. Recognizing this, the use of proper examination, evidence collection, and documentation procedures should be followed at all times.

Step 16 – Interrogation of peripheral devices: Flight controller, mobile device, etc.
- All secondary, ancillary storage devices, and methods in addition to the above (e.g., ground control station, tablet, remote controller, potential cloud services, recorders, sensors, cameras, etc.) should be included in the examination.

Step 17 – Extract removable data storage mediums (Destructive)
- In opposition to Steps 13 and 14 above, Step 17 should be the step of last recourse. If acquiring potential data via non-destructive means is no longer a viable option, then destructive (e.g., chip-off) procedures should be tried.

   The examiner should be aware, however, that destructive collection methods, while valid and potentially necessary, could potentially damage UAV components rendering the data stored on them, impossible to collect.

   Consideration should be given to what impact these destructive collection methods may have on potential data and the associated risks to the overall examination. Deciding if the risk warrants the approach or suspending the examination and researching alternative approaches.

   The examiner should also be prepared to document all destructive processes taken and their effect (if any) on the integrity of any data retrieved.

Analysis/Report

Step 18–Initial review of extracted data
- Here the examiner may have an advantage in the acquisition of possible evidential data, unintentionally left behind by the target of the investigation.

   Individuals unfamiliar with the deeper, inner workings of UAV file storage methods and both accessible and inaccessible data storage devices, located inside the UAV, may simply be unaware that they are leaving behind latent evidence.

   Evidence, in the form of .DAT files, GPS data, EXIF data, video and still images, etc. Evidence, which could potentially link the target of the investigation to the offense. Evidence that had the target known was there or how to get to it, could have been deleted, wiped, tampered with, or possibly removed and destroyed.

   It is also logical to suspect that the time between the offense and the acquisition of the UAV may not have afforded the target, time to alter, or destroy any data contained in the UAV.

Step 19 – Interpreting and translating of data – Into a human readable and evidential Format.
- Roder et al. define this examination stage as:
  - Data sifting is the process of reducing the data obtained through examination, to only case relevant data.
  - Data confirmation is the process of verifying the obtained data and confirming its accuracy
  - Data translation is the process of changing often complex datasets into a human readable format.[62]

Step 20 – Report/Statement
- The ability and necessity for the examiner to communicate succinctly and clearly the examination steps and processes, findings, limitations, and conclusions, related to the evidential data retrieved from the UAV is crucial. Clear, concise reports are essential to law enforcement and legal professionals and will have a direct impact on the ability for these individuals to successfully prosecute and/or defend, individuals suspected of criminal activity and the illegal use of the UAV.

As apparent, this proposed framework is extensive. Its application in the forensic examination of a UAV should be assessed on a case-by-case basis.

When selecting a framework, it may be best for the examiner to base the decision on the focus and strengths of the framework. In the end, the most reasonable and logical approach will be to blend frameworks into an examination approach that is most appropriate for the investigation.

Until a UAV examination standard and a standardized examination process has been tested, certified, and proven acceptable in a court of law, selecting the best features from among multiple frameworks may be the soundest approach.

UAV DATA PRESERVATION

In an effort to preserve potential evidential data that may exist on a UAV, it is best to review how data may be destroyed or altered, either intentionally by the target of the investigation or by examination processes.

Has the UAV been tampered with?

The UAV forensic investigator may wish to first determine if the UAV under examination has been 'unlocked.' Unlocking a UAV can be seen as the equivalent to rooting or ID jailbreaking your mobile device. Basically, unlocking the UAV gives the owner privileges to modify the software code on the UAV, which the manufacturer would not normally allow a user/owner to do.

To see how difficult (or rather easy) it would be to attempt to access data recorded and stored on the UAV and then attempt to erase it, just head over to Google and execute a search on 'unlock your drone.' As of this writing 31,600 entries popped up in 0.39 seconds. This has to tell you something about (a) the interest in doing this; (b) the potential ramification to the successful preservation of possible evidence; and (c) the resulting non-compliance and legal issues resulting in unlocking your drone.

By unlocking the UAV the owner/operator can make modifications such as the following:

- Disable or remove maximum altitude limit
- Fly beyond line of sight
- Change the UAV's serial number
- Increased Flying Range
- Override or remove no fly zones (NFZ)
- Increase the UAVs speed
- Upgrade or downgrade firmware

According to the folks at 911 Security, close to 1% of all drones detected and tracked with the company's drone detection platform have no specific city or state listed in their data. Meaning owners are manipulating their drone IDs and associated information to cloak or hide their identity.[63]

Data sources to be manipulated

Data, as we have discussed, may be recorded and found on the remote controller (RC), ground control system (GCS), or mobile device and the flight controller (FC).

Depending on UAV model type, configuration and payload, the examiner may expect to find two data recording media devices in the form of SD cards on the UAV – One easily accessible in the camera (if so equipped), the other not as easily accessible. Again, depending on the UAV model, this second SD card may be hidden and inaccessible, without opening and disassembling the UAV.

The internal SD card contains flight logs and diagnostic data of the aircraft every time the UAV is powered on.

Clark et al. conducted a digital forensic investigation of DJI Phantom 3 and noted that there were two primary sources for flight data. These include TXT files created by the DJI GO mobile application and stored on the mobile device and DAT files created by the drone itself and located on the drone's nonvolatile internal storage. Both files are encrypted and encoded using two different proprietary formats. After decrypting and decoding these files, data regarding the GPS, motors, remote control, flight status, and other information can be extracted. These files essentially serve as the electronic flight recorder for the drone.[64]

The .TXT files can be synced to the user's account on the manufacture's website. For example, a user piloting the DJI UAV would be able to sync the DJI Go app to his/her DJI account, as the flight files have been stored on the DJI servers, presumably in the cloud. This allows the user to access flight records on the existing device.

This raises two interesting points for consideration. First, if the UAV has been damaged and the examiner is unable to access or retrieve potential data from the UAV itself, the examiner may need to (a) examine the mobile device used to control the UAV (if available) and/or (b) examine files that have been stored on the manufacture's servers. Either step may require the execution of additional legal warrants to obtain the mobile device (if possible) and to access information stored on the manufacturer's servers.

Files stored on the manufacturer's app allow the user to easily transfer these files to a new UAV (same manufacturer) or delete the files stored on the mobile device.

The second kind of flight data, a .DAT file is as previously noted, is what the drone itself records to its nonvolatile internal storage. These files are not bound to a user's account and they cannot be wiped. If the examiner can locate and access, using forensically sound means, the UAV's internal SD card, access to these flight data files is possible.

Through an examination of the.TXT log file (on the mobile device) and the .DAT log file (from the UAV), the examiner would look to correlate the two log files, thus demonstrating that these log files are one to one match. This correlation and matching would be used as evidence linking the owner of the mobile device as the possible PIC of the UAV, while the UAV was operated and engaged in suspected illegal activity.

The examiner should be aware that these files are overwritten with new data as subsequent flights are taken. Thus, depending on when the UAV is acquired and examined, some data recorded to the internal SD card may have been overwritten.

The DJI Phantom 3, for example, will rotate logs as the internal SD Card fills up, deleting the oldest log to ensure there is available space. The SD Card may be reformatted at any time by the user by simply using the DJI GO App. Formatting will delete any and all remaining .DAT files and result in a clean SD Card.

Flight records are similar to an air traffic control tower's data. For DJI UAV models, Mavic Pro, Phantom 4, Phantom4 Pro, Inspire 2, Inspire 2 Pro, Matrice 100, Matrice 200, Matrice 600, and Spark, for example, the flight controller data refers to the data, including the working statuses of different modules, control, and navigation information, etc., generated by the flight controller after the DJI drone is powered on and stored in the internal memory until the drone is powered off.

One data file will be generated after the drone is powered on and off. The data files will be named in a sequential numbering order. The log file will be split when its file size exceeds 450 MB. Around 10 MB data will be generated after the drone has flown for one minute, so the log file size will reach 450 MB after the drone has flown for 45 minutes.

Each flight controller file has a DateTime and usually this can be used to determine the desired flight controller file(s) the examiner wishes to review. However, the DateTime comes from the remote controller. If the battery is turned on while the remote controller is not powered up the DateTime is set to the DateTime of the last flight controller file.

The result is two flight controller files with the same DateTime. Also, if the flight controller file reaches a size around 460 MBytes the UAV will end that flight controller file and start a new flight controller file. This new flight controller file will have the same DateTime as the first flight controller file.[65]

The size of the flight controller file can be used to estimate the length of the time the battery was on. The ratio is .11 MBytes/second. Be aware that the battery will be on longer than the length of the flight.

Once again, following approved examination protocols, if the UAV is off (which is highly likely), leave it off as you begin to search for and then examine the internal SD card. Note... examining a UAV when it is on and operational could be very dangerous and is not recommended. Always use safety precautions when examining any UAV.

If the UAV is still in-flight operating mode and 'on' when acquired, the examiner should note this and also be aware of any modifications to data that may occur as a result of turning the UAV off.

Data preservation

Presuming that you can actually find and get to the internal SD card, don't switch on the drone again until you take the SD out, any new image or video might accidentally flash into the memory, overwriting few last seconds of the unfinalized previous file. Potentially damaging evidence. As noted by Beckett et al., the drone should not be turned on as turning it on changes data on the drone by creating a new .DAT file, but may also delete stored data if the drone's internal storage is full.[66]

NIST and the UAV Computer Forensic Reference Datasets (CFReDS)

The National Institute of Standards and Technology (NIST) maintains a repository of images made from personal computers, mobile phones, tablets, hard drives, and other storage media. The images in NIST's Computer Forensic Reference Datasets, or CFReDS, contain simulated digital evidence and are available to download for free.

NIST has established a new section of CFReDS dedicated to UAVs, where forensic experts can find images popular makes and models. As of this writing there are 11 UAV manufacturers and 30 different UAV model's images available for download (see Table 3.9). The UAV images were created by VTO Labs, a Colorado-based digital forensic and cybersecurity firm.

Steve Watson, chief technology officer at VTO purchased three identical models of each UAV and flew them until they accumulated a baseline of data. He then extracted data from one while leaving it intact. He disassembled a second and extracted data from its circuit board and onboard cameras. With the third, he removed all the chips and extracted data from them directly. He also disassembled and extracted data from the pilot controls and other remotely connected devices. Various acquisition methods were applied across each set of drones, for example, logical, physical, chipoff, etc.

The images were created using industry standard data formats so that investigators can connect to them using forensic software tools and inspect their contents. The images for each model also come with step-by-step, photo-illustrated teardown instructions.

Table 3.9 UAV image files in the VTO database

AION	R1 UGV Rover
ArduPilot	DIY drone
DJI	Agras MG-1S
	Inspire 1
	Inspire 2
	Matrice 210
	Matrice 600 Pro
	Mavic 2 Enterprise
	Mavic 2 Zoom
	Mavic Air
	Mavic Pro
	Phantom 3
	Phantom 4
	Phantom 4 Pro V2
	Spark
	Spreading Wings S1000+
Intel	Falcon 8+
Parrot	Anafi
	Bebop 2 with Skycontroller
	Bluegrass
	Disco FPV
Qysea	Fifish P3
Ryze	Tello
SenseFly	Albris
	eBee
Sky Viper	v2450 GPS
Skydio	R1
Yuneec	H520
	Typhoon H
	Typhoon Q500 4K

Watson was able to retrieve serial numbers, flight paths, launch and landing locations, photos, and videos. On one model, he found a database that stores a user's credit card information.

The VTO developed image database is a valuable resource for UAV forensic examiners. The following information is included in the database for each UAV examined:

- Basic background and reference information about the drone model
- Detailed photos of the drone and its controller
- Instructions and accompanying photos for how the drone, controller and battery can be disassembled
- Component information and identification for the major chips/components on the drone, controller and battery circuit boards
- Information about the drone's operation/flight (when, where, flight duration, temperature, altitude) and the devices that were connected to it
- Information about and instructions for data acquisition methods that were used for the drone, identified data storage locations, and the results of the data acquisition methods that were applied
- A list of additional resources related to the drone model examined[67]

Investigators can use the images to practice recovering data, including deleted files.

Obtaining forensic evidence

A forensic image is a complete data extraction from a digital device.

The examiner may begin by taking a physical image of the SD card. Lock the SD card into 'read-only' mode and insert it into a write blocked/protected reader on a forensic workstation. Run an MD5 hash and stored the results. Use a utility such as disk dump (dd), dump the entire disk's contents to an image file. Hash the copied image file compare and verify. Duplicate the SD card, hash, verify, record, and store results. Secure and preserve the original SD card and hash documentation.

Extract flight controller files utilize a forensically validated tool, for example, FTK Imager (a data preview and imaging tool used to acquire data (evidence) in a forensically sound manner by creating copies of data without making changes to the original evidence). Use the most current release, at the time of this writing version 4.3.1.1.

It is good forensic practice to cross-validate data, findings That is producing the same result with two different tools. The examiner should, however, be acutely aware of one significant fact when utilizing cross-validation tools, as aptly stated by Beckett et al., 'These methods (cross validation of forensic tools), are sound methodologies in the context of Judicial or scientific reproducibility but, both have a major flaw inasmuch as they do not deal with what happens when the tools are incorrect.'[68]

Data are volatile and can easily be rendered inadmissible if not handled in a forensically sound and correct manner. The examiner should apply the same levels of data protection, loss prevention, data contamination, chain of custody, and examination procedures when examining UAV data and associated medium, as would be applied to any digital forensic examination. Examination should be carried out on the hash verified, duplicate copy of the backup copy of the original data.

UAV data may have different file formats, be encrypted, stored in the cloud, or hidden on hard to access media; however, it is still data. The examination of UAV data should not vary from accepted, accredited digital forensic examination procedures and protocols.

UAV DIGITAL EXAMINATION – QUESTIONS FOR MANAGEMENT

Because of the variety of data sources, digital forensic techniques can be used for many purposes, such as discussed in this chapter, the examination of unmanned aircraft systems and unmanned aerial vehicles. With the rapid increase in the use of UAVs for both legal and illegal purposes, both independent organizations and law enforcement agencies will eventually need and require need to have the capability to perform UAV digital forensic examinations.

The process for performing a digital forensic examination comprises the following basic phases:

- Collection: Identifying, labeling, recording, and acquiring data from the possible sources of relevant data, while following procedures that preserve the integrity of the data.
- Examination: Forensically processing collected data using a combination of automated and manual methods, and assessing and extracting data of particular interest, while preserving the integrity of the data.
- Analysis: analyzing the results of the examination, using legally justifiable methods and techniques, to derive useful information that addresses the questions that were the impetus for performing the collection and examination.
- Reporting: Reporting the results of the analysis, which may include describing the actions used, explaining how tools and procedures were selected, determining what

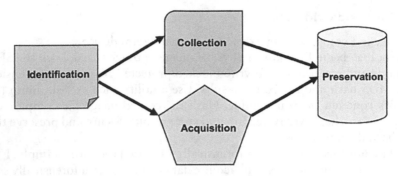

Figure 3.29 Evidence-handling processes according to ISO 27037[70]

other actions need to be performed (e.g., forensic examination of additional data sources, securing identified vulnerabilities, improving existing security controls) and providing recommendations for improvement to policies, procedures, tools, and other aspects of the forensic process.[69]

ISO/IEC 27037:2012 also provides guidelines for specific activities in the handling of digital evidence, which are identification, collection, acquisition, and preservation of potential digital evidence that can be of evidential value (see Figure 3.29).

The following is a series of questions, compiled to assist the reader/examiner in managing the UAV forensic examination process. This list of questions is by far not exhaustive. These questions are included here as a beginning, upon which to build a more comprehensive assessment tool for both pre- and post-forensic examination of a UAV.

UAV digital forensic examination questions

Administrative

1. Is the individual assigned to perform the examination, suitable trained to do so?
2. Are there any priorities, time scales by which results/responses are required?
3. Have all local police guidelines relating to the search and seizure of evidence been identified and followed throughout the acquisition of the UAV and associated components (e.g., GCS, flight controller, etc.)?
4. Are all conclusions derived from the data analysis written in a report that is concise and complete?
5. What are the potential sources of data internally to or tangentially associated with, the UAV?
6. Of the potential sources of data, which are the most likely to contain helpful information and why?
7. Which data source would be checked first and why?
8. Which forensic tools and techniques should be employed for an examination of the UAV?
9. Are there any additional tools and techniques might also be considered/used?
10. In addition to the forensic examiner assigned to perform the digital forensic analysis of the UAV, are there other individuals would probably be involved in the forensic activities?

11. What are their potential roles and responsibilities?
12. What communications with external parties might occur, if any? Why?
13. Who is responsible for managing these external communications?
14. Depending on the nature of the examination (internal, external, criminal) are there any privacy considerations which might affect the use of forensic tools and techniques?
15. Who should determine how much effort should be put into attempting to recover any data that may be encrypted? How would this be determined?

Operational

1. Have all appropriate approvals been obtained if the UAV examination will involve dismantling deconstructing or destroying the UAV, in the process of seeking possible digital evidence?
2. Have all steps been taken to assure, that wherever possible, no actions taken during the seizing of any evidential material would cause that material to be changed?
3. Where actions taken during seizure (e.g., shutting down UAV, GCS, flight controller, etc.) that may change data, have these actions been documented?
4. Are pre-printed forms used to complement the examiner's contemporaneous notes that will ensure that all the required information is recorded consistently?
5. Are all UAV evidential data and physical pieces and parts stored in a secure location throughout the examination process?
6. Are all digital examination tools selected from a library of tested and approved software?
7. Has the appropriate a chain of custody been established?
8. Have all UAV parts and associated airframe components been properly inventoried?

Procedural

1. Are there are any constraints (e.g., preservation of material for other purposes such as fingerprint examination, DNA, custody time limits, cost, etc.) that must be considered?
2. What procedures are in place that will mitigate improper handling of a mobile device during preservation and collection, to protect against the loss of digital data?
3. Are all activities relating to the seizure, access, storage or transfer of digital evidence fully documented, preserved and available for review?
4. Are traditional forensic processes, such as testing for fingerprints or DNA, performed in order to establish a link between a mobile device (flight controller) and its owner, user, PIC?
5. Is the recovery of digital evidence fully documented at all stages of the examination process, as appropriate?
6. Have appropriate anti-contamination precautions been established (and taken during the examination) to minimize any chance of accidental contamination of items which may subsequently be required for other laboratory examinations, (e.g., fingerprints which may be latent on the surfaces of the UAV's airframe, battery, flight controller, etc.)?
7. In most cases, the UAV must be switched off at the time of seizure and transportation in order to (a) safeguard personnel and (b) preserve the data (e.g., GPS, navigation equipment). Are procedures in place that will mitigate the possible loss or corruption (overwrite, destruction) of data when the UAV is switched back on?

8. Do appropriate procedures exist that document the collection of any digital devices (FPV goggles, camera, mobile device, remote controller (RC), etc.) associated with the UAV, in accordance with organizational guidelines and procedures?

9. What procedures are in place to disconnect mobile devices, used to control the UAV, from their networks to ensure data is not remotely modified or destroyed?

Technical

1. Has an assessment of the risk of contamination to electronic evidence been performed the examination commences?

2. Are appropriate and detailed procedures established that address the actions and steps to be taken and approvals required, if a UAV needs to be dismantled or the examination will cause the destruction of the device (e.g., desoldering memory chips from printed circuit board, chip-off).

3. Has a best effort attempt been made to obtain the following information, prior to the examination of the UAV?
   - PIN/Password access to flight controller (e.g., mobile device), third-party storage (e.g., cloud) storage services
   - Make/Model of all devices that generate any electronic data
   - Telecommunication data configuration (e.g., mobile carrier, configuration, etc.)
   - Operating system (OS) type, version
   - What is the suspected or known use of the UAV before, during and after the incident/ arrest
   - The person (PIC) or persons (flight ground crew) involved
   - Any known sequence or timings of events
   - Identification of the person or persons responsible for recovery of the UAV
   - The sequence and timing of events in the recovery of the UAV submitted for examination
   - Of the UAV items acquired for examination, has priority been given to identifying which parts and pieces of the UAV offer the best choice of target data in terms of evidential value

Post-Examination Hot Wash

1. From a forensic standpoint, what would be done differently, if the UAV would have had to be examined at a different physical location (in the field versus in a forensic lab)?

SUMMARY

This chapter has presented a review of the digital examination process for a commercial UAV (aka drone).

Figure 3.30 presents a model of the UAV forensic examination environment. This model includes the basic operational characteristics, features, and elements, which a forensic examiner will find, in the process of performing a UAV cyber forensic examination. The cyber forensic examiner must consider each of these factors when preparing to perform a UAV forensic examination.

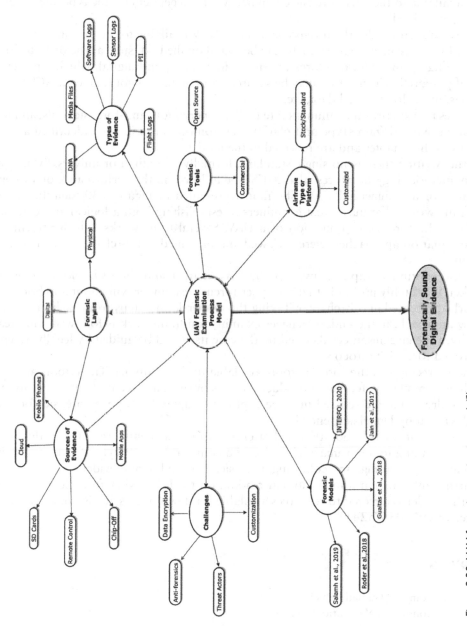

Figure 3.30 UAV forensic examination model[71]

Discussed in the chapter and represented in the model are the various and possible sources of UAV digital evidence and where this evidence may be obtained (e.g., SD cards, mobile phones, the remote controller, cloud storage, etc.). The model also includes the types of digital evidence that may be available and recovered from the UAV (e.g., flight-, software- and sensor-logs, operator/owner PII, media files, etc.). The sources, types, and location of digital evidence available to the UAV forensic examiner will be dependent on the type and model of UAV being examined.

As shown in Figure 3.30, the forensic layers of a UAV available to the examiner are represented as physical (e.g., evidence which may be found on the UAV such as the operator (PIC) or owner's fingerprints, photo, address, license number, etc.) and digital, which represents a breadth of potential evidence that may be secured from numerous sources (e.g., GCS, mobile phone, logs, files, cloud-stored data, etc.).

Challenges to a successful examination (e.g., data encryption, anti-forensic tools, and techniques) along with airframe type and platform customization have been identified and discussed within this chapter and are reflected in the model.

As of this writing there is no single standard, formally accredited or universally accepted approach, methodology, or process that MUST be followed in the performance of a forensic examination of a commercial UAV. The model presented in Figure 3.30 identifies several leading frameworks that may assist examiners in establishing both a logical and defensible approach to the forensic examination of a UAV. Several frameworks, which present very viable examination approaches, were reviewed and presented in this chapter for the reader's evaluation.

Given the continued, rapid growth of the UAV market and the technology supporting this market, it is highly probable that the cyber forensic examiner will adopt a hybrid of the frameworks presented in this chapter. Taking the most relevant steps from each or developing a new approach, in the end, whichever examination framework is followed will greatly depend upon the circumstances dictated by the case itself and be guided by legally accepted and approved forensic protocols.

The model recognizes the forensic tools available to the examiner. These tools are available both as commercial, fee-based products, and as open source technology. Cyber forensic tools are addressed and discussed in more depth in Chapter 10, 'Cyber Forensic Tools and Utilities,' written by Douglas Menendez.

The outcome of the collaboration and attention to the elements presented in this model will assist in leading to the primary objective of a commercial UAV cyber forensic examination…identifying, securing, and collecting, forensically sound digital evidence.

For further information and additional reading on the UAV cyber forensic examination process, the reader is directed to resources available for download as an eResource at https://routledge.com/9780367524180.

ACRONYMS

AGV   Automated Guided Vehicle
ANS   Autonomous Navigation System
AV   Autonomous Vehicle
C2   Command and Control
C2S   Command and Control System
CDL   Common Data Link
COTS   Commercial Off-The-Shelf

FDI	Fault Detection and Isolation
FLT CNTL	Flight Control
GEOS	Geosynchronous Earth Orbit Satellite
GIG	Global Information Grid
GPS	Global Positioning System
IMU	Inertial Measurement Unit
LADAR	Laser Detection and Ranging
LIDAR	Light Detection and Ranging
LOS	Line of Sight
MAE	Medium-Altitude and -Endurance
MCG&I	Mapping, Charting, Geodesy, and Imagery
MDR	Medium Data Rate
MMS	Mission Management System
MPM	Mission Payload Module
OCU	Operator Control Unit
OODA	Observe-Orient-Decide-Act
OTH	Over-The-Horizon
PFPS	Portable Flight Planning System
ROV	Remotely Operated Vehicle
SA	Situation Awareness
TCP	Transmission Control Protocol
UAV	Unmanned Aerial Vehicle
UCS	Unmanned Control System
VTOL	Vertical Takeoff and Landing

NOTES

1 The inclusion of an unmanned aircraft body, aircraft component by brand, model, name, manufacture, or unmanned aircraft software or associated tool, product, or website is not an endorsement of that aircraft, brand, model, software, tool, product, or website by the author and has been included herein as reference, research, or as an example, of which there may be many within the specified unmanned aircraft marketspace.

2 Anderson, P., (2014), "Electronic evidence – a basic guide for First Responders," European Union Agency for Network and Information Security, www.enisa.europa.eu, retrieved April 8, 2020.

3 (n.a.), (March 2020), "UAS by the Numbers," www.faa.gov/uas/resources/by_the_numbers, retrieved April 8, 2020.

4 Lukacs, M., Bhadra, D., Ekins, W.G., Teeter, J., (n.d.), "Unmanned Aircraft Systems," FAA Aerospace Forecast Fiscal Years 2020-2040 Full Forecast Document and Tables, www.faa.gov/data_research/aviation/aerospace_forecasts/media/unmanned_aircraft_systems.pdf, retrieved April 8, 2020.

5 (n.a.), (March 3, 2020), "Drone market outlook: industry growth trends, market stats and forecast," Business Insider Intelligence, www.businessinsider.com/drone-industry-analysis-market-trends-growth-forecasts, retrieved April 8, 2020.

6 (n.a.), (2011), "ICAO Cir 328, Unmanned Aircraft Systems (UAS)," International Civil Aviation Organization, 999 University Street, Montréal, Quebec, Canada H3C 5H7, www.icao.int/Meetings/UAS/Documents/Circular%20328_en.pdf, retrieved April 8, 2020.

7 Schaufele, R., (May 11, 2018), "Federal Aviation Administration (FAA) Aerospace Forecast for Fiscal Years 2018 to 2038," www.faa.gov/data_research/aviation/aerospace_forecasts/media/FY2018-38_FAA_Aerospace_Forecast.pdf, pg. 43, retrieved April 8, 2020.

8 U.S. Department of Transportation, Federal Aviation Administration, "Law Enforcement Guidance for Suspected Unauthorized UAS Operations," www.faa.gov/uas/resources/policy_library/media/FAA_UAS-PO_LEA_Guidance.pdf, Pg. 7, retrieved, April 10, 2020.

9 "UAS Sightings Report," www.faa.gov/uas/resources/public_records/uas_sightings_report, retrieved April 10, 2020.

10 This work, 'Figure 5, UAV Forensic Exam Flowchart,' is a derivative of 'RPAS Forensic Validation Analysis Towards a Technical Investigation Process: A Case Study of Yuneec Typhoon H, Figure 1, page 4' by Salamh, et al. used with permission, used under CC BY 4.0. 'Figure 5, UAV Forensic Exam Flowchart,' is licensed under CC BY by Albert J. Marcella.

11 United Kingdom's Association of Chief Police Officers (ACPO), Good Practice Guide for Digital Evidence, The Principles of Digital Evidence, www.digital-detective.net/digital-forensics-documents/ACPO_Good_Practice_Guide_for_Digital_Evidence_v5.pdf, pg. 6, retrieved April 12, 2020.

12 Compton, D., (April 20, 2009), "General Principles Applying to The Recovery of Digital Evidence," Guidelines for Best Practice in The Forensic Examination of Digital Technology, Best Practice Guide V6, the European Network of Forensic Science Institutes (ENFSI), https://cryptome.org/2014/03/forensic-digital-best-practice.pdf, pg. 18, retrieved April 12, 2020.

13 (n.a.), (n.d.), "What are Ground Control Stations?" Unmanned Systems Technology, www.unmannedsystemstechnology.com/category/supplier-directory/ground-control-systems/ground-control-stations-gcs, retrieved May 16, 2020.

14 Dual Screen UAV Ground Control Station, 12PCX HOTAS HD Portable GCS, Desert Rotor, 7302 E. Helm Drive, Suite #2004, Scottsdale, AZ 85260, +1 888.938.8470, info@desertrotor.com, www.unmannedsystemstechnology.com/company/desert-rotor, image used with permission, email communication with Debin Ray, Desert Rotor, May 16, 2020.

15 (n.a.), (February 11, 2013), "Scientific Working Group on Digital Evidence (SWGDE) Best Practices for Mobile Phone Forensics, Version: 2.0," www.swgde.org/documents/Current%20Documents/SWGDE%20Best%20Practices%20for%20Mobile%20Phone%20Forensics, pg. 6, retrieved April 12, 2020.

16 Ayers, R., Brothers, S., Jansen, W., (May 2014), "Guidelines on Mobile Device Forensics," NIST Special Publication 800-101, Revision 1, http://dx.doi.org/10.6028/NIST.SP.800-101r1, pp. 28, retrieved April 12, 2020.

17 Note: If data have been modified and all drone components (i.e., GCS and drone body) are present, the examiner should be able to identify and recover any potentially modified data. Dealing with known file systems such as FATs and EXTs, the examiner should be capable of recovering any modified data.

18 Roder, A., Choo, K.K.R., Le-Khac, N., (May 2018), "Unmanned 'Aerial Vehicle Forensic Investigation Process: DJI Phantom 3 Drone as A Case Study," https://commons.erau.edu/adfsl/2018/presentations/1/, retrieved April 14, 2020.

19 Bin Azhar, M., Barton, T., Islam, T., (March 31, 2018), "Drone Forensic Analysis Using Open Source Tools," Volume 13, Number 1 Article 6, www.rsmanuals.com/4991/definitive-technology-enfsi-bpm-fit-01/page-1/, pg. 8, retrieved April 14, 2020.

20 Clark, D., Meffert, C., Baggilli, I., Breitinger, F., (2017), "DROP (DRone Open source Parser) your drone: Forensic analysis of the DJI Phantom III," Elsevier Ltd, Proceedings of the Seventeenth Annual DFRWS USA, http://dx.doi.org/10.1016/j.diin.2017.06.013, retrieved April 16, 2020.

21 (n.a.) (March 2, 2020), "Is there a penalty for failing to register?" FAA Frequently Asked Questions, https://faa.custhelp.com/app/answers/detail/a_id/809/kw/Registration%20Questions/session/L3RpbWUvMTU4NjgxMTg2MC9zaWQvY0trR28%3D, retrieved April 13, 2020.

22 "Register Your Drone," Federal Aviation Administration, www.faa.gov/uas/getting_started/register_drone, April 13, 2020.

23 Adapted from pg. 99, Hartmann, K., Steup, C., (2013) "The vulnerability of UAVs to cyber-attacks - An approach to the risk assessment," Podins/Stinissen/Maybaum, 5th International Conference on Cyber Conflict, pp. 95–117, https://ccdcoe.org/uploads/2018/10/CyCon_2013_Proceedings.pdf, retrieved May 6, 2020.

24 Endianness refers to the order of storing bytes in computer memory. Words may be represented in big-endian or little-endian format. With big-endian the most-significant byte of a word is stored at the memory location with the lowest address and the least significant byte stored at the highest memory address.

With little-endian format the least-significant byte is stored at the lower memory address with the most significant byte is stored at the highest memory address.

Rajesh, (October 21, 2015) "Basics of Unicode and Encodings," My IT Learnings, http://myitlearnings.com/basics-of-unicode-and-encodings/, retrieved April 16, 2020.

25 Clark, D., Meffert, C., Baggilli, I., Breitinger, F., (2017), "DROP (DRone Open source Parser) your drone: Forensic analysis of the DJI Phantom III," Elsevier Ltd, Proceedings of the Seventeenth Annual DFRWS USA, http://dx.doi.org/10.1016/j.diin.2017.06.013, retrieved April 16, 2020.

26 Kovar, D., Bollo, J., (February 2018), "Drone Forensics," Digital Forensics Magazine, Issue 34, www.msab.com/download/case_studies/Digital_Forensics_Feb-2018-MSAB.pdf pg. 14–19, retrieved April 15, 2020.

27 The IMU serve as a supplement to GPS positioning systems, allowing the navigational device to continue with an estimated position and heading if it loses satellite connection. Arrow Electronics, (n.d.), "What is IMU? Inertial Measurement Unit Working & Applications," www.arrow.com/en/research-and-events/articles/imu-principles-and-applications, retrieved January 23, 2020.

28 Bin Azhar, M., Barton, T., Islam, T., (March 31, 2018), "Drone Forensic Analysis Using Open Source Tools," Volume 13, Number 1 Article 6, www.rsmanuals.com/4991/definitive-technology-enfsi-bpm-fit-01/page-1/, pg. 11, retrieved April 15, 2020.

29 Image provided courtesy of High Line Drones LLC, PO Box 441, Quinebaug, CT, 06262, 508-494-7018, www.highlinedrones.com, author correspondence with Elliot Webb, founder and chief pilot, High Line Drones, used with permission.

30 Ibid.

31 Additional sources for information on examining mobile devices:

- NIST Special Publication 800-101, Revision 1, Guidelines on Mobile Device Forensics, published in May 2014, but, be forewarned SP 800-101 isn't up-to-date with today's mobile technology and does not address the mobile components of a UAV, vital to a forensics examination.
- The Association of Chief Police Officers (APCO) Good Practice Guide for Digital Evidence, published in March 2012, does not provide UAV forensics guidance.
- The European Network of Forensic Science Institutes (ENFSI), Best Practice Manual (BPM) for the Forensic Examination of Digital Technology, published in 2015; however, this publication does not provide UAV forensics guidance.

32 (n.a.), (n.d.), "Drone Communication - Data Link: How do drones communicate with their operator?" www.911security.com/learn/airspace-security/drone-fundamentals/drone-communication-data-link, retrieved April 14, 2020.

33 Hall, T., (December 18, 2018), "DJI Follow Me – A Complete Guide," www.letusdrone.com/dji-follow-me-a-complete-guide, retrieved April 14, 2020.

34 DJI, (June 20, 2018), "6 Best Follow Me Drones (With Video Comparisons) 2020," https://store.dji.com/guides/camera-drone-that-follows-you/, retrieved April 14, 2020.

35 Corrigan, F., (November 24, 2019), "12 Best Follow Me Drones and Follow You Technology Reviewed," DroneZon, www.dronezon.com/drone-reviews/best-follow-me-gps-mode-drone-technology-reviewed, retrieved April 14, 2020.

36 DJI, (June 20, 2018), "6 Best Follow Me Drones (With Video Comparisons) 2020," https://store.dji.com/guides/camera-drone-that-follows-you/, retrieved April 14, 2020.

37 DJI, (December 18, 2017), "Film Like a Pro: DJI Drone" ActiveTrack" – With Video Tutorials," https://store.dji.com/guides/film-like-a-pro-with-activetrack/, retrieved April 15, 2020.

38 Adapted from pg. 99, Hartmann, K., Steup, C., (2013) "The vulnerability of UAVs to cyber-attacks – An approach to the risk assessment," Podins/Stinissen/Maybaum, 5th International Conference

on Cyber Conflict, pp. 95–117, https://ccdcoe.org/uploads/2018/10/CyCon_2013_Proceedings. pdf, retrieved May 6, 2020.

39 Prastya, S.E., Riadi, I., Luthfi, A., (March 2017), "Forensic Analysis of Unmanned Aerial Vehicle to Obtain GPS Log Data as Digital Evidence," International Journal of Computer Science and Information Security, 15(3):280–285, www.researchgate.net/publication/316220869_Forensic_ Analysis_of_Unmanned_Aerial_Vehicle_to_Obtain_GPS_Log_Data_as_Digital_Evidence/cita- tions, pg. 284, retrieved June 5, 2020.

40 (n.a.), (July 16, 2019), "SWGDE Best Practices for Mobile Device Evidence Collection & Preservation, Handling, and Acquisition," Scientific Working Group on Digital Evidence, https:// drive.google.com/file/d/1f8ayespt8uRuBI84lK_3fRqFZLCBUR2P/view, retrieved June 5, 2020.

41 DJI Forum, https://forum.dji.com/thread-35695-1-1.html, retrieved April 11, 2020.

42 DJI Drone Help Forum https://forum.phantomhelp.com/t/mavic-mini-teardown-by-dji-drone- service/4030, and https://datfile.net/ retrieved April 12, 2020.

43 Holybro is a registered trademark of Holybro, registered in the United States and other coun- tries. All Rights Reserved. PX4 is a registered Trademark of the Dronecode Foundation. All rights reserved. Pixhawk is a Registered Trademark of Lorenz Meier. All rights reserved. https:// shop.holybro.com/art/holybro_a0047.html. Images used with permission, email exchange with Holybro, May 15, 2020.

44 Ibid.

45 The gyroscopes will prevent any disturbance from external factors, but won't interfere with inputs from the pilot. The accelerometer tells the flight controller which direction is up and the flight controller spins the corresponding motors faster/slower to level itself. Reade, J., (January 28, 2018), "Flight Modes – Angle, Horizon, Acro what do they mean?", www.cnydrones.org/ flight-modes-angle-horizon-acro-what-do-they-mean, retrieved May 11, 2020.

46 Wren, C., (June 1, 2018), "What is the difference between autopilot and flight controller in drones?"   www.quora.com/What-is-the-difference-between-autopilot-and-flight-controller-in- drones, retrieved May 11, 2020.

47 Holybro is a registered trademark of Holybro, registered in the United States and other coun- tries. All Rights Reserved. PX4 is a registered Trademark of the Dronecode Foundation. All rights reserved. Pixhawk is a Registered Trademark of Lorenz Meier. All rights reserved. https:// shop.holybro.com/art/holybro_a0047.html. Images used with permission, email exchange with Holybro, May 15, 2020.

48 Drotek Electronics, https://electronics.drotek.com/autopilots/, ZA du ruisseau Route de Folcarde, 31290 Avignonet-Lauragais France, sales@drotek.com. Image used with permission, email exchange with Paul Thélu, Drotek Electronics, May 18, 2020.

49 Ibid.

50 Flight Data Recorder, UAV Navigation S. L., Avenida Pirineos, 7 - Bajo 11 - Edificio INBISA, 28703, San Sebastián de los Reyes – Madrid (España), +34 91 657 2723, www.uavnavigation. com/products/accessories/flight-data-recorder, image used with permission, email exchange with David Pinta, UAV Navigation S. L., May 15, 2020.

51 Zheng, Z., XIAO, G., (January 2019), "Evolution analysis of a UAV real-time operating system from a network perspective," Chinese Journal of Aeronautics, Volume 32, Issue 1, pp. 176-185, www.sciencedirect.com/science/article/pii/S1000936118301420, retrieved May 12, 2020.

52 Al-Zarouni, M., (December 3, 2007), "Introduction to Mobile Phone Flasher Devices and Considerations for their Use in Mobile Phone Forensics," Australian Digital Forensics Conference, https://ro.ecu.edu.au/cgi/viewcontent.cgi?article=1015&context=adf, retrieved April 12, 2020.

53 Gratzer, V. & Naccache, D. (2007) Cryptography, Law Enforcement, and Mobile Communications, http://info.computer.org/portal/site/security/menuitem.6f7b2414551cb84651286b108bc d45f3/index.jsp?&pName=security_level1_article&TheCat=1001&path=security/2006/ v4n6&file=crypto.xml, pp. 67–70, vol. 4, retrieved April 12, 2020.

54 Breeuwsma, Marcel & Jongh, Martien & Klaver, Coert & Knijff, Ronald & Roeloffs, Mark. (2007). Forensic Data Recovery from Flash Memory. Small Scale Digital Device Forensics Journal. 1,

www.researchgate.net/publication/252309999_Forensic_Data_Recovery_from_Flash_Memor, retrieved April 12, 2020.

55 Jain, U., Rogers, M., & Matson, E. T. (March 2017), "Drone forensic framework: Sensor and data identification and verification," In Sensors Applications Symposium (SAS), 2017 IEEE (pp. 1–6). IEEE, https://ieeexplore.ieee.org/document/7894059, email correspondence with E. Matson, June 12, 2020.

56 Gülataş, I, Baktir, S., (2018), "Unmanned Aerial Vehicle Digital Forensic Investigation Framework," Journal of Naval Sciences and Engineering, 14 (1), pg. 32–53. Retrieved from https://dergipark.org.tr/en/pub/jnse/issue/38665/420911, retrieved June 1, 2020.

57 (n.a.), (January 2020), "INTERPOL Framework for Responding to A Drone Incident - For First Responders and Digital Forensics Practitioners," Digital Forensics Laboratory of the INTERPOL Innovation Centre, Singapore, INTERPOL Global Complex for Innovation 18 Napier Road, Singapore 258510, www.interpol.int/es/content/download/15298/file/DFL_DroneIncident_Final_EN.pdf?inLanguage=eng-GB, retrieved August 25, 2020.

58 Roder, A., Choo, K.K.R., Le-Khac, N., (May 2018), "Unmanned Aerial Vehicle Forensic Investigation Process: DJI Phantom 3 Drone as A Case Study," https://commons.erau.edu/adfsl/2018/presentations/1/, retrieved May 15, 2020.

59 Note. For those operators in the United States, part 107 (commercial operations), all drones must be registered (regardless of weight). For the "Exception for Recreational Flyers" drones must be registered if they weigh 0.55 pounds (250 grams) or more.

60 (n.a.), (August 14, 2018), "Law Enforcement Guidance for Suspected Unauthorized UAS Operations," Version 5, The Federal Aviation Administration (FAA), www.faa.gov/uas/public_safety_gov/media/faa_uas-po_lea_guidance.pdf, retrieved June 12, 2020.

61 Roder, A., Choo, K.K.R., Le-Khac, N., (May 2018), "Unmanned Aerial Vehicle Forensic Investigation Process: DJI Phantom 3 Drone as A Case Study," https://commons.erau.edu/adfsl/2018/presentations/1/, retrieved May 15, 2020.

62 Ibid.

63 (n.a.), (n.d.), "How to unlock your drone and why we should not rely on UAS manufacturers for safety," 911 Security, 10874 Plano Road, Suite B, Bldg. 2 Dallas, TX 75238, www.911security.com/blog/how-to-unlock-your-drone-and-why-we-should-not-rely-on-uas-manufacturers-for-safety, retrieved June 19, 2020.

64 Clark, D., Merffert, C., Baggili, I., Bretinger, F. (2017), "DROP (DRone Open source Parser) your drone: Forensic analysis of the DJI Phantom III," DFRWS 2017 USA — Proceedings of the Seventeenth Annual DFRWS USA, www.sciencedirect.com/science/article/pii/S1742287617302001, retrieved June 18, 2020.

65 (n.a.), (n.d.), "Retrieving a V3 .DAT File from the AC," www.datfile.net/DatCon/retrieveV3Dat.html, retrieved June 16, 2020.

66 Beckett, J., Slay, J., (December 2006), "Digital Forensics: Validation and Verification in a Dynamic Work Environment," Proceedings of the 40th Hawaii International Conference on System Sciences – 2007, www.researchgate.net/publication/221177767_Digital_Forensics_Validation_and_Verification_in_a_Dynamic_Work_Environment, retrieved June 16, 2020.

67 The drone images, research results, and data discussed and displayed here were produced as part of the VTO Inc. Drone Forensics Program, sponsored by the United States Department of Homeland Security Science and Technology Directorate, Cyber Security Division. For more information visit www.droneforensics.com or contact VTO at droneforensics@vtoinc.com.

68 Beckett, J., Slay, J., (December 2006), "Digital Forensics: Validation and Verification in a Dynamic Work Environment," Proceedings of the 40th Hawaii International Conference on System Sciences – 2007, www.researchgate.net/publication/221177767_Digital_Forensics_Validation_and_Verification_in_a_Dynamic_Work_Environment, retrieved June 16, 2020.

69 Kent, K., Chevalier, S., Grance, T., Dang, H., (August 2006), "Guide to Integrating Forensic Techniques into Incident Response," Special Publication 800-86, National Institute of Standards

and Technology, https://nvlpubs.nist.gov/nistpubs/Legacy/SP/nistspecialpublication800-86.pdf, retrieved May 22, 2020.

70 Passia, K., (October 2012), "Guidelines for Identification, Collection, Acquisition and Preservation of Digital Evidence," ISO 27037, www.iso.org/standard/44381.html, retrieved May 24, 2020.

71 This work, 'Figure 3.30 UAV Forensic Examination Model,' is a derivative of 'A 3-Dimensional UAS Forensic Intelligence-Led Taxonomy (U-FIT), UAV forensic ontology, Figure 3.7' by Fahad Salamh, Ph.D., Purdue University. Unpublished doctoral dissertation. Used with permission. Used under CC BY 4.0. 'Figure 3.30 UAV Forensic Examination Model' is licensed under CC by Albert J. Marcella.

Chapter 4

Cloud forensics

Ronald L. Krutz

CONTENTS

As this chapter title implies, cloud forensics is the confluence of digital forensics technology and the cloud computing paradigm. In order to better address the topic, it is important to understand the fundamental characteristics of both subjects, their interactions, and complementary application to computing systems. The sections of this chapter will review the important relevant definitions of both areas and explore their integration into cloud forensic science.

CLOUD COMPUTING

The National Institute of Standards and Technology (NIST) defines Cloud Computing[1] as a model for enabling ubiquitous, convenient, on-demand network access to a shared pool of configurable computing resources (e.g., networks, servers, storage, applications, and services) that can be provisioned and released with minimal management effort or service provider interaction. This cloud model is composed of five essential characteristics, three service models, and four deployment models.[2]

ESSENTIAL CHARACTERISTICS

On-demand self-service. A consumer can unilaterally provision computing capabilities, such as server time and network storage, as needed automatically without requiring human interaction with each service provider.

Broad network access. Capabilities are available over the network and accessed through standard mechanisms that promote use by heterogeneous thin or thick client platforms (e.g., mobile phones, tablets, laptops, and workstations).

Resource pooling. The provider's computing resources are pooled to serve multiple consumers using a multi-tenant model, with different physical and virtual resources dynamically assigned and reassigned according to consumer demand. There is a sense of location independence in that the customer generally has no control or knowledge over the exact location of the provided resources but may be able to specify location at a higher level of abstraction (e.g., country, state, or datacenter). Examples of resources include storage, processing, memory, and network bandwidth.

Rapid elasticity. Capabilities can be elastically provisioned and released, in some cases automatically, to scale rapidly outward and inward commensurate with demand. To the consumer, the capabilities available for provisioning often appear to be unlimited and can be appropriated in any quantity at any time.

Measured service. Cloud systems automatically control and optimize resource use by leveraging a metering capability at some level of abstraction appropriate to the type of service (e.g., storage, processing, bandwidth, and active user accounts). Resource usage can be monitored, controlled, and reported, providing transparency for both the provider and consumer of the utilized service.

SERVICE MODELS

Software as a Service (SaaS). The capability provided to the consumer is to use the provider's applications running on a cloud infrastructure. The applications are accessible from various client devices through either a thin client interface, such as a web browser (e.g., web-based email), or a program interface. The consumer does not manage or control the underlying cloud infrastructure including network, servers, operating systems, storage, or even individual application capabilities, except for limited user specific application configuration settings.

Platform as a Service (PaaS). The capability provided to the consumer is to deploy onto the cloud infrastructure consumer-created or acquired applications created using programming languages, libraries, services, and tools supported by the provider. The consumer does not manage or control the underlying cloud infrastructure including network, servers, operating systems, or storage, but has control over the deployed applications and possibly configuration settings for the application-hosting environment.

Infrastructure as a Service (IaaS). The capability provided to the consumer is to provision processing, storage, networks, and other fundamental computing resources where the consumer can deploy and run arbitrary software, which can include operating systems and applications. The consumer does not manage or control the underlying cloud infrastructure but has control over operating systems, storage, and deployed applications; and possibly limited control of select networking components (e.g., host firewalls).

DEPLOYMENT MODELS

Private cloud. The cloud infrastructure is provisioned for exclusive use by a single organization comprising multiple consumers (e.g., business units). It may be owned, managed, and operated by the organization, a third party, or some combination of them, and it may exist on or off premises.

Community cloud. The cloud infrastructure is provisioned for exclusive use by a specific community of consumers from organizations that have shared concerns (e.g., mission, security requirements, policy, and compliance considerations). It may be owned, managed, and operated by one or more of the organizations in the community, a third party, or some combination of them, and it may exist on or off premises.

Public cloud. The cloud infrastructure is provisioned for open use by the general public. It may be owned, managed, and operated by a business, academic, or government organization, or some combination of them. It exists on the premises of the cloud provider.

Hybrid cloud. The cloud infrastructure is a composition of two or more distinct cloud infrastructures (private, community, or public) that remain unique entities, but are bound together by standardized or proprietary technology that enables data and application portability (e.g., cloud bursting for load balancing between clouds).

The cloud computing capabilities are made possible by using virtualization technology to run the associated software as discussed in the following section.

VIRTUALIZATION

Virtual machines (VMs) are software implementations of computers – and indistinguishable over a network from a physical computer. A VM is simply an environment, typically an operating system (OS) or a program, that is created within another environment.

A key concept here is that we are creating a virtual version of something (be it a server, application, storage, network, client) that can be separated from its underlying resources using an execution container, again usually an OS or a program. In some forms of virtualization, the underlying hardware layer is completely simulated, whereas in most implementations, this is not the case. In some cases, hardware may implement some virtualization support.

Another key concept is that virtualization is used in different areas, including server, storage, or network. Virtualization can mask complexity and enable resource sharing and utilization. Virtualization also can deliver a degree of isolation and insulation from the effect of some forms of vulnerability.

Supporting the operation of the VM is a hypervisor, which represents itself to the VM as the underlying hardware. The hypervisor is the part of a virtual machine that allows host resource sharing and enables VM/host isolation. Therefore, the ability of the hypervisor to provide the necessary isolation during intentional attack greatly determines how well the virtual machine can survive risk. One reason why the hypervisor is susceptible to risk is because it's a software program; risk increases as the volume and complexity of application code increases.

Ideally, software code operating within a defined VM would not be able to communicate or affect code running either on the physical host itself or within a different VM; but several issues, such as bugs in the software, or limitations to the virtualization implementation, may put this isolation at risk. In a normal virtualization scenario, the guest operating system (the operating system that is booted inside of a virtualized environment) runs like a traditional OS managing I/O to hardware and network traffic, even though it's managed by the hypervisor.

The hypervisor, therefore, has a great level of control over the system, not only in the VM but also on the host machine. If the hypervisor is compromised, it can pose serious security risks.

VIRTUALIZATION TYPES

Vendor implementations of virtualization will vary, but in general terms, there are several types of virtualization:

- **Type 1** also native or bare metal virtualization is implemented by a hypervisor that runs directly on bare hardware. Guest OSs run on top of the hypervisor. Examples include Microsoft Hyper-V, Oracle VM, LynxSecure, VMware ESX, and IBM z/VM.
- **Type 2** or hosted virtualization has a hypervisor running as an application within a host OS. VMs also run above the hypervisor. Examples include Oracle VirtualBox, Parallels, Virtual PC, VMware Fusion, VMware Server, Xen, and XenServer.
- **OS implemented virtualization** is implemented by the OS itself taking the place of the hypervisor. Examples of this type include Solaris Containers, BSDjails, OpenVZ, Linux-VServer, and Parallels Virtuozzo Containers.

Another area of concern with virtualization has to do with the nature of allocating and deal-locating resources such as the local storage associated with VMs. If during the deployment and operation of a VM, data are written to physical media – or to memory – and it is not cleared before those information resources are reallocated to the next VM, then there is a potential for information exposure.

A further area of concern with virtualization has to do with the potential for undetected network attacks between VMs that are co-located on a physical server. The problem is that unless the traffic from each VM can be monitored, you cannot verify that traffic is not possi-ble between VMs. One approach is to simply invoke OS-based traffic filtering or firewalling. One potential complication that can be faced by a customer who needs multiple communi-cating and cooperating VMs is that these VMs may be dynamically moved around by the service provider to load balance their cloud.

Because virtual machines can perform the same processes as actual systems, they are also able to track and record the activity trail of users. These capabilities are valuable tools in the quest to learn more about digital forensics, as they can produce evidence to be used to enhance understanding and application. For example, Virtual Forensic Computing (VFC)[3] software was first launched in 2007 and has become essential software for forensic investigators, as it allows for seamless recreation of a digital crime scene. It is widely used by law enforcement to extract images from a suspect's computer; launch a suspect machine in its native environment; take screenshots of key evidence, and view files and data in its natural state.[4]

DIGITAL FORENSICS

In its strictest connotation, digital forensics is the application of computer science and inves-tigative procedures involving the examination of digital evidence – following proper search authority, chain of custody, validation with mathematics, use of validated tools, repeatability, reporting, and possibly expert testimony.[5]

An alternative definition is 'The application of science to the identification, collection, examination, and analysis, of data while preserving the integrity of the information and maintaining a strict chain of custody for the data.'[6]

At the Digital Forensic Research Workshop (DFRWS), Palmer[7] defined digital forensics as

> the use of scientifically derived and proven methods toward the preservation, collection, validation, identification, analysis, interpretation and presentation of digital evidence derived from digital sources for the purpose of facilitating or furthering the reconstruction of events found to be criminal, or helping to anticipate unauthorized actions shown to be disruptive to planned operations.

The science of digital forensics was driven by the needs of law enforcement for a structured methodology for investigating computer crime. In 1984, the FBI, working with other law enforcement agencies, developed approaches to collecting and analyzing computer evidence. Because computer evidence is volatile and can be found in numerous components and locations, new paradigms were developed to address the acquisition, preservation, retrieval, and presentation of collected data. As an example, the FBI created the Computer Analysis and Response Team (CART) to analyze computer evidence.

In order to develop standards for computer forensic science, the FBI convened international conferences in 1995 in Baltimore, Maryland; in 1996 in Australia, and in the Netherlands in 1997. The result of these conferences was the establishment of the Scientific Working Group on Digital Evidence (SWGDE) to address digital forensics issues and standards.

In 1998, the U.S. National Institute of Justice (NIJ) established the Technical Working Group for Electronic Crime Scene Investigation (TWGECSI) with the assignment to 'identify, define, and establish basic criteria to assist agencies with electronic investigations and prosecutions.'

The working group comprised experts from federal, state, and local law enforcement agencies, prosecutors and district attorneys general, criminal justice agencies, commercial, academic, and professional organizations. As a result of the group's efforts, 'Electronic Crime Scene Investigation: A Guide for First Responders' [NIJ] was published in July 2001. This document was intended to be the first of a series addressing digital forensics methods. The latest version was published in 2008.[8] The document defines digital evidence as information and data of value to an investigation that is stored on, received, or transmitted by an electronic device.

This evidence is acquired when data or electronic devices are seized and secured for examination and has the following characteristics:

- Is latent, like fingerprints or DNA evidence.
- Crosses jurisdictional borders quickly and easily.
- Is easily altered, damaged, or destroyed.
- Can be time sensitive.

Before collecting evidence at a crime scene, first responders should ensure that legal authority exists to seize evidence, the scene has been secured and documented, and appropriate personal protective equipment is used.

It also recommends the following steps for the handling of digital evidence at an electronic crime scene:

- Recognize, identify, seize, and secure all digital evidence at the scene.
- Document the entire scene and the specific location of the evidence found.
- Collect, label, and preserve the digital evidence.
- Package and transport digital evidence in a secure manner.

The Guide also emphasizes that improperly accessing data stored on electronic devices may violate Federal laws. Investigators may need to obtain additional legal authority before they proceed.

Some of the relevant legislation and guidelines are summarized as follow:

- The Cable Communications Policy Act provides for discretionary use of PII by cable operators internally but imposes restrictions on disclosures to third parties.
- The Children's Online Privacy Protection Act (COPPA) is aimed at providing protection to children under the age of 13.
- Customer Proprietary Network Information Rules apply to telephone companies and restricts their use of customer information both internally and to third parties.
- The Electronic Communications Privacy Act protects exchanged information from being intercepted or disclosed by third parties, including law enforcement
- The 1973 U.S. Code of Fair Information Practices, which states:
  1. There must not be personal data record-keeping systems whose very existence is secret.
  2. There must be a way for a person to find out what information about them is in a record and how it is used.
  3. There must be a way for a person to prevent information about them, which was obtained for one purpose, from being used or made available for other purposes without their consent.
  4. Any organization creating, maintaining, using, or disseminating records of identifiable personal data must ensure the reliability of the data for their intended use and must take precautions to prevent misuses of that data
- The European Union (EU) has defined privacy principles that include the following:
  1. Data should be collected in accordance with the law.
  2. Information collected about an individual cannot be disclosed to other organizations or individuals unless authorized by law or by consent of the individual.
  3. Records kept on an individual should be accurate and up to date.
  4. Individuals have the right to correct errors contained in their personal data.
  5. Data should be used only for the purposes for which it was collected, and it should be used only for reasonable period.
  6. Individuals are entitled to receive a report on the information that is held about them.
  7. Transmission of personal information to locations where 'equivalent' personal data protection cannot be assured is prohibited.
- The Organization for Economic Cooperation and Development (OECD) has issued Guidelines that are summarized as follows:
  1. There should be limits to the collection of personal data and any such data should be obtained by lawful and fair means.
  2. Personal data should be relevant to the purposes for which they are to be used, and, to the extent necessary for those purposes, should be accurate, complete, and kept up to date.
  3. The purposes for which personal data are collected should be specified not later than at the time of data collection.
  4. Personal data should not be disclosed, made available, or otherwise used for purposes other than those specified.
  5. Personal data should be protected by reasonable computer forensics safeguards against such risks as loss or unauthorized access, destruction, use, modification, or disclosure of data.
  6. There should be a general policy of openness about developments, practices, and policies with respect to personal data.
  7. An individual should have the right to obtain from a data controller.
  8. A Member country should refrain from restricting transborder flows of personal data between itself and another Member country except where the latter does not

yet substantially observe these Guidelines or where the re-export of such data would circumvent its domestic privacy legislation.

9. A Member country may also impose restrictions in respect of certain categories of personal data for which its domestic privacy legislation includes specific basic principles and best practices in view of the nature of those data and for which the other Member country provides no equivalent protection.

Applying the aforementioned principles certainly enhances the ability to investigate computer crime and misuse. However, as will be discussed in this chapter, the application to cloud computing might make it difficult or, sometimes, impossible to satisfy all the recommended guidelines.

CLOUD FORENSICS

NIST defines cloud computing forensic science as 'the application of scientific principles, technological practices and derived and proven methods to reconstruct past cloud computing events through identification, collection, preservation, examination, interpretation and reporting of digital evidence.'[9]

Ruan et al.[10] propose a three-dimensional model to structure the complex domain of cloud forensics. It includes a technical dimension, organizational dimension and legal dimension.

TECHNICAL DIMENSION

The technical dimension includes data collection, live forensics, evidence segregation, virtualized environments, and proactive measures. Data collection is the process of identifying, labeling, recording and acquiring forensic data. The forensic data includes client-side artifacts that reside on client premises and provider-side artifacts that are in the provider infrastructure. The collection process should preserve the integrity of data with clearly defined segregation of duties between the client and provider. It should not breach laws or regulations in the jurisdictions where data are collected or compromise the confidentiality of other tenants that share the resource.

Ruan et al.[11] summarize the following important issues of the technical dimension:

- Because of cloud rapid elasticity, cloud forensic tools should include large-scale static and live forensic tools for data acquisition (including volatile data collection), data recovery, evidence examination, and evidence analysis.
- Procedures and tools must be developed to segregate forensic data among multiple tenants in various cloud deployment models and service models.
- Relative to cloud virtualization, hypervisor investigation procedures are practically non-existent.
- Procedures and tools must be developed to physically locate forensic data with specific timestamps while taking into consideration the jurisdictional issues.
- Proactive measures such as preserving regular snapshots of storage, continually tracking authentication and access control, and performing object-level auditing of all accesses, can enhance the investigative process.

Additional technical dimension concerns detailed by NIST[12] emphasize that the identification, collection, and preservation of media can be particularly challenging in a cloud computing environment given the following factors:

1. Identification of the cloud provider and its partners. This is needed to better understand the environment and thus address the factors below.
2. The ability to conclusively identify the proper accounts held within the cloud by a consumer, especially if different cyber personas are used.
3. The ability of the forensics examiner to gain access to the desired media.
4. Obtaining assistance of the cloud infrastructure/application provider service staff.
5. Understanding the topology, proprietary policies, and storage system within the cloud.
6. Once access is obtained, the examiner's ability to complete a forensically sound image of the media.
7. The sheer volume of the media.
8. The ability to respond in a timely fashion to more than one physical location if necessary.
9. E-discovery, log file collection and privacy rights given a multi-tenancy system. (How does one collect the set of log files applicable for this matter versus extraneous information with possible privacy rights protections?)
10. Validation of the forensic image.
11. The ability to perform analysis on encrypted data and the collector's ability to obtain keys for decryption.
12. The storage system no longer being local.
13. There is often no way to link given evidence to a particular suspect other than by relying on the cloud provider's word.

Traditional digital forensics tools have also found a place in cloud forensics due to improvements and updates. Some of the popular ones are summarized as follows:

- Access Data Forensic Toolkit (FTK) (www.accessdata.com/products-services/forensic-toolkit-ftk).
- EnCase (www.guidancesoftware.com/encase-forensic).
- F-Response (www.f-response.com).
- Magnet Forensics' Internet Evidence Finder (IEF) (www.magnetforensics.com/magnet-ief).

ORGANIZATIONAL DIMENSION

The organizational dimension requires auditors, carriers, cloud brokers, consumers, and providers to cooperate in obtaining digital evidence. Organizational policies and service level agreements (SLAs) can also support forensic activities.

To establish a cloud forensic capability, each cloud entity must provide internal staffing, provider–customer collaboration and external assistance that fulfill the following roles[13]:

Investigators – Investigators are responsible for examining allegations of misconduct and working with external law enforcement agencies as needed. They must have enough expertise to perform investigations of their own assets as well as interact with other parties in forensic investigations.

IT Professionals – IT professionals include system, network and security administrators, ethical hackers, cloud security architects, and technical and support staff. They provide expert knowledge in support of investigations, assist investigators in accessing crime scenes, and may perform data collection on behalf of investigators.

Incident Handlers – Incident handlers respond to security incidents such as unauthorized data access, accidental data leakage and loss, breach of tenant confidentiality, inappropriate system use, malicious code infections, insider attacks, and denial of service attacks. All cloud entities should have written plans that categorize security incidents for the different levels of the cloud and identify incident handlers with the appropriate expertise.

Legal Advisors – Legal advisors are familiar with multi-jurisdictional and multi-tenancy issues in the cloud. They ensure that forensic activities do not violate laws and regulations and maintain the confidentiality of other tenants that share the resources. SLAs must clarify the procedures that are followed in forensic investigations. Internal legal advisors should be involved in drafting the SLAs to cover all the jurisdictions in which a CSP operates. Internal legal advisors are also responsible for communicating and collaborating with external law enforcement agencies during forensic investigations.

External Assistance – It is prudent for a cloud entity to rely on internal staff as well as external parties to perform forensic tasks. It is important for a cloud entity to determine, in advance, the actions that should be performed by external parties, and ensure that the relevant policies, guidelines and agreements are transparent to customers and law enforcement agencies.

LEGAL DIMENSION

The legal dimension of cloud forensics requires the development of regulations and agreements to ensure that forensic activities do not breach laws and regulations in the jurisdictions where the data resides. Also, the confidentiality of other tenants that share the same infrastructure should be preserved. SLAs should define the terms of use between a CSP and its customers. The following terms regarding forensic investigations should be included in SLAs:

- The services provided, techniques supported, and access granted by the CSP to customers during forensic investigation.
- Trust boundaries, roles, and responsibilities between the CSP and customers regarding forensic investigations.
- The process for conducting investigations in multi-jurisdictional environments without violating the applicable laws, regulations, and customer confidentiality and privacy policies.
- Measurement of the effectiveness of incident resolution.
- Time of data delivery.
- Authentication method for eligible investigators.
- Type of data to be collected.

ADDITIONAL CONSIDERATIONS

There are number of additional considerations in establishing a cloud forensic capability that cover the technical, organizational, and legal dimensions.[14]

Forensic Data Collection – Access to forensic data varies considerably based on the cloud model that is implemented. IaaS customers enjoy relatively unfettered access to the data required for forensic investigations. On the other hand, SaaS customers may have little or no access to such data. Decreased access to forensic data means that cloud customers generally have little or no control – or even knowledge – of the physical locations of their data. In fact, they may only be able to specify location at a high level of abstraction, typically as an object or container.

In fact, CSPs intentionally hide data locations from customers to facilitate data movement and replication. In addition, cloud customers have very limited access to log files and metadata at all levels, as well as a limited ability to audit and conduct real-time monitoring on their own.

Static, Elastic, and Live Forensics – The proliferation of endpoints, especially mobile endpoints, is a challenge for data discovery and evidence collection. Because of the large number of resources connected to the cloud, the impact of a crime and the workload of an investigation can be massive. Constructing the timeline of an event requires accurate time synchronization.

Time synchronization is complicated because the data of interest resides on multiple physical machines in multiple geographical regions, or the data may be in flow between the cloud infrastructure and remote endpoint client. The use of disparate log formats is already a challenge in traditional digital forensics. The challenge is exacerbated in the cloud due to the sheer volume of data logs and the prevalence of proprietary log formats. Deleted data are an important source of evidence in traditional digital forensics. In the cloud, the customer who created a data volume often maintains the right to alter and delete the data.

Evidence Segregation – In the cloud, different instances running on a single physical machine are isolated from each other via virtualization. The neighbors of an instance have no more access to the instance than any other host on the Internet. Neighbors behave as if they are on separate hosts. Customer instances have no access to raw disk devices, instead they access virtualized disks. At the physical level, system audit logs of shared resources collect data from multiple tenants. Technologies used for provisioning and deprovisioning resources are being improved. However, it is a challenge for CSPs and law enforcement agencies to segregate resources during investigations without breaching the confidentiality of other tenants that share the infrastructure.

Multiple Data Copies – Cloud systems may generate multiple complete or partial copies of data across the cloud and on more than one local system. These data should be captured, and associated changes should be tracked and identified. Chen[15] proposes deep data analysis to correlate data from various locations such as locating different versions of the same file on cloud storage and local systems, synchronized among a group of clients who have shared the same document.

Virtualized Environments – Cloud computing provides data and computational redundancy by replicating and distributing resources. Most CSPs implement redundancy using virtualization. Instances of servers run as virtual machines that are monitored and provisioned by a hypervisor. A hypervisor is analogous to a kernel in a traditional operating system. Hypervisors are prime targets for attack, but there is an alarming lack of policies, procedures, and techniques for forensic investigations of hypervisors.

Data mirroring over multiple machines in different jurisdictions and the lack of transparent, real-time information about data locations introduces difficulties in forensic investigations. Investigators may unknowingly violate laws and regulations because they do not have clear information about data storage jurisdictions Additionally, a CSP cannot provide a precise physical location for a piece of data across all the geographical regions of the cloud. Finally, the distributed nature of cloud computing requires strong international cooperation – especially when the cloud resources to be confiscated are located around the world.

Internal Staffing – Most cloud forensic investigations are conducted by traditional digital forensic experts using conventional network forensic procedures and tools. A major challenge is posed by the paucity of technical and legal expertise with respect to cloud forensics. This is exacerbated by the fact that forensic research and laws and regulations are far behind the rapidly evolving cloud technologies. Cloud entities must ensure that they have enough trained staff to address the technical and legal challenges involved in cloud forensic investigations.

External Dependency Chains – CSPs and most cloud applications often have dependencies on other CSPs. For example, a CSP that provides an email application (SaaS) may depend on a third-party provider to host log files (i.e., PaaS), who in turn may rely on a partner who provides the infrastructure to store log files (IaaS). A cloud forensic investigation thus requires investigations of each individual link in the dependency chain. Correlation of the activities across CSPs is a major challenge. An interruption or even a lack of coordination between the parties involved can lead to problems. Procedures, policies, and agreements related to cross-provider forensic investigations are virtually nonexistent.

Service Level Agreements – Current SLAs omit important terms regarding forensic investigations. This is due to low customer awareness, limited CSP transparency and the lack of international regulation. Most cloud customers are unaware of the issues that may arise in a cloud forensic investigation and their significance. CSPs are generally unwilling to increase transparency because of inadequate expertise related to technical and legal issues, and the absence of regulations that mandate increased transparency.

Multiple Jurisdictions and Tenancy – Clearly, the presence of multiple jurisdictions and multi-tenancy in cloud computing pose significant challenges to forensic investigations. Each jurisdiction imposes different requirements regarding data access and retrieval, evidence recovery without breaching tenant rights, evidence admissibility, and chain of custody. The absence of a worldwide regulatory body or even a federation of national bodies significantly impacts the effectiveness of cloud forensic investigations.

NIST[16] identifies additional challenges as:

Architecture (e.g., diversity, complexity, provenance, multi-tenancy, and data segregation) – Architecture challenges in cloud forensics include dealing with variability in cloud architectures between providers; tenant data compartmentalization and isolation during resource provisioning; accurate and secure provenance for maintaining and preserving chain of custody; and infrastructure to support seizure of cloud resources without disrupting other tenants.

Data collection (e.g., data integrity, data recovery, data location, and imaging) – Data collection challenges in cloud forensics include locating forensic artifacts in large, distributed and dynamic systems; locating and collecting volatile data; data collection from virtual machines; data integrity in a multi-tenant environment where data are shared among multiple computers in multiple locations and accessible by multiple parties; and inability to image all the forensic artifacts in the cloud.

Local Computer Systems In addition to remote computer systems and storage, Zhang[17] emphasizes that local computer systems should be examined for digital evidence in the event that partial, deleted, or damaged files, data-communication logs, remote computer and server information, and digital certificates and public keys are still residing on them. Also, proxy servers and servers or systems running cloud computing security audits should be reviewed.

Forensic Triage An approach that is useful when dealing with large amounts of data in a cloud investigation is forensic triage, which is a screening process that typically happens at the initial stage of the investigation. Roussev[18] defines forensic triage as 'a partial forensic examination conducted under (significant) time and resource constraints.' Because investigators usually must search for and obtain relevant information from very large sources of data in a short amount of time, standard forensics triage methods and tools are valuable assets to have available.

Forensics as a Service (FaaS) A cloud-based method that supports handling large amounts of data associated with a computer forensic investigation. In this approach, a centralized system extracts indicators from the information for review by investigators in a shorter period of time. This capability is obtained by sharing interoperable forensic software and enabling investigators to customize forensic data-processing workflows.

Analysis (e.g., correlation, reconstruction, time synchronization, logs, metadata, and timelines) – Analysis challenges in cloud forensics include correlation of forensic artifacts across and within cloud providers; reconstruction of events from virtual images or storage; integrity of metadata; and timeline analysis of log data including synchronization of timestamps.

Anti-forensics (e.g., obfuscation, data hiding, and malware) – Anti-forensics are a set of techniques used specifically to prevent or mislead forensic analysis. Challenges in cloud forensics include the use of obfuscation, malware, data hiding, or other techniques to compromise the integrity of evidence.

Incident first responders (e.g., trustworthiness of cloud providers, response time, and reconstruction) – Incident first responder challenges in cloud forensics include confidence, competence, and trustworthiness of the cloud providers to act as first-responders and perform data collection; difficulty in performing initial triage; and processing a large volume of forensic artifacts collected.

Role management (e.g., data owners, identity management, users, and access control) – Role management challenges in cloud forensics include uniquely identifying the owner of an account; decoupling between cloud user credentials and physical users; ease of anonymity and creating fictitious identities online; determining exact ownership of data; and authentication and access control.

Standards (e.g., standard operating procedures, interoperability, testing, and validation) – Standards challenges in cloud forensics include lack of even minimum/basic SOPs, practices, and tools; lack of interoperability among cloud providers; and lack of test and validation procedures.

Training (e.g., forensic investigators, cloud providers, qualification, and certification) – Training challenges in cloud forensics include misuse of digital forensic training materials that are not applicable to cloud forensics; lack of cloud forensic training and expertise for both investigators and instructors; and limited knowledge by record-keeping personnel in cloud providers about evidence.

FORENSIC INVESTIGATION MODELS

There have been number of digital forensic and cloud forensic models aimed at quantifying the steps in their respective processes. The models vary in scope and detail but serve to illustrate different views of the steps in the forensic process.

A short list of these models as compiled by Simou et al.[19] is given as follows:

DIGITAL FORENSIC MODELS

The **Abstract Digital Forensic Model**[20] comprises the nine stages of identification, preparation, approach strategy, preservation, collection, examination, analysis, presentation, and returning evidence.

The Digital Forensic Evidence Processes Model[21] uses the stages of identification, collection, preservation, transportation, storage, analysis-interpretation and attribution, reconstruction, presentation, and destruction.

The DFRWS Model[22] defines the forensic processes of identification, preservation, collection, examination, analysis, presentation, and decision.

The Extended Model of Cyber-Crime Investigations[23] focuses on information flow description and includes the activities of awareness, authorization, planning, notification, search for and identify evidence, collection, transport, storage, examination, hypothesis, presentation, proof/defense, and dissemination of information.

The Harmonized Digital Forensic Investigation Process Model[24] is an iterative model with the phases of incident detection, first response, planning, preparation, incident scene documentation, identification, collection, transportation, storage, analysis, presentation, and conclusion.

The Integrated Digital Investigation Process (IDIP)[25] is divided into five groups: readiness, deployment, physical crime scene investigation, digital crime scene investigation, and review.

The Enhanced IDIP Model[26] is based on the IDIP model and consists of a primary crime scene phase, the trace back phase and the dynamite phase. The trace back phase identifies devices that were used, and the dynamite phase investigates objects that were discovered to possibly acquire additional evidence.

The Systematic Digital Forensic Investigation Model[27] focuses on phases of preparation, securing the scene, survey and recognition, documenting the scene, communication shielding, evidence collection, preservation, examination, analysis, presentation and, finally, result and review.

CLOUD FORENSIC MODELS

The Forensic Investigations Process[28] consists of four steps:

1. Determine the purpose of the forensics investigation
2. Identify the types of cloud services
3. Select the type of technology to be applied
4. Inspect the physical and logical locations

The Cloud Forensics Process[29] emphasizes the admissibility of the evidence through the following four phases:

1. Identify the purpose of the investigation
2. Determine the type of the cloud service,
3. Determine the type of cloud technology
4. Conduct the investigation

The Advanced Data Acquisition Model[30] comprises three stages of initial planning, the onsite survey, and the acquisition of electronic data.

The Integrated Conceptual Digital Forensic Framework for Cloud Computing[31] focuses on the four stages of identification and preservation, collection, examination and analysis, and reporting and presentation.

The Integrated Digital Forensic Process Model[32] is a comprehensive model comprising preparation, incident, incident response, physical investigation, digital investigation, presentation, and documentation.

The **Open Cloud Forensics Model**[33] comprises a preservation stage, identification, collection, organization, presentation, and verification.

An additional Cloud Forensics model with a different approach is:

The **Cloud Forensics Capability Maturity Model (CMM)**[34] that was developed by the Cloud Security Alliance 'to be used by both cloud consumers and Cloud Service Providers (CSPs) in assessing their process maturity for conducting digital forensic investigations in the cloud environment.'

The model was based on Carnegie Mellon University Software Engineering Institute's (SEI) Software Process Maturity Framework[35] which identifies five progressive levels of process maturity:

1. Initial – How are we ever going to do this?
2. Repeatable – Have we ever done this before?
3. Defined – What is our process for doing this?
4. Managed – What resources did this require?
5. Optimizing – How can we do this better

The SEI model is mapped to cloud forensics to provide high-level guidance per level and initially focuses on IaaS Cloud usage.

SUMMARY AND FUTURE RESEARCH

Reviewing current forensic techniques highlights that additional work must be done to address cloud forensics. One area that requires increased focus is the legal landscape. Jurisdictional and international cooperation issues still pose difficulties for cloud forensics processes.

Lopez et al.[36] have identified cloud forensic challenges and associated mitigation issues to be addressed as follows:

- **Extraterritorial jurisdiction** – Stronger international cooperation
- **Search warrant requirements** – Legal training
- **Lack of physical access** – Cloud provider cooperation
- **Competence and trustworthiness** – Ensure forensic procedures are followed and documented
- **Data location and collection** – Mobile forensics and data profiling
- **Multi-tenancy and resource sharing** – Cloud provider cooperation
- **Large and changing systems** – Cloud provider knowledge and live forensics
- **Massive volume of data** – Data Mining, Social Networks Forensics, Mobile forensics
- **Volatility** – Live Forensics
- **Chain of custody** – Training and legal advice
- **Make a forensic copy** – Snapshots
- **Data integrity** – Live forensic training
- **Recovery of deleted data** – Backups and repositories, snapshots, and mobile forensics
- **Cryptography** – Brute-force and mobile forensics
- **Data correlation issues** – Data mining and user profiling
- **Lack of interoperability** – Cloud provider cooperation
- **Partial Evidence** – Return to early stages of investigation
- **Investigation report** – Training
- **Choosing the right court** – Legal advice
- **Evidence return and secure deletion** – Legal training and legal advice

Likewise, NIST[37] has identified issues and areas, some similar, that would benefit from future research, listed as follows:

- Evidence correlation across multiple cloud providers
- Synchronization of timestamps
- No interoperability among providers
- No single point of failure for criminals
- Detection of the malicious act
- Intelligence processes for real-time investigation are often not possible in the cloud environment
- Malicious code may circumvent virtual machine isolation methods, and interfere with the hypervisor or other guest virtual machines
- Access to computer and network resources involve expanded scope and may involve more than one venue and geolocation
- Segregation of potential evidence in a multi-tenant system
- Decreased access and control of data at all levels by cloud consumers
- Chain of dependencies in multiple cloud systems
- Data associated with newly created virtual machine instances may only be available for a limited time
- Identifying storage media where artifacts, log files, and other evidence may be found
- Private and confidential details of cloud-based software/applications used to produce records are typically unavailable to the investigation
- Competence and trustworthiness of the cloud Provider as an effective, immediate first responder

Cloud forensics pose a number of challenges but, can provide valuable benefits in protecting cloud users and critical applications.

NOTES

1 Peter Mell, Timothy Grance, "NIST Special Publication 800-145," The NIST Definition of Cloud Computing, (Computer Security Division, Information Technology Laboratory, National Institute of Standards and Technology, Gaithersburg, MD, 2011).
2 Peter Mell, Timothy Grance, "NIST Special Publication 800-145," The NIST Definition of Cloud Computing.
3 Virtual Forensic Computing 5.0, MD5 Ltd., md5.uk.com.
4 J. Jeffers, "Computer Forensics: Forensic Issues with Virtual Systems," InfosecInstitute.com, 2020.
5 CNSSI No. 4009, "(CNSS) Glossary," (Committee on National Security Systems, 2015.)
6 Karen Kent, Suzanne Chevalie, Tim Grance, Hung Dang, "NIST Special Publication 800-86," Guide to Integrating Forensic Techniques into Incident Response, (Computer Security Division, Information Technology Laboratory, National Institute of Standards and Technology, Gaithersburg, MD, 2006).
7 G. Palmer (ed.), "A Road Map for Digital Forensic Research, "*Proceedings of the Digital Forensic Research Conference. DFRWS*, 2001.
8 Michael B. Mukasey, Jeffrey L. Sedgwick, David W. Hagy, Electronic Crime Scene Investigation: A Guide for First Responders, Second Edition, National Institute of Justice, 2008.
9 NIST Cloud Computing Forensic Science Working Group, Draft NISTIR 8006 NIST Cloud Computing Forensic Science Challenges, Information Technology Laboratory, 2014.
10 Keyun Ruan, Joe Carthy, Tahar Kechadi, Mark Crosbie. Cloud Forensics. 7th Digital Forensics (DF), Jan 2011, Orlando, FL, United States. pp. 35–46, 10.1007/978-3-642-24212-0_3. hal-01569563.

11 Keyun Ruan, Joe Carthy, Tahar Kechadi, Mark Crosbie. Cloud Forensics. 7th Digital Forensics (DF), Jan 2011.

12 Draft NISTIR 8006 NIST Cloud Computing Forensic Science Challenges, Information Technology Laboratory, 2014.

13 Keyun Ruan, Joe Carthy, Tahar Kechadi, Mark Crosbie. Cloud Forensics. 7th Digital Forensics (DF), Jan 2011.

14 Keyun Ruan, Joe Carthy, Tahar Kechadi, Mark Crosbie. Cloud Forensics. 7th Digital Forensics (DF), Jan 2011.

15 L. Chen, L. Xu, X. Yuan, et al., "Digital Forensics in Social Networks and The Cloud: Process, Approaches, Methods, Tools, and Challenges," *Proceedings of the 2015 IEEE International Conference on Computing, Networking and Communications (ICNC)*, IEEE, 2015.

16 Draft NISTIR 8006 NIST Cloud Computing Forensic Science Challenges, Information Technology Laboratory, 2014.

17 C. Zhang, "Cloud Calculative Environment Electronic Data Investigation and Evidence Collection," Netinfo Security, 2010.

18 Y. Roussev, C. Quates, and R. Martell, "Real-Time Digital Forensics and Triage," Digital Investigation, 2013.

19 S. Simou, C. Kalloniatis, S. Gritzalis, and H. Mouratidis, (2016) "A Survey on Cloud Forensics Challenges and Solutions," Security Comm. Networks, 9: 6285–6314. doi: 10.1002/sec.1688.

20 M. Reith, C. Carr, C. Gunsch, "An Examination of Digital Forensic Models," International *Journal of Digital Evidence* 2002.

21 FB. Cohen, "Fundamentals of Digital Forensic Evidence," *Handbook of Information and Communication Security*, Berlin Heidelberg, Springer, 2010.

22 G. Palmer, "A Road Map for Digital Forensic Research," The First Digital Forensic Research Workshop (DFRWS), Utica, New York, 2001.

23 SÓ, Ciardhuáin, "An Extended Model of Cybercrime Investigations," *International Journal of Digital Evidence*, 2004.

24 Valjarevic A, Venter HS, 'Harmonized Digital Forensic Investigation Process Model," Information Security for South Africa (ISSA), IEEE, Johannesburg, Gauteng, 2012.

25 B. Carrier, EH. Spafford, "Getting Physical with the Digital Investigation Process," *International Journal of Digital Evidence*, 2003.

26 V. Baryamureeba, F. Tushabe, "The Enhanced Digital Investigation Process Model," *The Proceedings of the Fourth Digital Forensic Research Workshop*, 2004.

27 A. Agarwal, M. Gupta, S. Gupta, SC. Gupta, "Systematic Digital Forensic Investigation Model," *International Journal of Computer Science and Security, (IJCSS)*, 2011.

28 H. Guo, B. Jin, T. Shang, "Forensic investigations in Cloud environments," *Computer Science and Information Processing (CSIP), 2012 International Conference on.* IEEE Xi'an, Shaanxi: 2012.

29 G. Chen, Y. Du, P. Qin, J. Du, "Suggestions to Digital Forensics in Cloud Computing," Network Infrastructure and Digital Content (IC-NIDC), *3rd IEEE International Conference*, IEEE, 2012.

30 R. Adams, "The Emergence of Cloud Storage and the Need for a New Digital Forensic Process Model," Cybercrime and Cloud Forensics: Applications for Investigation Processes, Ruan K. (ed), Hershey: IGI Global, 2013.

31 B. Martini, KKR Choo, "An Integrated Conceptual Digital Forensic Framework for Cloud Computing," Digital Investigation 2012.

32 MD Kohn, Mariki M Eloff, Jan Eloff, "Integrated Digital Forensic Process Model," *Computers & Security*, 2013.

33 S. Zawoad, R. Hasa, A. Skjellum. "An Open Cloud Forensics Model for Reliable Digital Forensics" *8th International Conference on Cloud Computing (CLOUD)*, IEEE, New York City, 2015.

34 Cloud Security Alliance, "Cloud Adoptions Practices & Priorities Survey Report" (2015).

35 Carnegie Mellon University Software Engineering Institute (1995), *The Capability Maturity Model: Guidelines for Improving the Software Process*, Addison-Wesley, pp. 15–17.

36 Erik Miranda Lopez, Seo Yeon Moon and Jong Hyuk Park, "Scenario-Based Digital Forensics Challenges in Cloud Computing," Symmetry, October 2016.

37 Draft NISTIR 8006 NIST Cloud Computing Forensic Science Challenges, Information Technology Laboratory, 2014.

Chapter 5

Forensics of the digital social triangle with an emphasis on Deepfakes[1]

James Curtis

CONTENTS

INTRODUCTION

Social Engineering has been an active method of manipulating, stealing from, and influencing others since two or more people gathered together. History is rife with victims of social engineering. Society has called them con men, grifters, and many other names. In today's society, the most common term used is scammer. While these people have always been dangerous, technology has vaulted the art of social engineering forward both in scope and potential impact as the use of computers, the internet, and applications have enabled the 'bad guys' to find, target, and successfully social engineer their victims.

Is social engineering a new mode of taking other people's stuff or getting people to do something they may not want to do? No. But, with the advancement of technology, especially the increasing ease of interactivity via the internet, it has exploded as a weapon and means for bad guys, and even some good guys, to utilize this skill toward their goals.

In this chapter, we discuss the structure of social media, networks, and engineering and the unique threats and challenges these techniques pose within the realm of cybersecurity forensic analysis. To do this, we will explore the challenges of conducting forensics within the vast parameters of social media, and present ground-breaking forensic analysis methods of Deepfakes – the newest social engineering technique as an example of how this aspect of the discipline is different than other cybersecurity threats. We will also evaluate several characteristics and methods we can use to conduct investigative forensics of the threat posed by social engineering across social media.

We first begin by presenting an overview of the relationship between the elements of this 'Digital Social Triangle' consisting of Social Media, Social Networks, and Social Engineering. The Digital Social Triangle (Figure 5.1) is the combined power of Social Media, Social Networking, and Social Engineering as applied by criminals, organizations, terrorists, or nation-states to achieve a goal.

It is important to understand that over the past decade, the art and science related to social engineering has been boosted significantly with the advent of social networking between people as they have become more comfortable with, accepting of, and often gullible to the exploding 'fake news' environment present on digital platforms; and then, when the art of social engineering is combined with this generally acceptable culture across social networks, society has added the final element of the triangle – social media. Specifically, internet applications such as Facebook, Twitter, Instagram, TikTok, and other social media have become a magnifier to make it easy for people to connect, share information, and to be manipulated by social engineers. The threat of the growing influence and power of the Digital Social Triangle has pushed this element of cybersecurity to the forefront of federal and state government concerns, especially regarding the rise of nation-state uses of the Digital Social Triangle.

The Russian social engineering attack on the United States of America's 2016 presidential election is one example of this explosion of use of the Digital Social Triangle as a weapon. And, terrorists groups such as ISIS have been very effective in their use of the triangle to achieve their goals of recruitment, intimidation, financial gain, and propagation of their message. This is an element of cybersecurity that is chilling in the possibilities of the growth in potential crime, nation-state influence, political disruption, and economic instability.

We will present several aspects of Online Social Networks (OSNs), specifically from the perspective of the Digital Social Triangle. And, we will conduct a detailed investigation into one of the newest technique called – *Deepfakes*. We will evaluate the approaches used to identify, specify, and counter Deepfake attacks through several rapidly maturing forensic methodologies.

It is also important to note that from a legal perspective the accumulation of forensic data from OSN is critical to identify, prosecute, and convict criminals who use the Digital Social

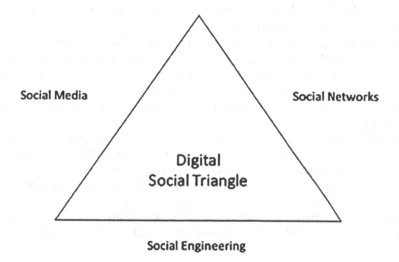

Figure 5.1 Digital Social Triangle[2]

Triangle. Digital tracing and forensic evaluation are critical to the process of successful conviction of criminals who use online networks to carry out their activities. The aspects of cloud services has continued to complicate these programs and more online services, enhanced memory storage options, as well as the enhanced personal user choices for encryption methods have significantly enhanced the 'bad guy's' tool kit to exploit, steal, hurt, and impact the lives of millions of innocent people, organizations, and nation-states.

WHY IS AMERICA (AND WESTERN CIVILIZATION) VULNERABLE?

To fully understand the threat of the Digital Social Triangle, it is essential to discuss the special vulnerabilities of open, democratic, societies, especially America. From our growing experience with digital social engineering, we are able to discern that America, as well as other western liberal democracies, is more vulnerable to social engineering attacks than other nations. Whether from nation-states such as Russia, China, or North Korea, or terrorist groups like ISIS, or the plain old online criminal, the United States of America has become target #1 for social engineering. To understand how this is true, it is important to examine several aspects of America that makes it an easy target for social engineers. The social linkages of societal norms, liberal laws, and an open business model, combine to be fertile ground for Americans as targets.

Russia – From a nation-state perspective, America's adversaries have learned quickly that using social media as the conduit for social engineering is inexpensive and quite effective. Russia is especially inclined to use the digital avenue to sow chaos, confusion, and influence American policies. In 2016, Russia was the source of the single largest, and most effective nation-state social engineering program in history – the attack on the U.S. presidential election campaign.

ISIS – The Islamic State of Iraq and the Levant, also known as ISIL, is the world's most famous terrorist group. It is also the first major terrorist group to use social media, propaganda, and social engineering to expand its influence, recruit, train, grow wealth, and other key elements of its survival. ISIS actually had an online magazine (Dabiq), chat rooms, marketing campaign, terror videos, and many other social media influencer programs. The effect across the western world was significant. The ability to influence new recruits in Europe was especially effective. In the ISIS study, it literally had its own technology department to manage its social media program. The primary focus was on intimidation of the enemy, recruitment of like-minded individuals, education of its goals and objectives, and any other online activities which would serve to advance its mission.

Plain Old Criminal – The combination of the technical advancements of the past two decades combined with the age-old art of the 'con man' served as a perfect storm for the modern day plain old criminal who was willing to learn how to take advantage of social media. The ability to trick, cheat, influence, con, or social engineer others to do something the criminal wants done has been magnified significantly with the melding of the art of the steal and technology.

'Western world' Cultural Vulnerabilities – Because the openness of the American and European democracies are inherently fertile ground for a well-organized, propaganda-based social engineering campaign. A lack of accountability, laws, and restrictions, combined with an open culture, makes these nations prime for manipulation and influence. In America, in particular, there is an inherent concept of individualism, choosing our own paths, an actual sense of pride and identity rejecting conformity that sets up the average American as a target for social media manipulation and social engineering victimization.

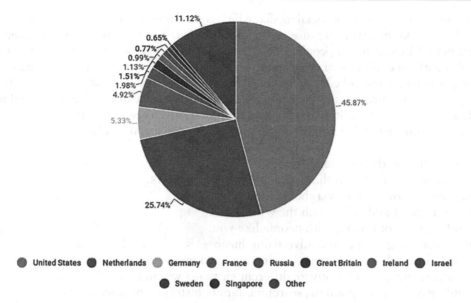

Figure 5.2 Web-based attack distribution by source country (Q2, 2018)[3]

Figure 5.2 displays the percentage of web-based cyber-attack distribution against representative nations. With just 4% of the world's population, the United States has almost 46% of the total attacks.

In his book, *Social Engineering: The Science of Human Hacking*, Christopher Hadnagy makes the claim that the first documented example of social engineering is from the Bible's Old Testament.[4] In the book of Genesis, there is the story about Jacob who tricks his father into believing he is actually his older brother Esau and giving him his brother's rightful blessing. Jacob used social engineering via deception by fastening goat skins to his arms, back, and neck. When his father, Isaac, who was almost blind, reached out to give his blessings to (Esau), it was actually Jacob deceiving him to believe he was Esau. And, it resulted in Jacob (who becomes the father of Israel) receiving the blessings of his dying father. This social engineering may not have had the internet, software, and hardware to help it to success, but it was an early example of social engineering.

UNDERSTANDING THE RELATIONSHIP BETWEEN SOCIAL NETWORKS, MEDIA, AND ENGINEERING

In order to comprehend the power of social engineering, which feeds into the difficulties of conducting successful forensic investigations, we provide an overview of key areas that enable social engineering to thrive as part of the digital social triad. First, Social Media.

Social media

Webster Dictionary defines **Social Media** as 'Forms of electronic communication (such as websites for social networking and microblogging) through which users create online communities to share information, ideas, personal messages, and other content (such as videos).'

Notice that this definition of social media, which is similar to other sources, puts an emphasis on social networking. That is because in our modern digitally centric world social media leads to social networking between people. Clearly, that has been a growth industry the past couple of decades with an explosion of this phenomenon really happening after the first smart phone enabled any person to be linked into the social network from anywhere at any time. And, while the dangers are ever-present and growing, we should pause to acknowledge that social media actually has some good qualities and positive aspects related to everyday living.

Below are some general examples of the positive uses of social media:

- Connection with family and friends.
- Listening and sharing feelings and empathy with others.
- Immediate connections via the internet.
- Being present and active with the world.
- Finding and connecting with people like you.
- Establishing, growing, and advertising businesses.
- Assists in teaching and learning using new and innovative means.
- Activate, engage, and motivate different elements within society.
- Finding a job, getting a date, searching family histories and ancestries.

Identifying online sites as social media

According to easymedia, a website dedicated to media students and media scholars, there are nine 'Key Features of Social Media Sites.' The author, Sunil Saxena, a media professional with experience in Social Media, presents the key features which support the ability of forensic analysts to utilize as they evaluate an online site as being part of social media.[5]

1. **Provide free web space:** Members of these sites don't need to own or share web servers. They can publish their content on the free space provided by these sites.

2. **Provide free web address:** Members are allotted a unique web address that becomes the web identity of an individual or a business. It can be used to identify, connect, and share content.

3. **Ask members to build profiles:** These sites require members to build their profiles. Information entered in the profiles is used to connect friends and contacts, and build networks that connect people with similar likes and interests across the world.

4. **Encourage members to upload content:** These sites allow members to upload text messages, photographs, audio and video files. All posts are published in in descending order with the last post coming first. Most important, all content is published in real time, and can be read, viewed, or shared instantly.

5. **Allow members to build conversations:** Members can browse content and comment upon it. By doing so, social media sites allow members to engage in conversations that increase engagement.

6. **Allow live chats:** Several social media sites have chat clients that enable members to chat with each other in real time.

7. **Direct Messaging facility:** Several social media sites provide direct messaging facility to their members. This allows members to send private messages, which can be read or viewed only by those for whom the message is intended.

8. **Provide tagging alerts:** Most social media sites alert members through e-mail or in site notifications whenever they are tagged in a message or in a photograph.

9. **Enable members to create unique pages:** On some social media sites, members can create theme-based pages. The pages can then be used to post articles or photographs related to a theme. The pages can also be used to promote businesses.

SOCIAL NETWORKING

Social Networking – Google's dictionary defines Social Networking as 'The use of dedicated websites and applications to interact with other users, or to find people with similar interests to oneself.'

Social Networking by itself gives the impression that you are trying to link up with someone or a group of people who share something in common.

John Scott suggests that contemporary social network analysis (SNA) draws on three lines of inquiry:[6]

1. Sociometric analysts in the United States during the 1930s, whose work had roots in Gestalt psychology, aimed to investigate how feelings of well-being are related to the structure of people's social lives. This movement is most closely associated with Jacob Moreno, who devised the sociogram, a visual diagram of people's relationship net-works in which individuals are represented as points and their connections to others as lines. Other major players in this research movement were Kurt Lewin, whose greatest legacy was his promotion of mathematical models of group relations, and Fritz Heider, who focused on people's perceptions about their relationships with others.

2. Also, in the 1930s, Harvard University researchers began focusing on cliques in social groups to identify cohesive subgroups (such as work, church, family, associations, and clubs) with-in social systems. This group was influenced by anthropologist Alfred Radcliffe Brown, whose work focused on factory and community life in the United States.

3. A group of anthropologists in Manchester, England, also drew on the work of Radcliffe Brown in the 1950s. John Barnes, a member of this group, is attributed with having coined the specific term 'social networks' in 1954. His work with Elizabeth Bott drew on the sociometric approach, but focused on people's informal social relationships rather than those associated with institutions and associations. In addition, their work focused on conflict and change in these networks. Clyde Mitchell extended the traditional sociometric approach with insights from the mathematics of graph theory to better deal with observations that were gathered.'

WHY IS SOCIAL NETWORKING SO POWERFUL?

Author J. Prier, writing in Strategic Studies Quarterly,[7] proffered that the propensity of social media being used as a weapon in information warfare continues to grow as both a capability and a concern for the U.S. Government.

- 72% of Americans get digital news primarily from a mobile device, and people now prefer online news sources to print sources by a two-to-one ratio.
- Homophily: 'Birds of a feather flock together' As social media usage has become more widespread, users have become ensconced within specific, self-selected groups, which means that news and views are shared nearly exclusively with like-minded users.
- Homophily within social media creates an aura of expertise and trustworthiness where those factors would not normally exist.
- Ultimately, this 'echo chamber' can promote the scenario in which your friend is 'just as much a source of insightful analysis on the nuances of U.S. foreign policy towards Iran as regional scholars, arms control experts, or journalists covering the State Department.'

Academic analysis conducted by political scientists are concluding that Americans are more tribal today than in many decades. We share with people who believe in what we believe, who like what we like, and we treat those who do not as the 'other.' This echo chamber type of communications via social networking using social media makes us extremely vulnerable to manipulation by the social engineer.

SOCIAL ENGINEERING

Now, let's turn to **Social Engineering**. What is it? We provide three definitions that are similar yet different in their nuances:

1. 'The art of gaining trust or acceptance in order to persuade someone to provide information or perform an action to benefit the attacker.'[8]
2. 'Getting people to do things they wouldn't ordinarily do for a stranger.'[9]
3. 'The act of manipulating a person to take an action that may or may not be in the target's best interest.'[10]

While the art of social engineering has been around as long as humans have existed, the advent of the digital realm has served as a tool for the social engineer to use in more impactful ways, with the ability to reach more potential 'victims,' combined with the capability to manipulate people in convincing fashion. The old, and often celebrated throughout history in literature and theater, con man skills are now on proverbial 'steroids' with the support of the digital social media environment. Combining social media, with the interactiveness of social networking, serves as the foundation for the social engineer to gain access, insight, and control over others.

In the end, social engineering is nothing more than the art of manipulating others to get them to do what you want; however, combine that skill with technology, and you end of with a very dangerous weapon which can do damage every bit as destructive as a kinetic weapon. Combining the use of social media with the intent of being active through social networking leads us directly to the modern-day version of social engineering or more accurately 'Digital Social Engineering.'

It is this combination of digital social engineering, social media, and social networking that form the Digital Social Triangle, resulting in elevating the online criminal threat to heights unimaginable just a decade or so ago.

Hadnagy's social engineering pyramid

To successfully conduct a forensic analysis, we must understand the attacks, specifically what type of pretext (planning) was used to develop the attack. Every social engineering attack can be broken down into one or more of these types of attacks. Hadnagy developed a concept called a Social Engineering Pyramid;[11] this concept can provide a forensic analyst a foundation for understanding and evaluating the standard process for a social engineering attack from inception to execution.

As presented in Figure 5.3, the pyramid is presented in several sections, and approaches social engineering from the perspective of a social engineer professional – that is, not one using social engineering for nefarious purposes but to help clients and customers. This is the perspective of a forensic evaluator. The following description of the social engineering pyramid enables the reader to gain insight into the process of a successful social engineering attack.

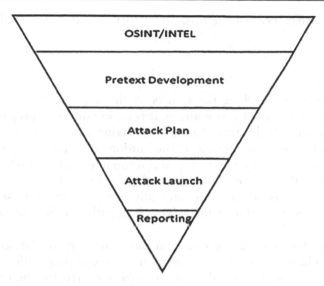

Figure 5.3 Hadnagy's social engineering pyramid (redrawn by author)[12]

OSINT/INTELLIGENCE

Open Source Intelligence (OSINT), is the life blood of every social engineering engagement. It is also the piece that should have the most time spent on it. Due to that, it occupies the first and largest piece of the pyramid. One piece of this part of the pyramid is rarely addressed: documentation. How will you document, save, and catalog all the information you find?

Hadnagy puts a special emphasis in documentation. Without a comprehensive set of events, an understanding of the 'who, what and how' of the attack, and a set of data to analyze, a forensic attempt will likely have little chance of success.

Pretext development

Based on all the findings from the OSINT period, the next logical step is to start to develop your pretexts. This is a crucial piece that's best done with OSINT in mind. During this phase, you see what changes or additions need to be made to ensure success. This is also when it becomes clear what props and/or tools are needed.

For the forensic analyst, understanding the pretext conceptualization, the logic behind the planning phase, and the general outline for a plan. This is similar to 'outlining' an idea before starting the more detailed planning.

Attack plan

Having a pretext in hand does not mean you are ready. The next stage is to plan out the three W's: what, when, and who. What is the plan? What is it we are going for and trying to achieve? What does the client want? These questions will help develop the next piece. When is the best time to launch the attack? Who needs to be available at a moment's notice for support or assistance?

Here, Hadnagy is putting the pretext into context. This is the most important phase. Similar to any other major activity whether it be project management, taking a vacation, or getting married, it is the planning stage that either makes it successful or a failure.

So, this is the phase that most of the time and effort should be invested in by the social engineer.

Attack launch

Now comes the fun part: launching the attacks. With the preparation done on the attack plan, you are prepared to go full steam ahead. It is important to be prepared but not to be so scripted that you can't be dynamic. I am all for having a written plan, and I think it can save you a ton of headaches down the road. The caution I have is that if you script out every word or action you feel needs to be taken, you can run into problems when the unexpected happens. Your brain realizes there is nothing on the script to help, and you begin to stutter, get nervous, and show signs of fear. This can really ruin your ability to succeed. Instead of scripting, I suggest using an outline, which gives you a path to follow but allows for artistic freedom.

While planning should be the phase where the most time is spent, the actual attack on the intended victim should be executed in a flexible manner with preparation for unknowns to encounter. Hadnagy advises the social engineer to be prepared for the unexpected and to adjust as the attack plan is implemented. There is an axiom in the American military regarding the art of war. It is essentially that 'no battle plan survives first contact.' Social engineers should operate under the same construct and be prepared to adjust the attack plan once the 'send button' has been pushed.

Reporting

Come back and read it. Yes, reporting is not fun, but you can think of it this way: Your customer just paid you x dollars to perform some services, and most likely, you were pretty darn successful in those attacks. But the customer didn't pay you just because they wanted to look cool. They paid you to understand what they can do to fix the problem. For that reason, the reporting phase is at the very tip of the pyramid, is the very pinnacle that the rest of the pyramid rests on. The five phases of this pyramid, if followed, will lead to your success not only as a social engineer, but as a professional who is offering social engineering services to your customers. The fact is that, with the exception of reporting, these steps are followed by the malicious social engineers in the world.

This final phase is where the forensic analyst finds their answers. It is the understanding of the social engineer's pretexting concepts, the detailed planning for the attack, and then how the execution of the attack actually unfolded leading to success or failure.

OSINT

Before presenting the specific types, traits, and techniques of social engineering that forensic analysts evaluate, we need to understand more about where the information comes from which social engineers use against intended victims. OSINT is the primary source for social engineers to gather the information they need to develop their attack plan.

The amount of data which can be turned into useful information to be developed into a digital weapon for the social engineer to use against their target grows by the day in the great digital divide. Combine the massive amounts of information with a pretext story, the social engineer is able to develop a plan to exploit the target and achieve their goal. Once the plan is ready, it can be executed and if done correctly, the social engineer achieves success. This can range from stealing money, to manipulating a person to do something illegal, immoral, or unethical, to influencing the actions of a nation-state in foreign policy or even acts of war.

In 2011, the United States Office of the Director of National Intelligence, defined OSINT as 'intelligence produced from publicly available information that is collected, exploited, and disseminated in a timely manner to an appropriate audience for the purpose of addressing a specific intelligence requirement.'[13]

And, the 1992 Intelligence Reorganization Act[14] determined key concepts of OSINT included two main concepts:

1. Must be objective intelligence free of bias.
2. Data must be available on public and non-public sources.

While OSINT is information collected from public sources such as those available on the internet, it is NOT limited to the internet. It also includes other information housed in libraries, periodicals, and similar sources. While the concept of OSINT has evolved significantly the past two decades of growth through OSN, as continues to be sourced by public sources in an expanding fashion.

There are many different ways to use OSINT when searching online. Some of these include:

1. Names of students, employees, colleagues.
2. Searching for corporate announcements, news releases, blog updates.
3. Addresses and telephone numbers.
4. Map identification of homes and offices.
5. Photograph identification for location and time/date.
6. Social network and media usage, announcements, events, birthdays, anniversaries, etc.
7. Search engine information gathering.

When diligent, an experienced social engineer can 'triangulate' information from multiple methods of OSINT to develop a profile of an intended target, establish patterns, obtain key personal identifiable information (PII), and even find out recent locations and social activities. All of this can then be used to develop and plan the social engineering attack against an individual or organization.

CATEGORIES AND TYPES OF SOCIAL ENGINEERING

Conheady[15] provides a list of six types of social engineering within three categories that represent the preponderance of the tactics utilized by social engineers. According to the author, the first category is **physical** social engineering which is defined as when an attacker tries to gain physical access to the targeted location; the second category is **remote** social engineering which is when an attacker gains access to information and/or resources via remote means such as the telephone or internet; and, the third category is a **combination** social engineering attack using one or more types from both of the other two categories. Table 5.1 outlines Conheady's categories and types of social engineering.

All of these types can be effective in meeting the social engineer's goal. And, sometimes the attacker may use one or more of them to achieve success. Is the person gullible enough to fall for a 'distraction' trick? Or, maybe their situation is a bit more sophisticated, thus requiring a longer term, more complex 'combination' method. This is the heart of social engineering – understanding the target, and then applying the correct method to achieve the social engineering results. Using today's technologies to include computers, transmission media, and

Table 5.1 Categories & types of social engineering[16]

Category	Type	Definition
Physical		Gaining access to a physical location
	Dumpster Diving	Literally combing through trash cans and dumpsters looking for sensitive materials, photos, old computers, etc., that may be used to gather sensitive information
	Distraction	Using a distraction to divert attention from the real attack target such as commotion, food delivery, or a physical altercation
Remote		Using electronic means that don't require real-time activities (you can wait for the victim)
	Email	Online scamming techniques to trick the receiver into clicking on a link or providing access to their system via electronic means
	Telephone	The verbal equivalent of email scamming attacks. Using the telephone to influence, coerce, or intimidate the victim
Combination		Combined physical, distraction, or electronic remote attacks to work in combination in an effort to confuse and overwhelm the victim
	'Boy who cried wolf'	Series of false attacks that tend to numb the victims so when the actual attack happens, they don't respond
	Road Apples	Physical object left in vicinity of target who is attracted to it and unknowingly loads it on their computer (classic bait-and-hook)

the internet, a modern social engineer can quickly become impactful against a target in the Digital Social Engineering era.

TRAITS OF SOCIAL ENGINEERING ATTACKS

It is important for the forensic analyst to understand the social engineers' traits. This is analogous to a criminologist work with fingerprints or other unique markers for a human being. The more the forensic team can understand about the traits of the social engineer, the higher the chances of identification. The key to defeating the plethora of OSN attacks is the process of Identification. Since OSN attacks are primarily time based (they lose impact the longer they are exposed) in their effect, the process of forensics is intertwined with the identification of the attack. The identification, using a thorough understanding of OSN social engineering traits, will aid in thwarting attacks and assist in the forensics analysis.

To conduct successful forensic analysis of an OSN attack, the analyst needs to understand the primary traits of an attack. We will primarily focus upon the main OSN attack technique (Phishing), but they apply to any OSN attack, especially ones that are similar to Phishing such as Spear Phishing, Whaling, and Baiting.

Additionally, while there are other traits, these ten stand out as the most common ones we need to be aware of as we work to counter an OSN attack.

1. The message contains a mismatched Uniform Resource Locator (URL) – If the address of the message is from one URL, but when you hover your cursor over the link and it is a different one, it is an indicator of a possible Phishing.
2. URLs contain a misleading domain name – It is important to fully understand the domain name. Look for misaligned or child domains that are not lined up correctly. As an example: the URL may have the company on the left side, and a different name further to the right. It is the one furthest to the right that will be the parent link.

3. The message contains poor spelling and grammar – This is one of the most common indicators of a Phishing attempt. Poor syntax, misspellings, bad grammar, and misaligned tenses of words are indicators. Most of the time, these are attacks from foreign entities. However, as the bad guys get better at this technique, they are learning to use software that assists them in correcting these mistakes.

4. The message asks for personal information – Always be on guard when asked for PII or banking, credit card, or related information.

5. The offer seems too good to be true – PT Barnum is credited with the famous saying 'there is a sucker born every minute.' – Phishers act in concert with the belief he was correct. If the offer seems amazing, it is likely not. Ask your friends and family to give you their assessment before committing to an offer that seems so good.

   Below is a typical example of both an offer too good to be true, one that will lead to asking for money, and is full of poor grammar and sentence structure:

Beloved,

I am writing this mail to you with heavy tears in my eyes and great sorrow in my heart. As I informed you earlier, I am (Mrs.) Sharon Suzuki from Japan and a widow to late Martin Suzuki, I am 63 years old, suffering from long time Cancer of the breast. From all indications my condition is really deteriorating and it's quite obvious that I won't live more than 2 months according to my doctors.

I have some funds I inherited from my late loving husband Mr. Martin Suzuki, the sum of ($2,000.000,00) which he deposited in a Bank. I need a very honest and God fearing person that can use these funds for Charity work, helping the Less Privileges, and 20% of this money will be for your time and expenses, while 80% goes to charities.

Please let me know if I can TRUST YOU ON THIS to carry out this favour for me. I look forward to your prompt reply for more details.

Yours sincerely
Mrs. Sharon Suzuki

6. You didn't initiate the action – often, you will receive a notice that you were the winner of the drawing, or selected to get a special deal. If you did not initiate the exchange, do not engage.

7. You're asked to send money to cover expenses – This is almost always a trap. No reputable organization will ask you to cover their expenses for something that benefits you. Do not send money without verifying the request through legitimate means.

8. The message makes unrealistic threats – Intimidation and threats get people nervous and then they make mistakes. If you receive a notice that you must comply with a demand for credit information, identification, or other PII within a certain period or you will be fined, arrested, turned over to a credit investigator, etc., it is likely a scam. Reputable companies use the postal service mail for these types of serious issues.

9. The message appears to be from a government agency – Similar to #8, the scammer is counting on the receiver to be respectful, concerned, and compliant with an official 'government' directive or request. The IRS, FBI, or other federal government agencies do not initiate official business with citizens via email. They use the United States Postal Service mail system.

10. Something just doesn't look right – Trust your gut. If it looks odd, feels strange, or seems out of sync with a normal communication, it likely is. Verify before committing to anything.

Even a seasoned cybersecurity expert can be lulled into victimization of a sophisticated OSN attack. And, while these traits are important for every person active online, successful Phishing attacks or related ones such as Spear Phishing will continue to mature and succeed. Therefore, it is logical that the forensics plan uses these same traits to conduct the analysis of the specific incident.

SOCIAL ENGINEERING LIFE CYCLE ATTACK AND OODA LOOP MODELS

From their analysis of social engineering attacks, Jamil et al. developed a Life Cycle of Social Engineering Attack model[17] to depict the social engineer's process for developing and executing an attack on their victims. Using this model, which also aligns with Hadnagy's pyramid concept, it is apparent that the life cycle of an attack is sequential and has a structural consistency regardless of which social engineering technique is applied in the execution of the plan. The research about the target using OSINT to determine operational, administrative, technical, or cultural weak points of entry is paramount to a successful attack. Once this is completed, the exploitation (plan in motion) is conducted. If the plan was sound, and the execution proper, the effects will likely meet the intent of the social engineer. Figure 5.4 depicts this Life Cycle Attack model in action.

The author's model shares a similarity to a United States Air Force concept called OODA Loop. OODA loop stands for Observe–Orient–Decide–Act. Even though it is many decades old, it has application to the realm of cybersecurity social engineering analytics and forensics. The OODA Loop concept (Figure 5.5)[19] was developed by Air Force Colonel John Boyd, and has been taught in Air Force officers' schools to train them to take in information, evaluate its values, form a plan, and then put the plan into action.

Figure 5.5 shows the OODA Loop process in action. It is the continuous process of learning, thinking, decision-making, and execution of a plan to keep the analyst in the moment regarding the threat. Taken together, the OODA Loop and Social Engineering Lifecycle serve as complimentary models for a forensic understanding of the complexities of the Digital Social Triangle.

While Colonel Boyd did not develop the OOPDA Loop concept for cybersecurity forensics, the process he outlines, especially when overlaid upon the social engineering attack life cycle,

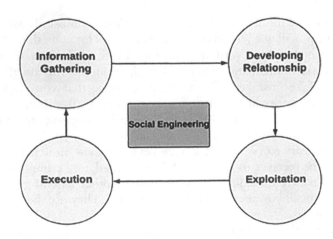

Figure 5.4 Life Cycle of Social Engineering Attacks[18]

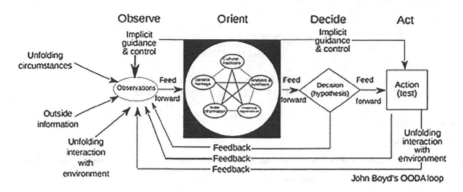

Figure 5.5 Boyd's OODA Loop concept

has a unique role in understanding how to approach a forensic evaluation of social engineering attacks.

The link is so clear that Richard Clarke and Robert Knake, leaders in national cybersecurity policies, and authors of *The Fifth Domain: Defending Our Country, Our Companies, and Ourselves in the Age of Cyber Threats*, believe Colonel Boyd's simple concept can be applied to cybersecurity analytics and forensics as a guide for understanding and countering the 'bad guys.'[20]

SOCIAL ENGINEERING TECHNIQUES

In this section, we present the primary and generally most impactful techniques used in social engineering to 'steal your stuff,' trick and hurt people, manipulate others, and to achieve the overall goals of a social engineering campaign against organizations to get others do something they would not normally do. You will notice the 'odd' names of these techniques with many of their names being derivatives of the original OSN method – Phishing. These older techniques maintain the family of aquatic names for the most part, but as we transition into the newer social engineering methods, we move away from the aquatic-linked names and go into new types of attacks such as Formjacking and Deepfakes.

Phishing (also known as spam phishing)

Phishing is the original digital social engineering technique, and one that continues to be the most prominent and successful to this day.

Most phishing scams share these characteristics:

- The 'Phish' seeks to gather PII. This could include names, addresses, birth dates, social security numbers, spouse, and children.
- The email or text has a highlighted or embed link that when clicked on sends the user to the intended Phishing websites.
- The attack has an air of authority, urgency, is threatening in nature or attempts to put the reader into a panic mode to react without thinking through the issue.

    Some Phishing emails are not as 'polished' or skillful as others. They are often written by individuals for whom English is not their primary language, and the text of the attack is filled with syntax, grammar, and spelling errors. One of the key areas to look for are 'tenses' because learning the proper use of tense in the English language can be

Webster
UNIVERSITY

MESSAGE *from the*
PRESIDENT *and* PROVOST

November 02, 2017

Dear Colleagues,

I have some important documents to share with you which requires all employee prompt attention.

Please _read_ through attached document.

It's of high importance all employee read through on what improves the welfare of our university.

Thank you for all you've done and for all you continue to do.

Sincerely,
Elizabeth (Beth) J. Stroble, Ph.D.
President

Julian Z. Schuster, Ph.D.
Provost, Senior Vice President & Chief Operating Officer

Webster University • 470 E. Lockwood Ave. • St. Louis, MO 63119-3194 U.S.A.

Preferences | Unsubscribe

Figure 5.6 Example of a standard phishing attack

a challenge. These straightforward types of Phishing attacks are almost always meant for the reader (victim) to click on the provided link which then takes them to malicious software or some other manor of gaining access to their computer data or a volunteering of information.

Figure 5.6 reflects an actual phishing attack against Webster University. Note the poor grammar ('...all employee read through on...'), as well as the misuse of singular tense versus plural for the sender(s). Also, hovering the cursor over the hyperlink 'read' shows the phishing source's link to gain access to the university's network – http: *pecl.pk.wallawalla/index.php*.

While this real example is not sophisticated, has several grammatical errors, and is fairly easy to discern that it is not credible, it is effective. Dozens of receivers clicked on the phishing link leading to corrective actions by the university's IT department. Unfortunately, the 'bad guys' are getting better with experience and are improving their presentation, use of the English language, and relational aspects of the phishing mode of attack.

The internet security firm Kaspersky identifies a sub-set group of methods used in phishing.[21] The list is a sub-set of the overall phishing concept targeted to a specific type of technology. They represent different modes of attack:

- **Voice phishing** (vishing) phone calls may be automated message systems recording all your inputs. Sometimes, a live person might speak with you to increase trust and urgency.
- **SMS phishing** (smishing) texts or mobile app messages might include a web link or a prompt to follow-up via a fraudulent email or phone number.

- **Email phishing** is the most traditional means of phishing, using an email urging you to reply or follow-up by other means. Web links, phone numbers, or malware attachments can be used.
- **Angler phishing** takes place on social media, where an attacker imitates a trusted company's customer service team. They intercept your communications with a brand to hijack and divert your conversation into private messages, where they then advance the attack.
- **Search engine phishing** attempt to place links to fake websites at the top of search results. These may be paid ads or use legitimate optimization methods to manipulate search rankings.
- **URL phishing** links tempt you to travel to phishing websites. These links are commonly delivered in emails, texts, social media messages, and online ads. Attacks hide links in hyperlinked text or buttons, using link-shortening tools, or deceptively spelled URLs.
- **In-session phishing** appears as an interruption to your normal web browsing. For example, you may see such as fake login pop-ups for pages you're currently visiting.

Spear phishing

Spear Phishing, in a sense, is almost the opposite attack mode of Phishing. Whereas Phishing can be thought of as dropping your line into the email pool and then wait to see who bites, with Spear phishing, the social engineer develops a specific line of attack focused upon a small, specific group or single person within an organization with a specific result as the goal.

Instead of sending out hundreds of emails to a large group, such as all the employees in a business, and then waiting to see which 'guppy' bites, a Spear Phishing attack takes more skill, requires more research, and demands more time and planning by the social engineer.

The social engineer will develop a plan of attack tailored to the high-profile target, and spend quite a bit of time in conducting OSINT to prepare a specifically tailored email that takes advantage of knowledge about the person's business, hobbies, personal activities, social life, etc. In these incidents, the knowledge of the individual target is paramount for success. And, having done their 'homework' the social engineer makes the detection of the attack very difficult to the average target.

Catfishing

Catfishing is probably the dastardliest and physically threatening of any social engineering technique. In a sense, it has been around for hundreds of years, but with modern technology giving it a boost, it is on a virtual 'steroid' trip and is so much more powerful as a means to trick innocent people than ever before.

Basically, a Catfish attack is when someone, known as the 'catfish,' creates a fake online identity and seeks out to establish a relationship with the victim. Generally, these attacks are motivated toward a sexual/romantic relationship. While they have taken actions to safeguard their clients, many online dating websites and cell phone dating apps are known phishing locations for a catfish. And, the Catfishers use other elements of the Digital Social Triangle to find, reel in, and attack their victims.

Catfishers use fake pictures, lies, false resumes, and other means to deceive the victim. One of the most effective uses is to play the role of a war veteran who has been hurt or has emotional damage as a result of combat. With the prevalence of war veterans in the nation the past two decades, this has been a particularly effective manner to Catfish.

Catfish warning signs

Like other OSN-based social engineering techniques, most Catfishers have a 'tell.' A 'tell' in gambling, primarily in card playing, is seen when the opponent gives a small clue regarding the type of hand they are playing. If we understand the 'tell,' we are at an advantage and can counter the attacker.

The following list contains some common features of the Catfisher's tells.

1. The Catfisher is evasive when asked about family, hometowns, friends, or other aspects of their history.
2. They seem to be available at inconsistent and odd hours of the day.
3. Their social media history and links to events, pictures, or activities are minimal.
4. They do not use American slang, or sometimes the English language properly.
5. They tell very imaginative stories about their life experiences, often based upon adventures traveling.
6. They seem worldly to include having careers and job responsibilities that has them traveling, and being unavailable for in-person meetings.
7. They do not want to conduct live online meetings such as Facetime sessions.
8. Their life story is too good to be real. They are war heroes, great athletes, inventors, etc.
9. They ask for 'favors' like paying bills or cash advances.
10. Something just doesn't seem 'normal.'

Whaling

Now, let's take Spear Phishing to the next level – Whaling.

Similar to spear phishing, whaling is a highly targeted attack vector that is designed to strike at an organization's 'big phish.' From popular Hollywood movies, we know that the big-time gamblers in Las Vegas are called Whales because they both have access to large sums of money and are susceptible to parting with it through gambling. In social engineering, a big phish is a high-value individual whose credentials or access to resources, if compromised, could endanger the entire business. Whaling attacks typically select targets specifically because of their position within the organization.

Similar to Spear Phishing, these attacks can be more difficult to detect because of their stealth and because they are generally sent on a one-time basis. Oftentimes, the social engineer will assume the role of the 'Whale' and attempt to persuade another individual within the organization to take an action which will transfer money or proprietary information to a seemingly legitimate person.

One of the most famous examples of Whaling is the victimization of the U.S.-based crypto-currency processor, *bitpay*. In this example, the company succumbed to a social engineering Whaling attack when the attacker successfully impersonated the company's CFO by convincing *bitpay*'s CEO to pay 5,000 bitcoins (approximately $650,000 in today's value) into the social engineer's account. In this case, two Whales were involved, the CEO and the CFO; and, this example serves to reconfirm that no one is insulated from a well planned and executed social engineering attack.

Baiting

A variation of Phishing is Baiting. In Baiting, the social engineer uses the enticement of the victim 'getting something' out of the exchange. It may be music CDs, or a flash drive, or a meal, even a gift card, but the goal of the social engineer is to entice the target into providing the information the social engineer wants as a result of obtaining the goods which serve as the enticement.

Baiting is a form of Phishing attacks, but it uses personal benefit and self-service as the primer for success. Quite often, the Baiter uses offers of personal benefit such as getting a free streaming video or digital song download just for the 'easy' exchange of access to information.

In the discussion of the Types of Social Engineering, we discussed the concept of 'road apples.' When operating within the confines of OSN, the use of road apples is through this type of online free discovery of something we want to use and think we are getting for free.

One study done in 2016 had researchers drop 297 USB drives around the campus of the University of Illinois. The drives contained files on them that linked to webpages owned by the researchers. The researchers were able to see how many of the drives had files on them opened, but not how many were inserted into a computer without having a file opened. Of the 297 drives that were dropped, 290 (98%) of them were picked up and 135 (45%) of them "called home."[22]

Vishing

We turn our attention back to a Phishing style attack but using telephones. In reality, Vishing is actually older than Phishing attacks because telephone scammers have been around since the day after the telephone was invented. Vishing, also commonly known as voice phishing or phone elicitation, is a rapidly growing social engineering attack vector. Vishing attempts are difficult to monitor and trace, and attackers are increasingly leveraging this mechanism to extract information and compromise organizations. Unfortunately, employees in customer service, sales and human resources departments are highly vulnerable to these types of attacks without a proper training and awareness program.

Like Phishing, the goal is to obtain usable information that assists the social engineer in their attack to 'steal our stuff' or influence, damage, or exploit people and organizations. As most people know today, the use of Caller ID is essentially worthless because the Vishers are able to spoof numbers. They use local numbers, fake organizations, and even legitimate government agencies on the Caller ID to trick the recipient into believing the call is legitimate.

A few of the most popular themes are acting as a government agency such as the IRS or FBI, being a Microsoft service representative, or acting as a helpdesk technician from the user's organization.

The example below is a real Vishing attack attempt that was viral in the Southern Illinois suburbs of St. Louis. In this case, the scam was set up as a call regarding the recipient missing a summons for jury duty. A plausible scenario – a jury summons is sent via the mail. Since it is true that sometimes we do not receive mail, there could have been a realistic situation where the 'recipient' did not know they were supposed to show up for jury duty.

As with other tactics, this one used a) authority to intimidate, b) threat of legal action or arrest, and c) a means to find a resolution. And, it didn't take too much to execute. A phone set-up, a script, a couple of 'actors' and start Vishing.

Example: Call from Sergeant Cooper from Saint Clair County Court Systems (#207-2058, ext. 3)

Vishing pretext:

- Subject: Summons for jury duty sent on February 17.
- Belleville Post Office has record of delivery of the summons to the citizen's residence.
- The citizen (target) did not show up on Monday for duty.
- Fines and possible jail for ignoring summons.

- Must submit fine via credit card or sheriff will execute warrant for arrest.
- Will discuss with Lieutenant in charge to determine if reduction may be made for not receiving the summons.

The day after the recipient received this Vishing attack, the local daily newspaper, *Belleville News Democrat*, published an article in reference to this Vishing attack which reiterated the same facts other local citizens were experiencing aligned to the attack. They cautioned readers that it was indeed a scam and recommended anyone who received the call to reach out to the Federal Trade Commission and report their experience. The Vishing criminals were never caught. And, the police were unable to discern how many actual victims existed.

Pretexting

Now, we turn to Pretexting. It is also a dominant and effective method of social engineer attacks. In Pretexting, there is generally an elaborate story for obtaining personal information or other information. Many times, the attacker takes on the fake profile of either a real person or an imaginary one so they can get access to an IT system or gain physical access to a building. There have been many books and movies which centered upon these types of criminals who have mastered the art of the steal by tricking innocents into believing the big lie.

The key to Pretexting is for the social engineer to develop a good story (the pretext) they then put into action and implement the con. They use many psychological traits such as friendliness, comradery, sharing of values, or other methods in order to get the victim to provide access to online information that serves the social engineer's purpose.

The primary element is the establishment of trust. The social engineer wants the victim to feel secure, and have a sense of trust so that they voluntarily and with ease provide the access to the targeted online systems. The planning for a successful Pretext is paramount as is the OSINT to obtain the intelligence required to profile the intended victim. The social engineer takes this information and develops a realistic story that they then put into play to build the rapport and trust with the victim.

One of the worst elements of this technique is the use of criminals to gain the trust of women, especially younger women and teenage girls to build a relationship online and then to convince the victim to send nude pictures of themselves. These can then be used for blackmail, online shaming, or other nefarious purposes.

Scareware (deception/fraudware software)

Scareware is exactly what it sounds like – the social engineer uses scare tactics to insert fear into the mind of the victim. They do this with fake online pop ups, banners and warnings to put a timely shock into the user resulting in a panicked reaction that fits in the social engineer's plan. The reader is tricked into believing there is something wrong with their software, operating system, online account, etc. One of the most effective tricks, is to actually tell the user that they are in the midst of an online attack, and that they are the 'legitimate authorized' person who will guide and assist them successfully through the attack.

An example email may state that they are '*Bill from IT security and we have an alert of a malware attack on your system. Please click on this link and provide your password so we can access your system and counter the attack.*' When people are intimidated, scared, or unsure, they will often follow the directions of the person who seems to know what they are doing. The victim will likely trust 'Bill' and provide their password or click the email link, and the attack succeeds.

Scammy Ads

Another newer social engineering technique is the Scammy Ad. While it may not actually be an attack trying to 'take your stuff,' it can be annoying and intrusive and sometimes deceptive.

Scammy Ads are anything that is and/or related to a scam online. It can be executed through an internet site, hard-copy items such as letters, ads, flyers, and any other media that uses their mode to let the intended victim think they are receiving something of value for free. Once the scam is in effect, the attack is built upon obtaining PII, insider information, or money.

Formjacking

One of the newest OSN attack techniques is Formjacking. This technique succeeds by using a Javascript-code injection to hack a website's online form (standardized) page. Essentially, the attack gathers PII and other information from the form as the victim fills it out thinking they are conducting a legitimate exchange with the company or service being provided. The victim believes they are just conducting a standard exchange, when in reality, they are voluntarily providing the social engineer whatever the form has as a completion field.

Formjacking was first diagnosed in 2018. The best way to understand this new social engineering technique is to think of it as an online version of the physical act of a gas station skimmer stealing credit card data from the unaware consumer. Just as insidious, Formjackers are able to invisibly steal the information a user submits via a form without anyone knowing.

Symantec's 2019 Internet Security Threat Report data show that 4,818 unique websites were compromised with Formjacking code every month in 2018. With data from a single credit card being sold for up to $45 on underground markets, just ten credit cards stolen from compromised websites could result in a yield of up to $2.2 million for cyber criminals each month.

The appeal of Formjacking for cyber criminals is clear. Symantec blocked more than 3.7 million Formjacking attempts in 2018, with more than one million of those blocks occurring in the last two months of the year alone. Formjacking activity occurred throughout 2018, with an anomalous spike in activity in May (556,000 attempts in that month alone), followed by a general upward trend in activity in the latter half of the year.[23]

Tailgating (piggybacking)

One of the 'oldie but goodie' techniques still prominent is Tailgating. This is more of a physical social engineering technique. Think of the *Ocean's Eleven* movie concept type technique often with an elaborate plan that includes confusion or distraction in order for the social engineer to gain physical access to a target site.

Tailgating (also known as Piggybacking) are types of attacks that involve someone who should not be in a specific location, and without any authentic identification, is able to trick or talk their way into the area. Oftentimes, a person impersonates a pizza, Grubhub, or DoorDash delivery driver and waits for the optimal opportunity to penetrate the security perimeter by deceptive methods such as a fake delivery and time-sensitive notification.

With modern physical access systems becoming a 'one pass – one entry' mode, this technique is becoming more of a challenge for the social engineer, and requires more planning to succeed.

Quid pro quo

A quid pro quo (from Latin, meaning 'a favor for a favor') is a promise a getting something of value (physical, assistance) in exchange for information – the benefit usually assumes as a form of service, whereas baiting frequently takes the form of a good. These benefits are generally not really worth much, but for the small 'favor' they are generally a worthwhile exchange to the victim because it is not until afterward that they may realize the mistake. It could be something as small as a music download, or a cheap carrying case or umbrella. It is more about getting something (the enticement) for doing the social engineer's favor.

This is one that your mother told you about – If it sounds too good to be true, it probably is a Quid Pro Quo social engineering attack. The victim enables the attacker to gain access to the company email, or disables their security software thinking they are helping the 'IT guy' fix something, but they didn't put much thought into it because they were more fixated on getting their end of the favor for a favor exchange.

Quid Pro Quo is one of the oldest human interactions in world history; so, why shouldn't it be a technique the social engineer uses to gain an advantage or achieve their ultimate goal?

Doxxing

Doxxing is another newer technique and is a growing issue, especially in social circles. Doxxing is searching for and publishing private or identifying information about (a particular individual) on the internet, typically with malicious intent, often as an intention of shaming or humiliation. Doxxing, or 'name-dropping,' is document (doxx) dropping. It's publicly exposing someone's real name or address on the internet.

This technique has gained fame for some exposures of Hollywood celebrities and other famous people who do not want their personal information available to the general population. It has even been used to try and intimidate or encourage people to use physical violence against elected officials or government employees by people who do not approve of their policies.

Different levels of the government are attempting to establish laws to hold Doxxers accountable for the results of their actions. However, if they find the information on OSINT, it is difficult to prove a crime based upon current legal statutes.

Deepfakes

The newest of the techniques presented; a Deepfake is a technique for human image synthesis based on artificial intelligence. It is used to combine and superimpose existing images and videos onto source images or videos using a machine learning technique called a 'generative adversarial network.' The combination of the existing and source videos results in a video that can depict a person or persons saying things or performing actions that never occurred in reality. Such fake videos can be created to, for example, show a person performing sexual acts they never took part in, or can be used to alter the words or gestures a politician uses to make it look like that person said something they never did.

Along with Formjacking, Deepfakes is evolving and maturing as a social engineering technique, and it is not an understatement to declare it may have be the most dangerous social engineering method ever devised, with nation-state threats and international political issues at its core.

AUTHORSHIP ATTRIBUTION

Evaluating the foundation of a forensic investigation into the Digital Social Triangle and the growing threat to OSN, it has become paramount to assign attribution to the originator (author) of the attack. This is a more complex and challenging task than in the realm of written word, audio, or other venues of communicating. Through their writing styles (Stylometry), forensic investigators attempt to identify the authors. The more evidence that can be collected on the targeted author, the higher the confidence in the forensic identification. Like so many other forensic techniques, this has been around for a long time; however, technology has enabled the attribution process to become more precise with higher levels of confidence in the designation of authorship.

According to the authors of 'Authorship Attribution for Social Media Forensic,'[24] the key considerations for forensic authorship attribution are:

1. No control over the testing set that predictions will be made from, which could be limited to a single sample.
2. No control over the quality of the training data used to create the attribution classifiers (in most circumstances). The training regime must be tolerant to some measure of noise, and a variable number of samples across known authors.
3. The need for a well-defined process. This is necessary for accurate and efficient algorithms, as well as legal consideration.
4. The determination of a well-defined error rate for an algorithm, before it is applied to a real-world problem. This is necessary to understand if the probability of making a mistake is too large for legal purposes.
5. The potential for adversaries. It is possible that the author under investigation is deliberately evading automated attribution methods.

As an example of these considerations, Rocha et al. provide the following comparison of two Twitter authors:

Notwithstanding, the underlying problem domains humanists and forensic examiners operate in can be rather different. The longer the text is, the easier it is to compute stylometric features, which become more reliable as more text is considered. For instance, a novel provides an investigator with a wealth of information from which to extract stylistic clues. But we face the opposite scenario when examining messages from social media, where texts are very short and therefore a smaller set of stylometric features is present in each one. In response to this, some researchers suggest joining the messages in a single document.

Even with preliminary results showing some improvement, this is not realistic since we may not have more than one message we wish to know the author of, and whenever dealing with anonymous messages we cannot guarantee that all of them belong to the same author. However, we know that there is often enough distinct information in even just a handful of sentences for a human reader to understand that they are from a common source.

For instance, consider the following tweets from two prominent Twitter personalities. These from Author A:

A.1: A beautiful reflection on mortality by a great man.
A.2: Unintentional reductio ad absurdum: 'Colleges Need Speech Codes Because Their Students Are Still Children'.
A.3: The great taboo in discussions of economic inequality: Permanent income has non-zero heritability.

And these from Author B:

B.1: Incredible wknd w/ @CASAofOC. Thx to all of the folks that helped raise $2.8 million to help young people in need.

B.2: Thx 4 having me. Great time w/ you all.
B.3: Great to meet you & thx for your support @ChrissieM10.

When reading these tweets, there can be no doubt that Author A and Author B are different people.

The first set of tweets comes from the experimental psychologist and popular science writer Steven Pinker. A notable stylist, Pinker's tweets tend to be well composed and include diverse vocabulary ('mortality,' 'reductio ad absurdum,' 'taboo,' 'heritability') even under the 140-character constraint.

The second set comes from Philadelphia Eagles quarterback Mark Sanchez, who deploys a more colloquial style of writing, including frequent use of abbreviation ('Thx,' 'wknd'). Such traits can be incorporated into a model that is learned from whatever limited information is at hand.

The continued research and application of technology to the art and science of Stylometry has much merit and promise. While the challenges are many because we know the message comprehension between sender and receiver are best received through first non-verbal, then verbal, and finally the written word and therefor this method of forensic identification is unto itself at a severe disadvantage at the outset of the investigation.

CENTRALITY

The forensic process to evaluate OSN attacks within the domain of the Digital Social Triangle can also benefit from the application of Centrality. In their introduction to social network methods course, Professors Robert Hanneman and Mark Riddle of the University of California, Riverside outlines several key aspects of Centrality.[25]

These professors present the actors as having constraints and opportunities in relational networks. If that actor has a lower number of constraints but a higher number of opportunities, then the actor is considered to be in a 'favorable structural position.' This, as is evident, puts the actor into a preferable position of being able to obtain better outcomes through bargaining as well as have more influence on the outcomes. It is a type of power position which not only provide more power to the actor but also provides the perception of power to other actors who have a less favorable position as the one in the favorable structural position.

While Hanneman and Riddle did not identify specific definitions for these constraints and opportunities, they do state that 'network analysis has made important contributions in providing precise definitions and concrete measures of several different approaches to the notion of the power that attaches to positions in structures of social relations.'

They used three types of graphs (star, line, and circle) to display these social networks. In the star network (Figure 5.7), actor A holds the position of being the most influential because of the increased opportunities for networking (literally with every other actor).

In the circle network (Figure 5.8), each actor can influence two other actors only.

Whereas in the line network (Figure 5.9) each actor (F-B) only has the ability to influence two other actors and actors G and A only one.

To understand the advantages to the actor with more opportunities, we need to assess the degree, closeness, and betweenness elements.

DEGREE

When considering options, it is clear in the star figure (Figure 5.7) that actor A has a plethora of options for opportunities. The degree is the relationship opportunity. Think of the modern

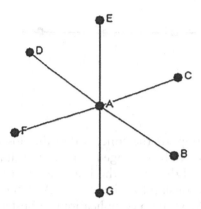

Figure 5.7 'Star' network

Bacon's Law that applies relationships to the actor Kevin Bacon. It is often called the Six Degrees of Kevin Bacon and alludes to linking relationships to six or fewer people. While this is a non-scientific game, it does hold some legitimacy from the perspective that the relational degrees of separation between actors is built upon the ability to have opportunities and options.

So, in the star figure, actor A has many opportunities and alternatives. While all of the other actors are limited to one. Actor A could be a conduit for actor B to interact with actor C, but that would require actor A to 'allow' the interaction and therefore control actor B's opportunities. Actor A clearly has all of the power and can disseminate allocations of opportunity to the other actors as actor A deems appropriate.

In Figure 5.8, this power position for actor A is eliminated. Now, each actor has the same amount of potential power, influence, and restrictions as any other actor. There are no positions of power or weakness. This alignment is the most equal and limiting of the three networks.

The line figure (Figure 5.9) reflects a somewhat hybrid presentation of power and opportunity. Actors A and G are on the ends with only one option of influence. While the other actors each have two and the possible ability to influence more actors throughout the network. From these three displays the key to Centrality is the fact that the more central in the location of the actor, the more opportunities for influence are available.

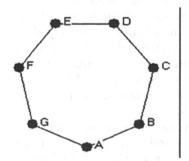

Figure 5.8 'Circle' network

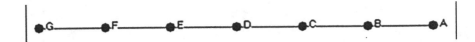

Figure 5.9 'Line' network

CLOSENESS

In a similar vein, Closeness supports the concept that the closer an actor is to other actors, the more influence and power can be leveraged. The relationships can be established, discussions occur, deals are made, and the actor who is at the center of this activity will be seen as the most powerful in the relationship to other actors. And, the direct linkage (from actor A to actor B) is short and specific with no complications or obstacles.

Looking at the star example, in Figure 5.7, it is clear that actor A has the closeness element with direct linkages to all other actors at equal distances and has the control of what each actor can do to interact with other actors. Hence, actor A is the power broker in the relationships.

In the circle example (Figure 5.8), each actor is the same. Same relationships, same distances, same ability to influence. So, the result is a total equivalency in terms of influence and power.

In the line example (Figure 5.9), actor D sits directly in the middle with equal actor opportunities to the left and right. Therefore, actor D has more opportunities for direct and indirect influence and power. Looking at actors A and G, we again see that these actors' options are extremely limited from a closeness concept and the ability to communication and influence another actor.

BETWEENNESS

Linked to Closeness between actors is the ability of an actor to influence the relationships of actors to each other. In the star example (Figure 5.7), actor A not only has many more relationships, but none of the other actors can interact with any actor except actor A without actor A's permission and linkage to the requested actor's subject. So, if actor B wants to interact with actor D, only actor A can make it happen. Like a traffic cop, actor A determines which actors go and which do not. This is the height of power in relationships. Actor A can demand service fees, favors, compensation, or whatever other acts deemed appropriate to grant another actor's request.

In the circle example (Figure 5.8), the network set up is such that only provides the same equal opportunity and pathway for each actor as another actor. No actor has a better leveraged position than a different actor. Actor B can influence or be influenced by two other actors, as can actor C. This example is one of total equality between actors.

In the line example (Figure 5.9), the end actors A and G are again at a disadvantage. They have but one option each. Whereas all other actors have more options for engagement based upon more possible first and second tier relationships. The key to power in a line relationship is being as close to the middle of the relationship structure as possible, thereby increasing the ability to influence other actors.

Acting within the forensic parameters of the Digital Social Triangle, we advise analysts to view these three different, yet related, approaches in terms of the centrality and ability to influence rather than the human perspective of wielding power. Viewing the relationship potential from this perspective enables a forensic analyst to better analyze possible social engineer's paths of influence as they conduct their analysis of an attack and the actors that were, or potentially could be, influenced. Hanneman and Riddle state that this is the better way to describe a network approach as measures of centrality rather than power. As should be evident, the more centralized the actor's position, the more potential to wield power and influence. That is the ultimate goal of the social engineer.

From a metric measurement for forensic analysts, the concept of Centrality is fundamental to evaluating the resilience of a network. In 2010, Piraveenan[26] provided a new metric to analyze the structural robustness of networks by measuring the change of size of the largest component with respect to the network node removals. He proffered that it would be a good method for measuring random and sequential node removals. The author showed how this metric could be applied toward evaluating actual network data. This research provides the ability for determining degree-based ordering, betweenness centrality based ordering, and closeness centrality based ordering; and, provides evidence that it can be an important tool in choosing a strategy to attack or defend a network. Degree-based attack analysis is performed by removing the highest order node in a network whereas random attack analysis is performed by removing a node in random degree order.

SOCIAL NETWORK ANALYSIS (SNA)

SNA is a method of using network and graph theories in order to investigate and analyze social structures for different circumstances. SNA analyzes various communion social structures looking at the social ties and how they are intertwined. The foundation of conducting SNA is a focus on the relationships with the belief that they have a bearing on individual's belief systems, the ability to influence behavior and other social impacts such as following a leader or idea. When looking at how to use SNA in a forensic application, it is important to focus on the relationships (centralities) between the actors in the social engineering attack. This concept uses modeling, predictive techniques and software application tools to aid in finding and exposing these social structures. Of particular interest in this concept could be application to large-scale social engineering campaigns (related multiple attacks) by organized crime families, terrorists, or nation-states.

Algarni et al.[27] examined social engineering threats on social networking sites and asked the questions 'which entities exist and how do they effect social engineering in social networking sites?' Their findings state that for social engineering attacks to be successful they are impacted by four entities which include the OSN (the environment), the social engineer (the attacker), the plan and technique (the trick), and the OSN user (the victim). Their findings align with Hadnagy's Social Engineering Pyramid presented previously in this chapter.

The key facet of countering social engineering attacks relies upon the ability to detect the deception emanating from the OSN. From their research at the University of Illinois, Alowibdi, Buy, Yu, and Stenneth[28] determined that there are three general approaches for detecting deception in OSNs. Depending on how one uses information from profile characteristics. Here are some examples.

1. Detecting deception by comparing different characteristics for each user in a data set obtained from a single OSN (e.g., first names and colors in a given OSN).
2. Detecting deception by comparing characteristics from different OSNs (e.g., Twitter and Facebook) for the same user.
3. Detecting deception by comparing a combination of characteristics from a user's profile in a given OSN (e.g., first name, user name, and colors in a Twitter profile) with a ground truth obtained from external source.

SOCIAL NETWORK INVESTIGATIONS IN DIGITAL FORENSICS

From their work in A Survey of Social Network Forensics,[29] the authors provided an overview of what best defines digital evidence. This evidence is 'data stored or transmitted using a computer that supports or refutes a theory on how an event occurred.' As in other digital forensic areas, social engineering forensics relies upon the importance of evidence based 'bits and bytes' that can be directly linked to the social engineer's attack. This includes the online pretext, the emails, application and system log files, even the hardware (cell phone, laptop, desktop, tablet) devices that were in use by the victim.

The core of the Digital Social Triangle forensic initiative is the credibility of the digital evidence. Verifying the validity of the social engineer's pretext plan is paramount to any successful forensic analysis of an attack. Anne David et al.[30] proposed a 'two-stage model for identifying and contextualizing features from artifacts created as a result of social networking activity. This technique can be useful in digital investigations and is based on understanding and the deconstruction of the processes that take place prior to, during and after user activity; this includes corroborating artifacts.'

This two-stage model for investigations of social network activities presents a logical method to understanding how to identify and understand artifacts based upon investigations into social engineering attacks.

Stage 1: URL feature extraction

The first stage of this model involves the identification and recovery of URLs from disk and the extraction of features from the URLs. The URL in this instance is the main/core source of features for online activity. For example, the social network site visited, or the actions performed by the user (search, follow).

It is important to note that URLs are not platform dependent, so this approach can be applied to any platform (i.e., OS or browser). Features are extracted using a combination of RegEx and the sqlite3 module in Python. Artifacts recovered are stored in CSV files containing the dates and times of activity, the full URL, and extracted feature(s) which can be used to infer user activity or allude to the user's intent.

Stage 2: Corroborating evidence

Corroborating artifacts validate each piece of evidence found during an investigation. In the context of this paper, corroborating artifacts provide both confirmation and supplementary information about the artifacts recovered during the URL feature extraction stage. This stage of the model involves the identification and recovery of artifacts that validate what a URL feature indicates. These types of artifacts provide context to the features extracted from a URL.

For example, the HTTP header information in the cache may show a URL that contains '/settings/account' was created as a result of clicking on the 'account' link in the page 'settings' causing the browser to respond by updating the URL and rendering the requested content. In addition to information derived from HTTP headers found in the browser cache (or unallocated space), metadata from the web page HTML could be useful in understanding the user's interaction with the social networking site.

This stage also involves the recovery of core OS artifacts that backup what has been inferred of the user's activity. For example, downloaded files associated with the recovered URLs; a local copy of uploaded data (associated with a 'www.domain-name/ upload' URL); artifacts indicating that a downloaded application was installed and run 'X' number of times

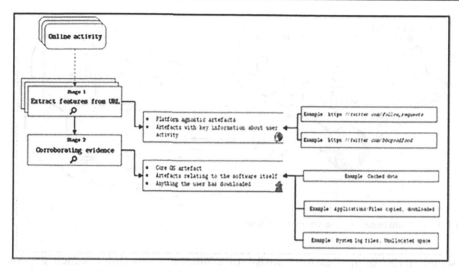

Figure 5.10 Two-stage model for investigations of social networks[31]

including the physical path of the application; artifacts verifying application paths. The proposed model is useful for both the recovery of actionable intelligence and for focusing and ensuring a structured investigation. Having a 'URL feature extraction' stage takes the bulk of URL artifacts and extracts meaningful information from them.

This is useful because the digital forensic investigator needs a clear understanding of the URL structure in order to extract usable information from it. When artifacts from the 'URL' stage have been extracted, corroborating (supplementary) artifacts are used to contextualize events and help digital evidence meet the requirement to be beyond reasonable doubt.

Figure 5.10 is a reproduction of the schematic illustration of the proposed two-stage model for the investigations of social network activity as presented in the paper 'A Two-Stage Model for Social Network Investigations in Digital Forensics,' by David et al.

SOCIAL SNAPSHOT FRAMEWORK

In their work, Digital Forensics for Online Social Networks,[32] the authors developed a Social Snapshot Framework for security research in digital forensics. Their work is applied to the Digital Social Triangle in a method that 'enables an investigator to snapshot a given online social network account including meta-information, a method we termed "social snapshot".'

Figure 5.11 shows the core framework of our social snapshot application. (1) The social snapshot client is initialized by providing the target user's credentials or cookie. Our tool then starts the automated browser with the given authentication mechanism. (2) The automated browser adds our social snapshot application to the target user's profile and sends the shared application programming interface (API) secret to our application server. (3) The social snapshot application responds with the target's contact list. (4) The automated web browser requests specific web pages of the user's profile and her contact list. (5) The received crawler data are parsed and stored. (6) While the automated browser requests specific web pages, our social snapshot application gathers personal information via the OSN API. (7) Finally, the social data collected via the third-party application is stored on the social snapshot application server.

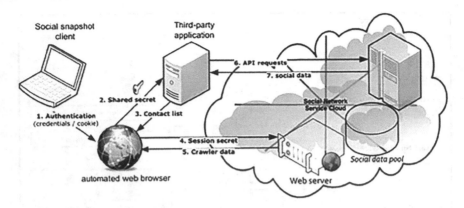

Figure 5.11 Core framework of the social snapshot[33]

Our social snapshot application consists of a number of modules, which we describe in the following. The core modules are the automated web browser and our custom third-party application as outlined in Figure 5.11.

Social snapshot client

The social snapshot client module initializes the data gathering process with a given user's credentials or cookies. Once started, the client first authenticates itself against the target OSN. In the following, the client automatically adds our custom third-party application with the highest possible permissions to the target's account. Information that cannot be retrieved through our third-party application is crawled and parsed by the client. Once all information has been retrieved, the client removes the third-party application and logs out of the given social networking account. The interaction with the social network as well as web-crawling is performed by the Selenium framework, which we describe in the following. We implemented the social snapshot client in Java and the module offers a command line interface.

Automated web browser

The browser module is responsible for the basic interaction with the target OSN. We used the Selenium testing framework to automate the Mozilla Firefox browser. Selenium comes with a command line server that receives Selenium commands. Therefore, we can use the framework to script the behavior of an average user using her Firefox web browser to surf a social networking website. We had to overcome one initial obstacle though: cookie authentication with Selenium which was not supported out-of-the-box. We finally patched the original Java source code of the command line server to be able to correctly set HTTP cookies for the cookie authentication mode.

Third-Party social snapshot application

Our OSN social snapshot application is a third-party application, which sole purpose consists of gathering all possible account data through the target OSN's API. The main design goal of our third-party OSN application is performance, thus multiple program threads are used to gather information as quickly as possible. The third-party application can be configured to prioritize specific account data and to download only a predefined set of account artifacts (social snapshot depth).

Hijack

The hijack module is a network sniffer module that collects valid OSN HTTP authentication cookies from sources such as LAN or WiFi connections. We built our hijack module on the basis of Mike Perry's modified libpkt library, which works out of the box with LAN, unencrypted WiFi, and WEP encrypted WiFi connections. The hijack module offers a command line interface and is implemented in Python.

Digital image forensics

The digital image forensics module matches image files gathered from OSNs with their original source. The goal is to find the pristine image of a compressed picture extracted through our social snapshot application. All images are initially clustered according to their color histograms, rescaled and compressed to the target picture size, and finally matched with pattern recognition techniques. As social networks typically remove meta (EXIF) information of uploaded images this module is helpful in finding the source of collected pictures from OSNs and thus restore information such as the original image creation time and camera.

Analysis

The analysis module is a parser for the results gathered with the data collection modules of our application. It parses the crawled data as well as the information collected through the OSN's API. Furthermore, the analysis module fetches additional content such as photos that are openly available by knowing the URI from OSNs. Finally, it generates a report on the social snapshot data. The analysis module can be used to generate exact timelines of communication, metadata summaries, for example, of pictures, a weighted graph from the network of friends, or their online communication.

DATA TRACING FOR FORENSICS AND OTHER OSN METHODS

The Social Snapshots concept was intended to collect social network data with the aim of harvesting more data compared to the other available tools such that it can be utilized more effectively for searching and analyzing online evidence. While most OSNs have APIs that assist with data collection, they are not always capable of providing the detailed information required for a full forensic evaluation of the OSN. The researchers designed a hybrid consisting of web browsers with an OSN mediator application consisting of six modules.

Other methodologies utilized for OSN forensics include profile matching, categorization of users, authorship analysis, and other characteristic matching processes such as gender, native language identification, and other emerging profiling techniques. Some of the emerging investigative analytical concepts include credibility analysis of emails, texts, and other OSN attacks and taking an offensive versus defensive tact toward them by using these scoring algorithms to initially automatically block them from entering into an organization's network. An isolate, triage, evaluate, and inform concept could be applied where the suspected attack is first discovered based upon the algorithm credibility assessment, then a triage of the attack (a first look by an IT professional) is conducted. If the subject is deemed suspect, it is then sent along to be evaluated by a forensic analyst. If found to be an OSN attack, the intended victim and appropriate authorities are informed.

We focus the remainder of this chapter on a representative forensic evaluation discussion of the newest social engineering technique – **Deepfakes**. Deepfakes represent the advancements of technology merged with the traditions of social engineering and are considered to be the fastest rising threat to nation-states.

DEEPFAKE FORENSICS

Deepfake social engineering techniques get the name from merging the terms 'deep learning' and 'fake.' The best estimate for the origination of the term comes from a Reddit user who actually gave himself the name 'Deepfakes.' He had used open source technology to create pornographic videos. He used the origins of face swapping of famous people with actual participants in the videos. And, since it is a form of artificial intelligence in practice, it has the potential for growth in both use and its ability to project reality. Using artificial images and algorithms, a Deepfake can be generated by almost anyone. As software becomes more readily available, individuals are able to manipulate video and images.

The process of deepfake creation

There are two primary methods for making a Deepfake. First, GAN. GAN stands for Generative Adversarial Network. It uses two separate neural networks, based upon algorithms that recognize different patterns that then 'learn' the attributes of real people within pictures or videos and interact with each other through labeling, clustering, and classifying. While one of the networks produces an image, the other one is learning what image is fake and what is real. Through this learning process, GAN then enables itself to produce fake photos of the intended victim. Those photos can be used standalone or produced in a streaming video.

The first part is called a Generator. It basically produces and image based upon the dataset. The second part of GAN is the Discriminator. It determines if the generated image belongs to the dataset. The Generator and Discriminator have a somewhat of an attack-defend relationship with each other in an iterative manner as the datasets are sent by the Generator with the goal of deceiving the Discriminator to believe the 'fake' image is an actual one. Essentially, the Discriminator acts as the 'check' on the Generator, so the Generator continuously sends images to be screened until they are modified to the point of acceptance.

The second method is one that is continuously expanding as technology evolves. It is the use of Artificial Intelligence encoding (an algorithmic process) to replace real faces with a fake imposter (a real person or a digital facsimile). This method uses many (hundreds or more) photographs from different angles, perspectives, and lighting, and looks for similarities between the real and faker's faces. Once this is completed, a decoder is used to take the fake pictures and carefully overlay (replace) upon the body of the victim.

Tools and skills to generate deepfakes

The availability of inexpensive software and hardware to the general population makes Deepfakes fairly easy to produce. A modern, gaming desktop computer is a good example of a machine that possesses the computing power needed. And, graphic software is maturing at a rate that a Deepfake can be developed by a novice user. Some of the software available include open source free software or inexpensive software. Python-based software such as DeepFaceLab is both free and easily accessible on several operating systems. There are similar software tools available with the numbers are growing at a fast pace.

Identifying a Deepfake is a challenging process, and becomes more difficult as technology and techniques mature. However, there are some general 'giveaways' to look for when a Deepfake is suspected.

These include:

1. Out of aligned body positioning.
2. Odd wrinkles, colors, or shapes with the person's clothing.
3. Hair that seems unnatural or misaligned.
4. Words and lips not in sync.
5. Inconsistent skin color (i.e., face and hands are different).
6. Voice and body seem disconnected or unrealistic.
7. Odd or electronic sounds.
8. A 'Max Headroom' effect (jittery image).
9. Wide-eyed video with slow eye blinking or fast movement of eyes.

Norton, a leading online security firm, produced an overview of what the U.S. government, as well as private organizations, are doing to combat Deepfakes. According to Norton,[34] as of 2020 the following activities are ongoing regarding technology, rules, and legislation.

1. Social media rules. Social media platforms like Twitter have policies that outlaw Deepfake technology. Twitter's policies involve tagging any Deepfake videos that aren't removed. YouTube has banned Deepfake videos related to the 2020 U.S. Census, as well as election and voting procedures.
2. Research lab technologies. Research labs are using watermarks and blockchain technologies in an effort to detect Deepfake technology, but technology designed to outsmart Deepfake detectors is constantly evolving.
3. Filtering programs. Programs like Deeptrace are helping to provide protection. Deeptrace is a combination of antivirus and spam filters that monitor incoming media and quarantine suspicious content.
4. Corporate best practices. Companies are preparing themselves with consistent communication structures and distribution plans. The planning includes implementing centralized monitoring and reporting, along with effective detection practices.
5. U.S. legislation. While the U.S. government is making efforts to combat nefarious uses of Deepfake technology with bills that are pending, three states have taken their own steps. Virginia was the first state to impose criminal penalties on nonconsensual Deepfake pornography. Texas was the first state to prohibit Deepfake videos designed to sway political elections. And California passed two laws that allow victims of Deepfake videos – either pornographic or related to elections – to sue for damages.

In the U.S. government's January 2019 Worldwide Threat Assessment report,[35] the Director of National Intelligence, Dan Coats, wrote this about the threat of Deepfakes to the United States of America.

Adversaries and strategic competitors probably will attempt to use Deepfakes or similar machine-learning technologies to create convincing—but false—image, audio, and video files to augment influence campaigns directed against the United States and our allies and partners.

To consistently identify and counter Deepfakes, at this early stage in the evolution of this growing threat, there is not a lot we can do. However, the four primary areas of focus are:

1. Educate everyone we can on what a Deepfake is, how they can be used, and what we as consumers should be doing to watch for them. This is especially important for social

media users who rely primarily upon this mode for their news. And, it is key to educate children who are most vulnerable to a Deepfake social engineering attack;

2. It is imperative that government, at all state and federal levels, develop a tool kit of laws to assist with this emerging technology, organizations, states, and companies will have some support against the Deepfake attacks. As a result of the national intelligence and federal law enforcement warnings, the U.S. Congress is working with the intelligence community to determine what policy and legal options are available to counter the threat from criminals, nation-states, and terrorist organizations;

3. Working in partnership, private industry, academia, and governments need to team with each other to develop technology countermeasures against Deepfake attacks and sources. By developing fast-reaction methods to identify a Deepfake or to conduct the forensics of a Deepfake product, we can assist law enforcement and the public in understanding a real versus fake video;

4. Finally, it is critical to be able to rapidly expand the capabilities of Authentication verification services to enable a video to be instantly verified as authentic. Some of the ideas being considered include an authentication process to add a digital 'watermark' to commercially produced video and pictures. However, for the average private individual, this is a significant obstacle. There are a lot of other challenges to this concept to include management, time commitments, and authentication protocols; however, it is a promising area of investigation to discern reality from fake.

Aligned with the Director of National Intelligence warning, in 2020 President Donald Trump signed the National Defense Authorization Act[36] which provided three goals for addressing Deepfakes:

1. Initiate a required reporting and documentation of foreign Deepfake weaponization.
2. Notification to Congressional of Deepfake disinformation that targets U.S. elections.
3. Establish a Deepfake competition program (to be determined) to encourage the creation of detection technologies.

THE PROCESS OF PRODUCING A DEEPFAKE

We know from Hollywood movies that using technology to change facial features has been around for a long time, but computer technology has taken this to a new level of sophistication. In this example, we present the movie *Rogue One: A Star Wars Story* which used an actress who was digitally altered through the process of Computer-Generated Imagery (CGI) to reproduce Princess Leia (Carrie Fisher) as a young Princess 40 years younger than she actually was. Figure 5.12 shows the process of using the CGI.

In the picture on the left, note the dots used by computers to 'map' her face to Fisher. (Note: technically, this CGI movie process is not a Deepfake because there is no deception intended but the process is what would be used in a comprehensive social engineering Deepfake).

In another example from the same article published by the online site, KDnuggets (a leading site on AI, Analytics, Big Data, Data Mining, Data Science, and Machine Learning), Gaurav Oberoi, Allen Institute for AI,[38] showed how easy it is to develop a Deepfake.

Oberoi, wrote a script to use with YouTube videos of two late night comedians. Figure 5.13 displays the script (faceit.py) which allowed him to specify a list of YouTube videos for each person. Next, he ran commands 'to preprocess (download videos from YouTube, extract

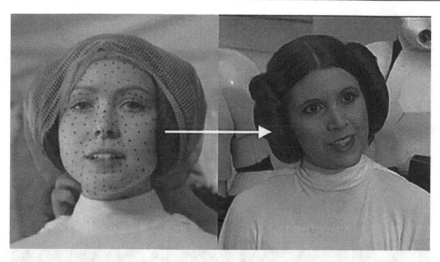

Figure 5.12 CGI process for the movie *Rogue One: A Star Wars Story*[37]

```
faceit = Faceit('fallon_to_oliver', 'fallon', 'oliver')
faceit.add_video('oliver', 'oliver_trumpcard.mp4', 'https://www.youtube.com/watch?v=J1xQ3IUVT0I')
faceit.add_video('oliver', 'oliver_tmrreform.mp4', 'https://www.youtube.com/watch?v=g23v7VP5xU0')
faceit.add_video('oliver', 'oliver_zazu.mp4', 'https://www.youtube.com/watch?v=Y0lUPv15Qqg')
faceit.add_video('oliver', 'oliver_pastor.mp4', 'https://www.youtube.com/watch?v=aUndipbufkg')
faceit.add_video('oliver', 'oliver_cookie.mp4', 'https://www.youtube.com/watch?v=H316EVndP_A')
faceit.add_video('oliver', 'oliver_lorelai.mp4', 'https://www.youtube.com/watch?v=Gl>P2f1_1Jg')
faceit.add_video('fallon', 'fallon_eom.mp4', 'https://www.youtube.com/watch?v=g1Irs20-te4')
faceit.add_video('fallon', 'fallon_charlollesville.mp4', 'https://www.youtube.com/watch?v=L97JJsv67OmE')
faceit.add_video('fallon', 'fallon_dakota.mp4', 'https://www.youtube.com/watch?v=tPtMP-NAM20')
faceit.add_video('fallon', 'fallon_single.mp4', 'https://www.youtube.com/watch?v=xfFVuiN0FSI')
faceit.add_video('fallon', 'fallon_sesamestreet.mp4', 'https://www.youtube.com/watch?v=tHocr7ol1_M')
faceit.add_video('fallon', 'fallon_tombstone.mp4', 'https://www.youtube.com/watch?v=PLB5oC_2IYo')
faceit.add_video('fallon', 'fallon_xfinity.mp4', 'https://www.youtube.com/watch?v=7Jv88ZRLgxM')
faceit.add_video('fallon', 'fallon_bank.mp4', 'https://www.youtube.com/watch?v=q-8hmTKVYgE')
faceit.add_model(faceit)
```

Figure 5.13 Oberoi's Deepfake faceit.py Script[39]

frames, find faces), train, and convert videos with audio and options to resize, show side-by-side, etc.' The code is available online and can be utilized by individuals with little software programming skills.

The result of this script resulted in two famous television late night comic hosts' images being transposed via a Deepfake process. Jimmy Fallon's legitimate video is manipulated to show John Oliver's head on Jimmy Fallon's body, Figure 5.14.

Oberoi fully explains the process as follows:

> At the core of the Deepfakes code is an autoencoder, a deep neural network that learns how to take an input, compress it down into a small representation or encoding, and then to regenerate the original input from this encoding, Figure 5.15.

In this standard autoencoder setup, the network is trying to learn how to create an encoding (the bits in the middle), from which it can regenerate the original image. With enough data, it will learn how to do this. Putting a bottleneck in the middle forces the network to recreate these images instead of just returning what it sees. The encodings help it capture broader patterns, hypothetically, like how and where to draw Jimmy Fallon's eyebrow.

Deepfakes goes further by having one encoder to compress a face into an encoding, and two decoders, one to turn it back into person A (Fallon), and the other to person B (Oliver). It's easier to understand with a diagram. See Figure 5.16.

Figure 5.14 Deepfake result using Oliver for Fallon[40]

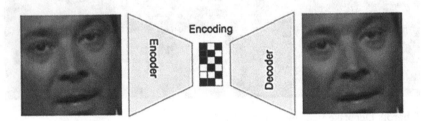

Figure 5.15 Autoencoder learning to regenerate original image[41]

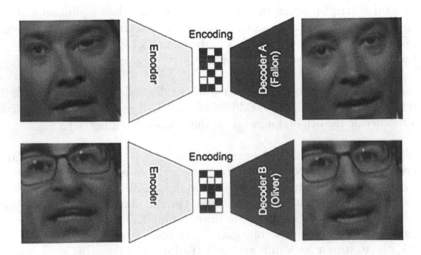

Figure 5.16 Encoding with double decoding process[42]

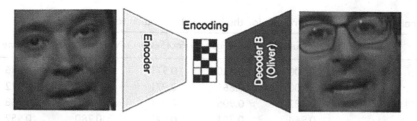

Figure 5.17 Encoder pass to Decoder B (instead of Decoder A)[43]

There is only one encoder that is shared between the Fallon and Oliver cases, but the decoders are different. During training, the input faces are warped, to simulate the notion of 'we want a face kind of like this.' In the above, we're showing how these three components get trained:

- We pass in a warped image of Fallon to the encoder and try to reconstruct Fallon's face with Decoder A. This forces Decoder A to learn how to create Fallon's face from a noisy input.
- Then, using the same encoder, we encode a warped version of Oliver's face and try to reconstruct it using Decoder B.
- We keep doing this over and over again until the two decoders can create their respective faces, and the encoder has learned how to 'capture the essence of a face' whether it be Fallon's or Oliver's.
- Once training is complete, we can perform a clever trick: pass in an image of Fallon into the encoder, and then instead of trying to reconstruct Fallon from the encoding, we now pass it to Decoder B to reconstruct Oliver, Figure 5.17.
- This is how we run the model to generate images. The encoder captures the essence of Fallon's face and gives it to Decoder B, which says 'ah, another noisy input, but I've learned how to turn this into Oliver... voila!'

It's remarkable to think that the algorithm can learn how to generate these images just by seeing thousands of examples, but that's exactly what has happened here, and with fairly decent results.

DEEPFAKE DETECTION TOOLS

There are experts and groups making great strides in the detection arena. One of those is Matthias Niessner of Germany's Technical University of Munich. Niessner is part of a team that's been studying a large data set of manipulated videos and images to develop detection tools. On March 14, 2019, his group released a 'faceforensics benchmark' where, he told Mashable via email, 'people can test their approaches on various forgery methods in an objective measure.'[44]

What is ground-breaking about this work is that Niessner and his colleague's tool is able to analyze the growing body of detection software offerings to determine their accuracy in identifying Deepfakes using readily available software.

According to their published results, 'FaceForensics is a video dataset consisting of more than 500,000 frames containing faces from 1004 videos that can be used to study image or video forgeries. To create these videos, we use an automated version of the state of the art Face2Face approach. All videos are downloaded from YouTube and are cut down to short

Table 5.2 FaceForensics benchmark results for the binary classification scenario[45]

Method	Info	Deepfakes	Face2Face	FaceSwap	NeuralTextures	Pristine	Total
ZAntiFakeBio		1.000	0.920	0.971	0.907	0.936	0.940
Leo		1.000	0.861	0.971	0.853	0.922	0.917
NoSenseAtAll		0.982	0.905	0.951	0.827	0.908	0.908
Cancer		0.964	0.781	0.942	0.780	0.952	0.903
RobustForensics		0.991	0.891	0.951	0.807	0.904	0.902
Aquarius		1.000	0.854	0.971	0.807	0.884	0.890
PredictFake		0.973	0.847	0.913	0.820	0.894	0.887
StableForensics		0.991	0.847	0.951	0.787	0.884	0.883
FAKEDET		0.964	0.832	0.922	0.687	0.918	0.877
ATDETECTOR		0.955	0.796	0.922	0.780	0.898	0.875

continuous clips that contain mostly frontal faces.' To do this, FaceForensics provides two types of datasets.

- Source-to-Target: In this type, the tool can 'reenact' videos extracted from the original and apply new facial expressions. Those results are then used to train and educate forensic analysts in identifying Deepfakes.
- Selfreenactment: A similar, but subtle difference from the Source-to-Target type in that they use Face2Face to reenact the facial expressions of people from original videos and then use that individual's own facial changes and movements to pair with the original and train 'supervised generative refinement models.'

Table 5.2 is an abbreviated representative snapshot of ten (as of October, 2020) of FaceForensics benchmark results for the Binary Classification scenario. In this table, the ten detection software methods have been applied toward five different forgery methods. A score of 1.0 is considered to be a one-for-one match in identifying the Deepfake forgery. Scores less than 1.0 are a percentage of success less than a 100% rate.

LEVELS OF FORENSIC TECHNIQUES

High Level – The example discussed with a focus on eye blinks is considered a high level forensic area. In addition to eye blinks, other areas may include physiological signals, mannerisms, and head and mouth movements.

Low Level – This level primarily focuses upon the digital artifacts of pixel detection. We consider this a low level due to the fact that this type of forensic evaluation is more prone to being manipulated by transcoding or resizing. There is a benefit to low-level forensics in the fact that they are able to better detect artifacts that may not be readily visible via a high-level assessment.

REPRESENTATIVE SAMPLING OF DEEP FAKE FORENSIC METHODS

Because of its rapid popularity, significant concern by governments, and the potential for damage to individuals and organizations, a world-wide scientific investment in Deepfake

Table 5.3 Representative sampling of Deepfake forensic methods

Deepfake Method	Title	Author(s)
Eye Blinking	In Ictu Oculi: Exposing AI Generated Fake Face Videos by Detecting Eye Blinking	Yuezun Li, Ming-Ching Chang and Siwei Lyu
Biological Photoplethysmography	FakeCatcher: Detection of Synthetic Portrait Videos using Biological Signals	Umur Aybars Ciftci, Ilke Demir, and Lijun Yin
Recurrent Neural Networks	Deepfake Video Detection Using Recurrent Neural Networks	David Güera and Edward J. Delp
White Box and Black Box Detectors	Evading Deepfake-Image Detectors with White- and Black-Box Attacks	Nicholas Carlini and Hany Farid
Forensic Transfer	Forensic Transfer: Weakly supervised Domain Adaptation for Forgery Detection	Davide Cozzolino, Justus Thies, Andreas Rossler, Christian Riess, Matthias Nießner, Luisa Verdoliva
Fake Image Properties	What makes fake images detectable? Understanding properties that generalize	Lucy Chai, David Bau, Ser-Nam Lim, and Phillip Isola
Co-motion Pattern Detection	Exposing Deepfake Videos by Anomalous Co-motion Pattern Detection	Gengxing Wang, Jiahuan Zhou, and Ying Wu

forensics solutions is like the Wild West of technology. As of 2020, new studies are being presented focused upon both high- and low-level testing results, often with great potential for conducting future Deepfake forensic evaluations. We have selected a representative sampling of some of the recent studies which reflect the progress and dialogue of researchers as we move forward with this nascent social engineering technique.

The Deepfake forensic methods reviewed in this chapter are represented in Table 5.3. In general, the more traditional type of methods using pattern detection, image artifacts, and digital properties were effective in identifying most Deepfakes generated by using commercially available software. We find the methods with the potential for breakthrough success investigated the biological and neural aspects of a Deepfake.

With these methods, the propensity to improve the percentages of Deepfake identification rates a high confidence assessment, and as technology and research continues in these areas, we believe new techniques will evolve and faster and more accurate forensic identification will be reality.

IN THE BLINK OF AN EYE

To help uncover a Deepfake, one of the areas technology has yet to fully integrate in a seamless fashion is eye blinking. So, when a person is watching OSN, they can look for the sign of natural eye blinking as one possible indicator of a Deepfake technique such as Face Swap or Face2Face.

According to a 2018 study published by the University of Albany, SUNY,[46] that focused on how the act of eye blinking is poorly represented in Deepfake videos, there was evidence that without a significant number of video/photographs of the targeted subject with their eyes closed, it is very difficult to present a natural imaging Deepfake.

On average, people blink at a pace approximately every 2 and 10 seconds. A blink's duration has lasts about one-tenth to four-tenths of a second. In an original video, these characteristics of s standard blinking process are evident. The researchers investigated the differences between a real blinking process and how this biological action is displayed in a Deepfake.

For OSN forensic professionals, the SUNY study holds particular promise and merits further evaluation and support. The team was able to present a new method to expose fake face videos generated with neural networks based upon the detection of eye blinking in the videos. And, they proffered that eye blinking is a physiological signal that is not well presented in the synthesized fake videos (not enough captured video/pictures). They used a method to test benchmarks of eye-blinking detection datasets which resulted in positive signs of being able to detect videos generated using the Deepfake social engineering technique.

BIOLOGICAL SIGNALS

The robust research from Ciftci et al.,[47] using biological signals within pictures has significant potential for forensic applications. This approach has potential as a forensic tool, FakeCatcher, as a fake portrait video detector using biological signals. Primarily through the use of three sections (left, middle, and right) of facial biologicals, their 'Deepfake Detector' is capable of detecting synthetic content in videos. To achieve this, the researchers focus their work on biological signals hidden in portrait videos. These hidden signals are described as 'implicit descriptors of authenticity because they are neither spatially nor temporally preserved in fake content.'

Relying upon Photoplethysmography, an optical technique used to detect volumetric changes in blood in peripheral circulation, this evaluation of Deepfake videos moves into a new realm, different from the predominant focus on reproductive distortions, artifact compressions, and image quality. FakeCatcher is a low cost and non-invasive method that makes measurements at the surface of the skin.

The researcher's Deepfake Detector process with the FakeCatcher tool consisted of these steps:

1. Engage several signal transformations to address the pairwise separation problem
2. Utilize those findings to formulate a generalized classifier for fake content; this is achieved by analyzing proposed signal transformations and corresponding feature sets
3. Generate novel signal maps and employ a Convolutional Neural Network (CNN) to improve the traditional classifier for detecting synthetic content
4. Release an 'in the wild' dataset of fake portrait videos collected as a part of the evaluation process.

Essentially, this is the continuation of evidence-based assessments of human biological traits used by law enforcement to identify an individual. Fingerprints have been used for a millennium and in recent decades, we have matured the use of DNA even to the point of convicting or proving the innocence of people charged with crimes. And, continued maturation of the use of eye retina and iris distinctions are growing, and even the uniqueness of eye colors is a growing area of discernment between people. This research's Deepfake Detector is consistent with these historical forensic and identification techniques.

NEURAL NETWORKS

In their study titled Neural Networks,[48] the authors used a CNN to extract frame-level features which were then utilized as a training tool with a recurrent neural network (RNN).

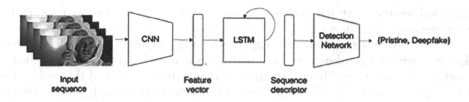

Figure 5.18 Overview of temporal aware detection system[49]

Doing this exercise resulted in the RNN that has learned to determine whether or not a video has been manipulated.

Figure 5.18 provides an overview of this detection system. The system consists of a convolutional Long Short-Term Memory (LSTM) apparatus that processes the video frames. Using CNN for frame feature extraction combined with the LSTM temporal sequence analysis resulting in an estimate (probability) of the sequence being identified as a legitimate video or a Deepfake.

This Neural Network system uses an end-to-end manner to learn about the subject video sequence; and then, using this process it produces a probability of the video being a Deepfake. The authors describe their three-step methodology as:

1. Obtain a set of features for each frame produced by the CNN.
2. Concatenate the features of multiple consecutive frames and transfer to the LSTM for analysis.
3. Produce an estimate of the likelihood of the sequence identified as a Deepfake or an actual original video.

WHITE AND BLACK-BOX ATTACKS

Carlini et al.[50] have developed a classification methodology called White Box and Black Box in which they investigate the forensic classifiers robustness by evaluating the access (white being access is granted and black being it is not) to the classifier's parameters. They developed this methodology out of the work from the adversarial machine learning research areas. An overview of their work with the White Box/Black Box concept states:

> ...we evaluate the robustness of forensic classifiers to an attacker who has complete access to the classifier. This attacker is therefore able to compute the gradient of the input with respect to the classifier output, a so-called white-box threat model. We apply three attacks that have previously been studied in the adversarial example literature, and then develop our own attack that subverts forensic classifiers by modifying the generator's latent space... black-box threat model corresponds to the situation when the adversary does not have access to the exact details of the forensic classifier, but is aware what type of classifier is in place. Defenses are only interesting if they remain secure against an adversary who is aware they are present; 'security through obscurity' is not a valid defense. As such, the black-box threat model assumes that the adversary knows that there is a defense in place, and similarly knows the general strategy of the defense.

Adversarial modeling is particularly important in the types of data-driven, machine-learning based techniques described here. We have shown that these techniques, are highly vulnerable to attack because the same power and flexibility of the underlying neural-network classifiers that leads to high classification accuracies, can also be easily manipulated to create adversarial images that easily subvert detection. This subversion takes the form of white-box attacks in which it is assumed that the details of the forensic classifier are known, and black-box attacks in which it is assumed that only a forensic classifier, of unknown detail, exists. These attacks can imperceptibly modify fake images so that they are misclassified as real, or imperceptibly modify real images so that they are misclassified as fake.

FORENSICTRANSFER

Another forensic technique with excellent potential is the result of research by Cozollini et al.[51] Their research focused upon ForensicTransfer (FT). The authors emphasized the effectiveness of FT which increased accuracy of transferability analysis to 95%. According to the authors:

> Our approach disentangles the information necessary to make the real/fake decision in the source domain from a faithful latent-space representation of the image, which may be exploited in new target domains. To prevent the net from discarding precious information during training, we rely on autoencoder-based representation learning by which the latent space is constrained to preserve all the data necessary to reconstruct the image in compact form. Therefore, the latent space holds both the image representation and the data used for the real/fake decision, but these pieces of information live in orthogonal spaces, and do not interfere with one another. This is obtained by dividing the latent space in two parts, one activated exclusively by real samples, and the other by fake samples. Since the network has to reproduce the image anyway, all relevant information on the input image is stored in both parts. Thus, the features of the learned forensic embedding keep all desired information, useful for diverse forensic tasks, and easily adapted to new domains based on a small number of new training samples.

This research, while conducted on a small scale, has the potential for identifying, storing, sorting, and providing the forensic analyst key elements of the Deepfake to aid in them in their work.

UNDERSTANDING PROPERTIES OF FAKE IMAGES

Related to the FaceForensics tool, is another study focusing on several automating detection techniques utilized to understand the properties of fake images. From their research, Chai et al.[52] were able to compile several initiatives to predict metadata or other low-level artifacts, similarity embeddings and graphs, training classifiers, CNN extract features over image patches, and other property factors.

In particular, they identified manipulation and expression transfers by

> using the FaceForensics++ dataset which includes methods for identity manipulation and expression transfer. Identity manipulation approaches, such as FaceSwap, paste a source face onto a target background; specifically, FaceSwap fits detected facial landmarks to 3D model and then projects the face onto the target scene.

The deep learning analogue to FaceSwap is the Deepfake technique, which uses a pair of autoencoders with a shared encoder to swap the source and target faces. On the other hand, expression transfer maps the expression of a source actor onto the face of a target. Face2Face achieves this by tracking expression parameters of the face in a source video and applying them to a target sequence. Neural Textures uses deep networks to learn a texture map and a neural renderer to modify the expression of the target face.

Through the use of heat maps laid over the images, this process is capable of predicting an image as being fake. By using patch-based classifiers over sliding patches of an image, they are able to draw the heat maps for analysis.

In Figure 5.19, the authors used classifiers trained on CelebA-HQ PGAN images depicting heat map predictions. The results of their work include an averaged heat map where the colorization (red) reflects the results that are likely the correct classification.

The results from the study also

> ...show an averaged heat map over the 100 most real and most fake images, The average heat maps highlight predominately hair and background areas, indicating that these are the regions that patch-wise models rely on when classifying images from unseen test sources.

Their approach included equalizing the preprocessing of two classes of images for the purpose of looking at the difference between actual images from a camera to the fake image. Similar to other researchers, they put emphasis on mouths, eyes, hair, and unique background effects. The heat map resulting from the classification of these areas (datasets, training parameters, generators). The results of the study's findings are promising. The authors completed their investigation with this summary:

> We show a technique to exaggerate the detectable artifacts of the fake images, and demonstrate that image generators can still be imperfect in certain patches despite fine tuning against a given classifier. While progress on detecting fake images inevitably creates a

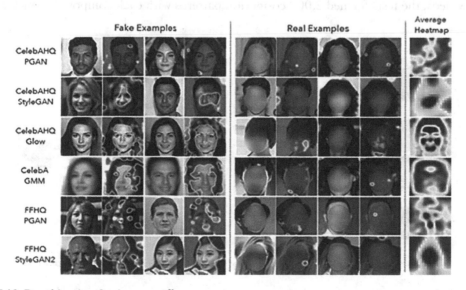

Figure 5.19 Deepfake classifier heat maps[53]

cat-and-mouse problem of using these results to create even better generators, we hope that understanding these detectors and visualizing what they look for can help people anticipate where manipulations may occur in a facial image and better navigate potentially falsified content in today's media.

CO-MOTION PATTERN DETECTION

Wang et al.[54] developed a video forensic method which models the conjoint patterns of local motion in real videos to assist in identifying the abnormal (Deepfake) video motions through the comparison of the extracted motion patterns versus the actual patterns. From their research, we are able to discern significant possibilities for the forensic use of Co-motion pattern detection (Figure 5.20).

To enhance generalizability on videos with various content, we model the temporal motion of multiple specific spatial locations in the videos to extract a robust and reliable representation, called co-motion pattern. Such kind of conjoint pattern is mined across local motion features which is independent of the video contents so that the instance-wise variation can also be largely alleviated. More importantly, our proposed co-motion pattern possesses both superior interpretability and sufficient robustness against data compression for deepfake videos.

The pipeline of the study's proposed co-motion pattern extraction method. The first step estimates the motion of corresponding key points which are then grouped for analysis. Additionally, this study constructed a co-motion pattern as a compact representation to describe the relationship between motion features.

Table 5.4 presents the results of the study method against four of the primary forgery databases in common use. Each of the four forgery databases were presented via a binary classification task compared to actual videos. The authors evaluated their theory using the previously discussed FaceForensics++ dataset consisting of four sub-databases that produce face forgery via different methods. The evaluation used Deepfake, FaceSwap, Face2Face, and NeuralTexture forgery databases, as well as externally sourced data to demonstrate the similarity of co-motion patterns based upon real videos. Because each sub-database contained 1,000 videos, the tests formed 2,000 co-motion patterns with each composed of picking N ρ

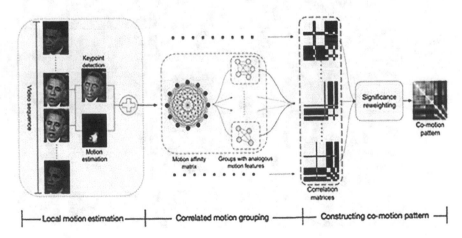

Figure 5.20 Co-motion pattern detection[55]

Table 5.4 Co-motion pattern detection forgery database results[56]

Method/Dataset	Deepfakes	FaceSwap	Face2Face	NeuralTexture	Combined
Xception [48]	93.46%	92.72%	89.80%	N/A	95.73%
R-CNN [49]	96.90%	96.30%	94.35%	N/A	N/A
Optical Flow + CNN [3]	N/A	N/A	81.61%	N/A	N/A
FacenetLSTM [53]	89%	90%	87%	N/A	N/A
N = 1 (Ours)	63.65%	61.90%	56.50%	56.65%	57.05%
N = 10 (Ours)	82.80%	81.95%	72.30%	68.50%	71.30%
N = 35 (Ours)	95.95%	93.60%	85.35%	83.00%	88.25%
N = 70 (Ours)	99.10%	98.30%	93.25%	90.45%	94.55%

matrices for training and testing, respectively. The percentages displayed are the rate of detection for each method as applied toward the four forgery databases.

The study's authors provide four contributions resulting from their work:

1. *Co-motion pattern*: a descriptor of consecutive image pairs that can be used to effectively describe local motion consistency and correlation.
2. *Attributes of the co-motion pattern*: co-motion pattern is explainable, robust to video compression/pixel noises and generalizes well.
3. *Proof through Experimentation*: Using experiments under both classification and anomaly detection settings, the process showed the co-motion pattern is able to accurately reveal the motion-consistency level of given videos.
4. *Robust and Transferability*: The method was used with datasets with different quality and forgery methods. The purpose of these aspects was to demonstrate the robustness and transferability of the method.

SUMMARY AND FUTURE FOCUS AREAS

The Digital Social Triangle continues to expand across the world as the primary vehicle for the bad guys to conduct OSN social engineering attacks against individuals, organizations, and nation-states. Within OSN, the newest technique with the propensity to be extraordinarily damaging is Deepfakes. Throughout this chapter, we have focused on presenting the aspects of the Digital Social Triangle, types and traits of social engineering, an overview of the social engineering lifecycle, and social media authorship authentication. These areas served as the foundation of social engineering attack techniques. After reviewing the primary social engineering techniques, we presented the important concepts related to influence and relationships through the theory of centrality. Completing that discussion, we moved into the more specific social engineering forensics with focus upon the newest technique – Deepfakes.

Based upon this forensic focus upon Deepfakes, we close this chapter with recommended areas to invest in; and, which will provide both new challenges and opportunities for forensic analysis. While traditional OSN methods do generally apply, the uniqueness of a Deepfake makes new forensic methods important.

As it is such a new social engineering technique, there is a fast growing yet untested investment by professionals to develop modes of forensic science. We expect this area to continue to rapidly expand as OSN evaluations and testing occurs.

Recommended areas to support future forensic evaluation may include:

- Unique Face Identifiers: Increased focus on the more complex areas of a face to include facial hair and eye blinking. These are aspects of humans which are quite challenging to emulate and could hold future possibilities for forensic determination.
- Improvement of online tools: This area has significant potential. The automation analysis toolsets available today are reacting to the online Deepfake software sets. As more investigations occur, the possibilities of more intrusive and accurate detection tools will be key to forensic evaluations. Like all other social engineering attacks, the offensive side (bad guys) have the advantage as they generate new and improved methods of leveraging technology. However, to identify and analyze Deepfake technology requires a defensive capability equal to the task.
- Bot management: Bots are not only a danger for other social engineering techniques, OSN propaganda, and online manipulation, but they may become another method for Deepfake manipulation as AI matures.
- Establishing partnerships between industry, academia, and government: Deepfakes are so new, with the potential to cause great damage across nation-states, that there should be a world-wide effort with leadership by the United Nations, leading IT companies, and major research universities to partner in finding identification and forensic methods for Deepfakes.
- Government leadership: Similar to the technology partnership, governments across the world must establish laws and policies to address Deepfakes. Deterrence and accountability should be the primary focus of these public policies. Criminal penalties should be strong and consistent across nation-states.
- Authentication Methods: Deepfakes are particularly insidious from a deception perspective. Developing faster, more accurate, and available authentication methods is paramount to countering this threat. Similar to online virus protection software, an inexpensive, readily available to users, and easy to manage authentication software suite of tools will greatly aid in defending against the threat.
- Education and Training: Like all social engineering techniques, the most important element of any counter to the attack is awareness, understanding, and acting to keep the attack from succeeding. Deepfake training in organizations should be integrated with other information security training programs. And, universities, community colleges, and K-12 schools should develop education programs for students.

For additional, recommended reading resources and publications, which accompany this chapter and are provided as a downloadable eResource, readers are encouraged to visit the Publisher's website at https://routledge.com/9780367524180.

FIVE MANAGEMENT AWARENESS DISCUSSION POINTS FOR MANAGERS

It should be clear from this chapter that social engineering is a paramount cybersecurity attack vector of concern to leaders and managers in any organization, small or large.

To assist leaders and managers in preparing their organizations to educate, train, and defend against this mode of attack, we offer the following discussion points.

1. **Employees Need to be Able to Recognize Social Engineering Techniques**. Many organizations do not invest the appropriate time and resources into assisting their employees to both recognize and defend against social engineering attacks. To do this, the people

need to a) become aware of the threat, b) be able to recognize attempted attacks, and c) know what to do when they are an intended victim of an attack. Organizations need to invest in their employees to ensure they understand their individual roles in protecting the organization's information.

2. **Training Programs.** Most information security awareness training programs are based upon an annual, online and short presentation consisting of slides or short videos. The employee is required to watch the presentation and then take a 'test' at the end. However, if they fail the test, they are only required to revisit that particular area of the training and then answer the missed question(s) again. This is not only an ineffective manner to train employees, but it misses the opportunity to engage employees regarding actual threats to the organization's networks. Managers should consider investing in innovative, hands-on, in-person training. Bringing in 'white hat' social engineers to conduct the training can be an effective mode to impact employees. Having the ability to interact with experts who can answer questions, discuss social engineering techniques, and actually demonstrate them in action, can be an impactful and effective training experience.

3. **Policies and Processes.** No organization can have an effective information security protection program without a solid foundation built upon policies and processes that utilize best practices, rely upon government guidance such as the NIST standards, and are both current and available. We recommend, as part of the employee-training program, an overview of these policies is included. Key policies such as password management, access control, and physical security are examples that should be included in all information security orientations and recurring training programs. Additionally, it is important to include organization processes such as wearing identification badges and access to sensitive areas, in a continuous and consistent cultural environment of expected adherence to appropriate processes.

4. **Budgeting for Information Security.** While recent trends have shown an increase for organizational information security budgets, more is needed. As the social engineering threats increase, the potential damage from a successful attack is magnified manifold. Organizations need to invest more funding into the information technology department's security programs, the employee training and awareness programs, and external expertise to stress and test the security of the OSN.

5. **Exercises, External Assessments, and Auditing.** Few organizations use white hat, internal security unit, or other means to stress their OSN and conduct risk analysis of defense against social engineering attacks. This is an area of cybersecurity that has been gaining recognition for the potential benefits of return on investment costs. Experts who can conduct objective risk analysis, threat assessments, and provide other services such as penetration testing, can be extremely valuable in relationship to the costs of their services.

SOCIAL ENGINEERING FORENSIC DISCUSSION QUESTIONS

1. Why is it important for forensic analysts to understand the relationship of the Digital Social Triangle to the process of social engineering attacks?
2. How does the Social Media mode make the cybersecurity forensic evaluation process more challenging than other modes?
3. What role does OSINT play in the forensic evaluation of a social engineering attack?
4. Why and how are the Traits of Social Engineering important to understand for a forensic analyst as they approach an investigation?
5. How do the Social Engineering Life Cycle and the OODA Loop Concept support a forensic approach of social engineering attacks?

6. What are some common elements found in the different social engineering techniques?
7. How is the process of Authorship Attribution a key factor in social engineering forensics?
8. Why do social engineering forensic analysts need to understand the theory of Centrality?
9. What is the role of the Social Snapshot Framework within digital forensic research?
10. Why is it important to understand and evaluate the General Giveaways of a Deepfake at the outset of an investigation?
11. What are some of the promising forensic advancements in the area of biological markers for Deepfake investigations?
12. How do Deepfake detection tools such as FaceForensics aid in the process of forensic analysis?

NOTES

1 The inclusion of any social engineering procedure, specifically related to Deepfakes, methods, software, processes, or techniques, is not an endorsement of that software, tool, product, process, procedure or website by the author or publisher and have been included herein as reference, research or as an example, of which there may be many within the social engineering space.
2 Figure 5.1 drawn by author.
3 European Union Agency for Network and Information Security (ENISA): ENISA Threat Landscape Report 2018. 15 Top Cyberthreats and Trends (Final Version), 1.0, ETL 2018, JANUARY 2019, retrieved October 28, 2020.
4 Hadnagy, C., Social Engineering: The Science of Human Hacking. Pages 11-13. Wiley, 2018 ISBN:9781119433385, retrieved June 1, 2019. Reproduced with permission of Copyright Clearance Center LICENSE #: 4945991438057.
5 Saxena, S., "What Is Social Media and what are its Main Features?" easymedia, December 24, 2017. www.easymedia.in/social-media-main-features/#:~:text=%20Here%20are%20some%20key%20features%20of%20social,sites%20require%20members%20to%20build%20their...%20More%20, Reproduced with permission of Professor Sunil Saxena, owner, easymedia site and blog, via email on December 6, 2020.
6 Scott, J., (2012), Social Network Analysis, SAGE Publications Ltd, ISBN-139781446209042, reproduced with permission via Copyright Clearance Center, License # ID1075933-1, March 17, 2020.
7 Prier, J., (2017), "Commanding the Trend: Social Media as Information Warfare," Strategic Studies Quarterly, 11(4), 50-85, reproduced with permission of W. Michael Guillot, Editor, Strategic Studies Quarterly, via email on November 9, 2020.
8 Gardner, B., & Thomas, V. (2014). Building an information security awareness program Defending against social engineering and technical threats (1st ed.). page: 45, retrieved September 14, 2019.
9 Mitnick, Kevin & Simon, William. (2002). The Art of Deception: Controlling the Human Element of Security. Page: xii, retrieved April 10, 2018.
10 Hadnagy, C., Social Engineering: The Science of Human Hacking. Pages 11–13. Wiley, 2018 ISBN:9781119433385, retrieved June 1, 2019. Reproduced with permission of Copyright Clearance Center LICENSE #: 4945991438057
11 Ibid.
12 Ibid.
13 Office of the Director of National Intelligence, U.S. National Intelligence: An Overview 2011, Washington, D.C., 2011. As of November 10, 2017: https://www.dni.gov/files/documents/IC_Consumers_Guide_2011.pdf, retrieved May 12, 2020.
14 Intelligence Reorganization Act of 1992. S.2198. 102nd Congress (1991–1992), retrieved May 12, 2020.
15 Conheady, S., Social Engineering in IT Security: Tools, Tactics, and Techniques, McGraw-Hill Education; 1st edition (2014). ISBN-139780071818469, Pages: 2–7, reproduced with permission via Copyright Clearance Center, License # ID1076563-1 retrieved October 11, 2018.

16 Ibid.

17 Jamil, A., Asif, K., Ghulam, Z., Nazir, K. M., Mudassar Alam, S., & Ashraf, R., "MPMPA: A Mitigation and Prevention Model for Social Engineering Based Phishing attacks on Facebook," 2018 IEEE International Conference on Big Data (Big Data), Seattle, WA, USA, 2018, pp. 5040-5048, doi: 10.1109/BigData.2018.8622505, Reproduced with permission via Copyright Clearance Center, License # 4944840892935, retrieved January 17, 2020.

18 Ibid.

19 Richards, Chet., Boyd's OODA Loop (It's Not What You Think). (2012) J. Addams & Partners, Inc. Retrieved from https://fasttransients.files.wordpress.com/2012/03/boydsrealooda_loop.pdf, retrieved July 17, 2020.

20 Clarke, Richard., & Knake, Robert. The Fifth Domain: Defending Our Country, Our Companies, and Ourselves in the Age of Cyber Threats. Penguin Press (2019), retrieved August 19, 2020.

21 Kaspersky Resource Center, "What is Social Engineering?" (2020) Retrieved from https://usa.kaspersky.com/resource-center/definitions/what-is-social-engineering, reproduced with permission of Sarah Kitsos, Kaspersky, Head of Corporate Communications, Kaspersky North America on December 4, 2020.

22 Tischer, Matthew., Durumeric, Zakir., Foster, Sam., Duan, Sunny., Mori, Alec., Bursztein, Elie., & Bailey, Michael. (2016) Users Really Do Plug in USB Drives They Find. University of Illinois. doi:10.1109/SP.2016.26, retrieved September 14, 2020.

23 Symantec Internet Security Threat Report. (February 2019) Volume 24. Retrieved from: https://www.lifelockbusinesssolutions.com/wp-content/uploads/2019/05/ISTR_24_2019_en.pdf, retrieved July 22, 2020.

24 Rocha, A., Scheirer, W., Forstall, C., Cavalcante, T., Theophilo, A., Shen, B., Carvalho, A., & Stamatatos, Efstathios. (2016). "Authorship Attribution for Social Media Forensics," in IEEE Transactions on Information Forensics and Security, vol. 12, no. 1, pp. 5-33, Jan. 2017, doi: 10.1109/TIFS.2016.2603960, Reproduced with permission of (Copyright Holder) via Copyright Clearance Center License # 4944901351025, retrieved September 28, 2020.

25 Hanneman, Robert A. & Riddle, Mark. (2005) Introduction to social network methods. Riverside, CA: University of California, Riverside. Pages: Online Chapter 10, retrieved March 16, 2020.

26 Karabiyik, Umit; Canbaz, Muhammed Abdullah; Aksoy, Ahmet; Tuna, Tayfun; Akbas, Esra; Gonen, Bilal; and Aygun, Ramazan S. (2016), "A Survey of Social Network Forensics," Journal of Digital Forensics, Security and Law: Vol. 11: No. 4, Article 8. doi: https://doi.org/10.15394/jdfsl.2016.1430, reproduced with permission via http://creativecommons.org/licenses/by-nc/4.0/, retrieved September 11, 2020.

27 Algarni, A., Xu, Y., Chan, T., & Tian, Y.-C. (2013) Social engineering in social networking sites: Affect-based model. In Internet technology and secured transactions (icitst), 2013 8th international conference for (pp. 508–515), retrieved July 17, 2020.

28 Alowibdi, J. S., Buy, U., Yu, P. S., Stenneth, L., et al. (2014) Detecting deception in online social networks. In Advances in social networks analysis and mining (asonam), 2014 ieeejacm international conference on (pp. 383–390), retrieved July 17, 2020.

29 Karabiyik, Umit; Canbaz, Muhammed Abdullah; Aksoy, Ahmet; Tuna, Tayfun; Akbas, Esra; Gonen, Bilal; and Aygun, Ramazan S. (2016) "A Survey of Social Network Forensics," Journal of Digital Forensics, Security and Law: Vol. 11: No. 4, Article 8. doi: https://doi.org/10.15394/jdfsl.2016.1430, Reproduced with permission via http://creativecommons.org/licenses/by-nc/4.0/, retrieved July 19, 2020.

30 David, Anne; Morris, Sarah; and Appleby-Thomas, Gareth. (2020) "A Two-Stage Model for Social Network Investigations in Digital Forensics," Journal of Digital Forensics, Security and Law: Vol. 15, Article 1. Available at: https://commons.erau.edu/jdfsl/vol15/iss2/1, reproduced with permission via http://creativecommons.org/licenses/by-nc/4.0/ retrieved October 1, 2020.

31 Ibid.

32 Huber, M., Mulazzani, M., Leithner, M., Schrittwieser, S., Wondracek, G., & Weippl, E. (2011). Social snapshots: Digital forensics for online social networks. In Proceedings of the 27th annual computer security applications conference (pp. 113–122). New York, NY, USA: ACM, ISBN-13978-1-4503-0672-0, Retrieved from http://doi.acm.org/10.1145/ 2076732. 20767 48 doi: 10.1145/2076732.20767 48, reproduced with permission of (Copyright Holder) via Copyright Clearance Center, License # ID1076792-1, retrieved August 2, 2020.

33 Ibid.

34 Johansen, G., July 24, 2020, "Deepfakes: What they are and why they're threatening," Norton Security Center, https://us.norton.com/internetsecurity-emerging-threats-what-are-deepfakes.html, reproduced with permission via email exchange with Norton Lifelock, NortonLifeLock copyrights@nortonlifelock.com on November 16, 2020.

35 Office of the Director of National Intelligence, Worldwide Threat Assessment of the US Intelligence Community by Dan Coats, Director of National Intelligence as presented to the Senate Select Committee on Intelligence. (January 29, 2019) https://www.dni.gov/files/ODNI/documents/2019-ATA-SFR---SSCI.pdf, retrieved July 3, 2020.

36 National Defense Authorization Act for Fiscal Year 2020, H.R. 2500, 116th Congress (2019) https://www.congress.gov/bill/116th-congress/house-bill/2500/text, retrieved July 3, 2020.

37 Oberoi, Gaurav. "Exploring DeepFakes" KDnuggets, News 18:n13, Mar 28, 2018. https://www.kdnuggets.com/2018/03/exploring-deepfakes.html, Reproduced with permission of Gregory Piatetsky-Shapiro, Ph.D., President, KDnuggets via email on November 3, 2020.

38 Ibid.

39 Ibid.

40 Ibid.

41 Ibid.

42 Ibid.

43 Ibid.

44 Rossler, Andreas., Cozzolino, Davide., Verdoliva, Luisa., Riess, Christian., Thies, Justus., & Nießner, Matthias. Faceforensics++: Learning to detect manipulated facial images. In IEEE International Conference on Computer Vision, pages 1–11, 2019, TEXT AND TABLE 5.2 reproduced with permission of Andreas Rossler via email on October 15, 2020

45 Ibid.

46 Li, Yuezun., Chang, Ming-Ching., & and Lyu, Siwei. (2018) In Ictu Oculi: Exposing AI Created Fake Videos by Detecting Eye Blinking. 1-7. 10.1109/WIFS.2018.8630787. http://www.cs.albany.edu/~lsw/papers/wifs18.pdf, retrieved September 25, 2020.

47 Ciftci, U. A., Demir, I., and Yin, L. "FakeCatcher: Detection of Synthetic Portrait Videos using Biological Signals," in IEEE Transactions on Pattern Analysis and Machine Intelligence, doi: 10.1109/TPAMI.2020.3009287, ISSN1939-3539, Reproduced with permission via Copyright Clearance Center, License # ID1076278-1, retrieved September 27, 2020.

48 Guera, David., & Delp, J. Edward. "Deepfake Video Detection Using Recurrent Neural Networks" IEEE, 10.1109/AVSS.2018.8639163, November 2018, https://engineering.purdue.edu/~dgueraco/content/deepfake.pdf, reproduced with permission via Copyright Clearance Center, LICENSE #: 4944861368446, retrieved September 25, 2020.

49 Ibid.

50 Carlini, N., Farid, H., "Evading Deepfake-Image Detectors with White- and Black-Box Attacks," 2020 IEEE/CVF Conference on Computer Vision and Pattern Recognition Workshops, 2020, arXiv:2004.00622v1, reproduced with permission via Copyright Clearance Center, License # 4944870048636, retrieved October 11, 2020.

51 Cozzolino, D., Thies, J., Rössler, A., Riess, C., Nießner, M., Verdoliva, L., (2018), "ForensicTransfer: Weakly-supervised Domain Adaptation for Forgery Detection," arXiv:1812.02510v2, Reproduced with permission from Andreas Rossler via email on November 20, 2020.

52 Chai, L., Baul, D., Lim, Ser-Nam., & Isolal, Phillip. (2020) "What makes fake images detectable? Understanding properties that generalize," https://arxiv.org/ arXiv:2008.10588, reproduced with permission via Copyright Clearance Center License #: 4954260999323, retrieved October 2, 2020.

53 Ibid.

54 Wang, G., Zhou, J., & Wu, Y., (2020), "Exposing Deepfake Videos by Anomalous Co-motion Pattern Detection," Northwestern University, arXiv:2008.04848v1, reproduced with permission of Gengxing Wang, Northwestern University, via email on November 10, 2020.

55 Ibid.

56 Ibid.

Chapter 6

Operational technology, industrial control systems, and cyber forensics

Albert J. Marcella

CONTENTS

PREFACE

Operational Technology

> Programmable systems or devices that interact with the physical environment (or manage devices that interact with the physical environment). These systems/devices detect or cause a direct change through the monitoring and/or control of devices, processes, and events. Examples include industrial control systems, building management systems, fire control systems, and physical access control mechanisms.

Just how important are industrial control systems (ICSs) and the critical infrastructure which they control? VERY important.

The Cybersecurity and Infrastructure Security Agency defines National Critical Functions (NCF) as:

> The functions of government and the private sector so vital to the United States that their disruption, corruption, or dysfunction would have a debilitating effect on security, national economic security, national public health or safety, or any combination thereof.

These NCF are organized as four areas – supply, distribute, manage, and connect, see Table 6.1.

Table 6.1 National critical functions organized by function area

Supply
• Exploration, Extraction, Refining, and Processing of Fuels
• Generate Electricity and Supply Water
• Produce and Provide Agriculture, Human, and Animal Food Products and Services
• Provide:
Metals and Materials and Manufacture Equipment
Information Technology Products and Services
Materiel and Operational Support to Defense
Research and Development
Housing
Distribute
• Transmit and Distribute Electricity
• Maintain Supply Chains
• Transport
Cargo and Passengers by Air, Rail, Road, and Vessel
Passengers by Mass Transit
Materials by Pipeline
Manage
• Develop and Maintain Public Works and Services, Educate and Train, Enforce Law
• Perform Cyber Incident Management Capabilities
• Preserve Constitutional Rights, Operate Government, and Conduct Elections
• Protect Sensitive Information, Identity Management, and Associated Trust Support Services
• Store Fuel and Maintain Reserves, Manage Hazardous Materials and Wastewater
• Provide and Maintain Infrastructure
• Provide:
Capital Markets and Investment Activities, Funding, and Liquidity Services
Consumer and Commercial Banking, Payment, Clearing, and Settlement Services
Insurance Services
Provide Medical Care, Maintain Access to Medical Records and Support Community Health
Public Safety and Prepare for and Manage Emergencies
Connect
• Operate Core Network
• Provide:
Internet-Based Content, Information, and Communication, Routing, Access, and Connection Services
Positioning, Navigation, and Timing Services
Radio Broadcast, Satellite, Wireless, and Cable Access Network Services

What exposures and risks and resulting impacts does a nation state and her citizens face from compromise of ICSs and the critical infrastructure, within which these NCF operate?

The following is a sampling, over a 7-year period, of global incidents targeting ICSs and corresponding critical infrastructure.

2014

- An Iranian cyber campaign targeted government agencies and critical infrastructure companies in the United States (U.S.), Canada, Europe, the Middle East, and Asia.

2015

- Security researchers say that power outages in Western Ukraine were the result of a coordinated attack on several regional distribution power companies. SCADA systems and system host networks were targeted and damaged. Malware was used to probe for network vulnerabilities, establish command and control, and wipe SCADA servers to delay restoration.

2017

- Department of Homeland Security (DHS) and Federal Bureau of Investigation (FBI) reports warn of Russia-linked hackers targeting ICSs at U.S. energy companies and other critical infrastructure organizations.

2018

- The U.K.'s National Cyber Security Centre released an advisory note warning that Russian state actors were targeting U.K. critical infrastructure by infiltrating supply chains.
- The FBI and DHS issued a joint technical alert to warn of Russian cyber-attacks against U.S. critical infrastructure. Targets included energy, nuclear, water, aviation, and manufacturing facilities.

2019

- The U.S. Department of Justice announced an operation to disrupt a North Korean botnet that had been used to target companies in the media, aerospace, financial, and critical infrastructure sectors.
- Networks at several Bahraini government agencies and critical infrastructure providers were infiltrated by hackers linked to Iran.

2020

- Government and energy sector entities in Azerbaijan were targeted by an unknown group focused on the SCADA systems of wind turbines
- Suspected Iranian hackers unsuccessfully targeted the command and control systems of water treatment plants, pumping stations, and sewage in Israel[1]
- Cyber-attack on BlueScope Steel halted global operations
- Australian logistics Toll Group saw their operations shut down
- Honda suffered a Snake attack[2]

The National Security Agency (NSA) along with the Cybersecurity and Infrastructure Security Agency (CISA), Alert (AA20-205A) indicated that in 2020, cyber actors have demonstrated their continued willingness to conduct malicious cyber activity against critical infrastructure (CI) by exploiting internet-accessible operational technology (OT) assets.[3]

2021 (January – June)

- A hacker altered the amount of sodium hydroxide (lye) added to the water supply for Oldsmar, Florida. The hacker gained access via an unnamed remote software program that allows employees to troubleshoot IT problems. The operator who first noticed the intrusion initially suspected the remote access belonged to another worker. A plant operator noticed the change and reversed it before the tainted water entered the municipality's water supply.[4]
- Colonial Pipeline, a critical infrastructure company in the energy sector, fell victim to a ransomware attack. The company, which operates a major pipeline system that transports 45% of the East Coast's fuel (2.5 million barrels per day of gasoline, diesel, jet fuel, and other products), was forced to shut down a massive 5,500-mile pipeline.
- A cyberattack against Scripps Health disrupted the organization's electronic medical records, radiology, patient portal, and other systems.
- Cyberattacks launched against transportation systems...the New York Metropolitan Transportation Authority, North America's largest transit system and Massachusetts-based Martha's Vineyard Ferry.
- The world's largest meat processing company, Brazil-based meat processor JBS SA, attacked. The attack targeted servers supporting JBS's operations in North America and Australia.

The Stuxnet attack, first discovered in 2010, disrupted Iranian nuclear facilities through a series of events: the malware infiltrated Windows systems through USB drives, then autonomously spread to Programmable Logic Controllers (PLCs) that ultimately destroyed 984 uranium enrichment centrifuges. Stuxnet showcases an early case of successfully targeting ICS, and illustrates how a cyber-attack can have very serious physical consequences.[5]

While there may appear to be many similarities, subtle and important differences exist between the more well referenced and studied information technology (IT) systems and operational technology (OT) systems.

This chapter focuses on operational technology and the cyber forensic examination of such technology. Operational technology...the technology that runs ICSs, which in turn are installed in and responsible for, operating and controlling substantial pieces of a nation's critical infrastructure.

As shown in Table 6.2, there and many different types of Energy, Installation, and Energy (EI&E) and Facility-Related Control Systems (FRCSs).

This chapter explores the use of cyber forensic analysis as one means to analyze ICS incidents, which could include human error, unexpected environmental conditions, failure of equipment, communication related issues, or the result of direct exploitation of inherent ICSs vulnerabilities (e.g., legacy applications and unsecured networks).

In the past, ICSs operated in closed, networked environments, utilizing proprietary operating systems, inherently self-protected and secure. Traditional ICSs can have 30-year lifecycles and are purpose-built, stand-alone systems designed for reliability rather than security. The greatest risk being physical security and access control to operations facilities.

Today, with the demand for greater reliability and improved monitoring, ICSs are now found connected to business IT systems and those systems, to the global Internet. ICSs vulnerabilities can now be exploited by unseen threat actors plying the Internet.

Table 6.2 Different Types of EI&E FRCSs[6]

Advanced Metering Infrastructure	Fire Sprinkler System
Building Automation System	Interior Lighting Control System
Building Management Control System	Intrusion Detection Systems
Carbon Monoxide/Dioxide Monitoring	Physical Access Control System
Digital Signage Systems	Public Safety/Land Mobile Radios
Electronic Security System	Renewable Energy Geothermal Systems
Emergency Management System	Renewable Energy Photo Voltaic Systems
Energy Management System	Shade Control System
Exterior Lighting Control Systems	Smoke and Purge Systems
Fire Alarm System	Vertical Transport System (Elevators and Escalators)
	Closed Circuit Television (CCTV) Surveillance System

Research findings from a 2019 study, released by the Ponemon Institute and Siemens disclosed that 56 percent of the over 1,700 respondents, reported at least one attack involving a loss of private information or an outage in the OT environment in the past 12 months.

In that same study, the researchers noted that the risk that cyber-attacks pose to the OT environment is increasing in frequency and potency as malicious actors' ability to accurately target critical infrastructure assets improves, causing even greater consequences for utility sector operators, managers, and executives. Fifty-four percent of respondents expect an attack on critical infrastructure in the next 12 months.[7]

Due to the critical nature of ICSs and the role which they play, we begin the chapter with an overview of ICS, components, functionality and application. Then a review of what type of data an examiner may find within ICSs and the system's associated components, the challenges associated with examining ICSs and finally present questions which management should ask at the onset of a cyber forensic examination of ICS.

INDUSTRIAL CONTROL SYSTEMS (ICSs)

ICSs which include Supervisory Control and Data Acquisition (SCADA) systems, distributed control systems (DCSs), and other control system configurations such as PLCs are often found in the industrial control sectors.

An ICS consists of combinations of control components (e.g., electrical, mechanical, hydraulic, pneumatic) that act together to achieve an industrial objective (e.g., manufacturing, transportation of matter or energy).

The part of the system primarily concerned with producing the output is referred to as the process. The control part of the system includes the specification of the desired output or performance. Control can be fully automated or may include a human in the loop. The part of the system primarily concerned with maintaining conformance with specifications is referred to as the controller (or control).

Systems can be configured to operate open-loop, closed-loop, and manual mode. In open-loop control systems, the output is controlled by established settings. In closed-loop control systems, the output has an effect on the input in such a way as to maintain the desired objective. In manual mode, the system is controlled completely by humans.

Most manufacturing environments fit into one of five general categories: repetitive, discrete, job shop, process (batch), and process (continuous). Repetitive processing has dedicated production lines that produce the same or similar items consistently without change.

It requires minimal setup or changeover, so it can be accelerated, slowed down, or another production line added. Job shop processing has production areas, rather than production lines. One or a number of product versions are assembled in the areas. If demand deems necessary, the job shop operation is converted to a discrete processing environment with automated equipment.[8]

Batch manufacturing involves multiple discrete steps. After each step in the process, production typically stops so samples can be tested offline for quality. Sometimes during these 'hold times' between steps, the material may be stored in containers or shipped to other facilities around the world to complete the manufacturing process. Batch production process manufacturing examples include such products as: chemicals, gasoline, beverages, baked goods; clothing; computer chips; die- or mold-making; electrical goods; jet engine production; machine tool manufacturing.

Continuous-flow manufacturing, describes a manufacturing method in which the materials (dry bulk or fluids) that are being processed are continuously in motion, undergoing mechanical, thermal, and/or chemical treatment. Synonyms include: continuous manufacturing, continuous processing, continuous production, and continuous flow process. Materials processed using continuous manufacturing are moved nonstop within the same facility, eliminating hold times between steps. Material is fed through an assembly line of fully integrated components. Some examples of continuous processes are gases, liquids, powders, or slurries. Or in areas like mining, they can be granular materials.

Discrete manufacturing produces finished products that can be recognized as distinct physical units via serial numbers or other labeling methods. Discrete processing is also an assembly or production line process, but it is highly diverse, with a wide variation of setups and changeover frequencies. The variation is based on whether the products being produced are alike or very disparate.

ICSs are used to control geographically dispersed assets, often scattered over thousands of square kilometers. ICSs are typically used in electrical power grids, water and wastewater distribution systems, oil and natural gas pipelines, chemical, transportation, pharmaceutical, pulp and paper, food and beverage, agricultural irrigation systems, and discrete manufacturing (e.g., automotive, aerospace, air traffic control, Postal Service mail handling, and durable goods) industries.[9]

Across the U.S., the private sector owns and operates a vast majority of the nation's critical infrastructure, so partnerships between the public and private sectors that foster integrated, collaborative engagement and interaction are essential to maintaining critical infrastructure security and resilience. The 16 U.S. critical infrastructures and the U.S. government agencies responsible for oversight of these infrastructures are shown in Table 6.3.

According to DHS, because the private sector owns approximately 85 percent of the nation's Critical Infrastructure and Key Resources (CIKR) – banking and financial institutions, telecommunications networks, and energy production and transmission facilities, among others – it is vital that the public and private sectors work together to protect these assets.[10]

The private sector is the part of the economy that is run by individuals and companies for profit and is not state controlled. Therefore, it encompasses all for-profit businesses that are not owned or operated by the government. Companies and corporations that are government run are part of what is known as the public sector, while charities and other nonprofit organizations are part of the voluntary sector.[11]

Examining the CIKR energy sector, for example, publicly owned utilities account for 59.1 percent of the market, while federal (U.S. government) owned power agencies make up only three-tenths (0.3) of a percent of the market (See Figure 6.1).

The key control components of an ICS, including the control loop, the Human–Machine Interface (HMI), and remote diagnostics and maintenance utilities, supported by an array of network protocols, is shown in Figure 6.2.

Table 6.3 United States 16 critical infrastructure sectors and responsible sector by specific agency

Responsible sector-specific agency (SSA)	U.S. critical infrastructure sector
Department of Homeland Security (DHS)	Chemical
	Commercial Facilities
	Critical Manufacturing
	Dams
	Emergency Services
	Information Technology
	Nuclear Reactors, Materials, and Waste
	Transportation Systems
	Government Facilities
Department of Transportation (DoT)	Transportation Systems*
General Services Administration (GSA)	Government Facilities**
Department of Agriculture (DoA)	Food and Agriculture
Department of Health and Human Services (DHHS)	Healthcare and Public Health
	Food and Agriculture***
Department of Defense (DoD)	Defense Industrial Base
Department of Energy (DoE)	Energy
Department of the Treasury (DoTR)	Financial Services
Environmental Protection Agency (EPA)	Water and Wastewater Systems

* DHS and DoT share oversight responsibilities
** DHS and GSA share oversight responsibilities
*** DHHS and DoA share oversight responsibilities

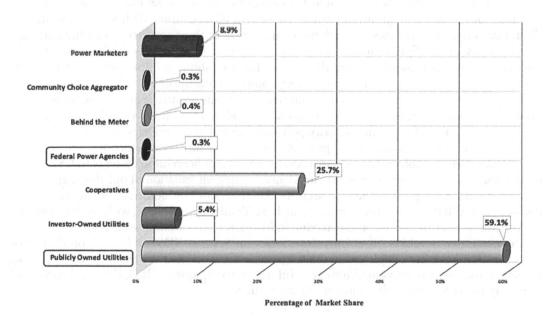

Figure 6.1 Energy sector – percentage of market by provider type[12]

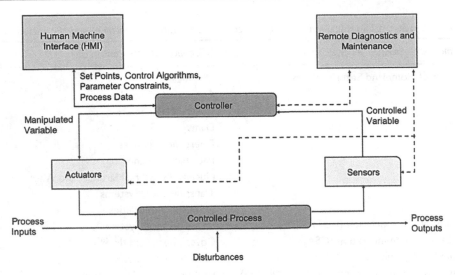

Figure 6.2 Key control components of an industrial control system[13]

The cyber forensic examiner should also be concerned with what, if any, prevailing internal controls or security features are present within the ICS under examination. Such internal controls, in the form of appropriate security devices (e.g., firewalls), may provide useful, digital forensic evidence of an ICS-related incident/event.

Firewalls are used to maintain security across the system, keeping unwanted/unauthorized external (e.g., Internet) accesses from the enterprise network. Multiple firewalls are most commonly used to segregate networks of different sensitivity levels. The primary system firewall should be placed between the Internet and the enterprise network. A second firewall should be positioned between the enterprise network and the ICS's control network, to thwart unwanted access form inside the corporate network.

Several questions that the cyber forensic examiner should consider include but, are not limited to: Does such a configuration exist in the system being examined? If so, what sources of digital data may be available through the examination of these security device (i.e., firewall) logs? If no such configuration exists, the examiner may have serious doubts regarding the integrity of any data collected from the system. Data residing on unprotected systems could potentially be subject to unauthorized access and/or modification.

Intrusion detection is the process of monitoring the events occurring in a computer system or network and analyzing them for signs of possible incidents, which are violations or imminent threats of violation of computer security policies, acceptable use policies, or standard security practices. Intrusion prevention is the process of performing intrusion detection and attempting to stop detected possible incidents. Intrusion detection and prevention systems (IDPS) are primarily focused on identifying possible incidents, logging information about them, attempting to stop them, and reporting them to security administrators. In addition, organizations may consider using IDPSs for other purposes, such as identifying problems with security policies, documenting existing threats, and deterring individuals from violating security policies.

The examiner should verify if Intrusion Detection Systems (IDS) and Intrusion Protection Systems (IPS) have been deployed in the ICS control network and the SCADA server in order to detect and log any unwanted/unsuccessful access to the system. IDS, IPS, or, if used IDPS logs may provide a valuable source of digital evidence.

ICSs and their associated, component parts are susceptible to misuse and attack. Systems that were once air-gapped and separated from general, external communications are now increasingly being exposed to externally accessed technology (e.g., cloud environments, Internet of Things [IoT]), which leave these systems vulnerable to attack by external threat actors. In a report issued by the Kaspersky Lab Industrial Control Systems Cyber Emergency Response Team, exploitation of vulnerabilities in various ICS components by attackers can lead to arbitrary code execution, unauthorized control of industrial equipment and that equipment's Denial of Service (DoS). Importantly, most vulnerabilities can be exploited remotely without authentication and exploiting them does not require the attacker to have any specialized knowledge or superior skills.

The largest number of vulnerabilities, as shown in Figure 6.3, were identified in:

- Engineering software
- SCADA/HMI components
- Networking devices designed for industrial environments
- PLCs[14]

Vulnerable components also include industrial computers and servers, industrial video surveillance systems, various field level devices, and protection relays.

These ICSs are vital to the operation of the U.S. and most developed and developing nation's critical infrastructures. These systems are often highly interconnected and mutually dependent systems.

The digital forensic examination of communications, logs, files, and data created and processed by ICSs could potentially lead to the source and or identity of threat actors and their sponsors. The cyber forensic examiner should be familiar with ICSs in general and what data may be available for examination, if called upon to perform an examination of an organization's ICSs.

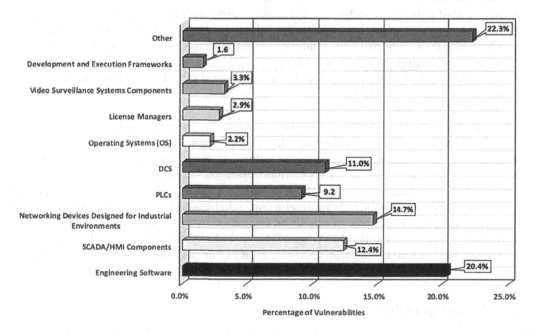

Figure 6.3 Distribution of vulnerabilities identified by ICS components[15]

SUPERVISORY CONTROL AND DATA ACQUISITION (SCADA) SYSTEMS

The SCADA server is the interface between the control center and the many processing facilities for which the control center has oversight responsibility. SCADA systems integrate data acquisition systems with data transmission systems and HMI software to provide a centralized monitoring and control system for numerous process inputs and outputs.

The software used by operators in the 'control center' of a SCADA system is referred to as HMI software. The HMI software serves a dual purpose of presenting the data acquired from various elements of the SCADA system and allowing the operator to manipulate parameters of the system that are under the operator's supervisory control. HMI software is often configured to mimic the look and feel of a tangible control panel, with elements like switches, dials, sliders, and readouts.[16]

SCADA systems are designed to collect field information, transfer it to a central computer facility, carry out any necessary analysis and control and display the information to the operator graphically or textually, thereby allowing the operator to monitor or control an entire system from a central location in near real time. Based on the sophistication and setup of the individual system, control of any individual system, operation, or task can be automatic, or it can be performed by operator commands.[17]

A SCADA system consists of a number of remote terminal units (RTUs) collecting field data and sending that data back to a master station, via a communication system. The master station displays the acquired data and allows the operator to perform remote control tasks.

An RTU is an electronic device that is controlled by a microprocessor. The device interfaces with physical objects to a DCS or SCADA system by transmitting telemetry data to the system.[18]

RTU performs the following functions:

- The connection with supervised equipment;
- Reading of equipment status (such as open/closed position of the valve or relay);
- Acquisition of measured signals, such as the pressure, flow, voltage, or current;
- The control of equipment by sending command signals, such as the closing of a valve or relay or setting the speed of a pump;
- Reading the digital or analog signals, and sending the commands using both
- digital or analog signals.[19]

The Master Terminal Unit (MTU) in SCADA systems is a device that issues the commands to the RTU, which is located remotely from the control center. The MTU gathers the required data, stores the information, and process the information and display the information in the form of pictures, curves, and tables to human interface and assists the operator in making control decisions.[20] The accurate and timely data allows for optimization of the plant operation and process. Other benefits include more efficient, reliable and most importantly, safer operations. This results in a lower cost of operation compared to earlier non-automated systems.[21]

Communication between the MTU and RTU is bidirectional; however, the major difference is RTU cannot initiate the conversation; an RTU simply collects the data from the field and stores the data.

SCADA configuration

Typical hardware includes a control server placed at a control center, communications equipment (e.g., radio, telephone line, cable, or satellite), and one or more geographically distributed field sites consisting of RTUs and/or Programmable Logic Controllers (PLCs), which controls actuators and/or monitors sensors.

Generally, all SCADA units will comprise of a combination of common categories of hardware components:

- Audio components (microphone and speaker)
- Battery and charging unit
- Digital signal processor (DSP)
- Human input interface (such as a keypad, keyboard, or touch screen)
- Main board
- Measurement devices and sensors
- Microprocessor
- Radio module and antenna
- Random access memory (RAM)
- Read only memory (ROM)
- Visual display unit (this may be solely a function of the HMI)[22]

The control server stores and processes the information from RTU inputs and outputs, while the RTU or PLC controls the local process. The communications hardware allows the transfer of information and data back and forth between the control server and the RTUs or PLCs.

The software is programmed to tell the system what and when to monitor, what parameter ranges are acceptable, and what response to initiate when parameters change outside acceptable values.

An Intelligent Electronic Device (IED), such as a protective relay, may communicate directly to the control server, or a local RTU may poll the IEDs to collect the data and pass it to the control server. IEDs provide a direct interface to control and monitor equipment and sensors. IEDs may be directly polled and controlled by the control server and in most cases have local programming that allows for the IED to act without direct instructions from the control center.

SCADA systems are usually designed to be fault-tolerant systems with significant redundancy built into the system. Redundancy may not be a sufficient countermeasure in the face of malicious attack.

The typical configuration of a SCADA system consists of a control center, which houses a control server and the communications routers. Other control center components include the HMI, engineering workstations, and the data historian which are all connected by a local area network (LAN).

The historian is a crucial component of the SCADA system, responsible for storing and logging site data. The historian is responsible for storing and logging all of the data that the SCADA system aggregates. It allows operators and stakeholders to look at historical data for the plant. A historian can also have reporting capability. It can generate manual or automated reports containing different sets of data and show what happened at the plant over specified periods of time.

Manufacturing operates in real time, requiring very fast data collection for optimal analyses. A plant-wide historian provides 10–20 times faster read/write performance over a relational database and 1-millisecond resolution for true real-time data. Additionally, the plant-wide historian is optimized for 'time series' data, while a relational database is built to manage relationships. For example, relational databases are great at answering a question such as: What customer ordered the largest shipment? A plant-wide historian, on the other hand, excels at answering questions such as: What was today's hourly unit production standard deviation?[23]

The historian helps track problems that build up over periods of time. By monitoring changes in value for certain data points over days, weeks, and months, operators can see if the correlating systems and devices are developing problems that need maintenance attention. If a device malfunctions suddenly, its historical data may offer visibility and insight into resolving the problem.[24]

A Historian system is composed of three primary components:

1. Data collectors for interfacing with the data sources such as a PLC, networked devices, Object Process Control (OPC) servers, files, and other data sources.
2. Server software that processes the data from the data collectors, stores the data, and serves the data to client applications. The server software components can also provide other services such as a calculation engine, alarm management, and subsystems to provide context for the data.
3. Client applications for data reporting, charting, and analysis.[25]

Data collected, stored and processed by the Historian could provide valuable information in the forensic examination of an ICS incident. The examiner should strive to preserve and collect relevant data from the Historian.

The control center collects and logs information gathered by the field sites, displays information to the HMI, and may generate actions based upon detected events. The control center is also responsible for centralized alarming, trend analyses, and reporting. The field site performs local control of actuators and monitors sensors.

Field sites are often equipped with a remote access capability to allow operators to perform remote diagnostics and repairs usually over a separate dial up modem (legacy systems) or WAN connection. Standard and proprietary communication protocols running over serial and network communications are used to transport information between the control center and field sites using telemetry techniques such as telephone line, cable, fiber, and radio frequency such as broadcast, microwave, and satellite.[26]

The complexity and breadth of the system or process being monitored will dictate whether the SCADA system is a simple configuration or a very complex one. SCADA systems operate in real-time using a database system referred to as Real Time Data Base (RTDB). This a database, which operates based upon real-time processing to handle workloads whose state is constantly changing. This type of processing differs from the traditional information system databases, whose content is typically persistent data and which typically does not have a temporal dependency.

The Supervisory Station (SCADA Server) layer or MTU contains OPC software and HMI applications. OPC software is a communication standard based on Object Linking and Embedding (OLE) technology provided by Microsoft Windows that provides an industrial standard exchange mechanism between plant floor devices (e.g., RTUs, PLCs) and client applications (e.g., HMI).[27] Figure 6.4 shows the components and general configuration of a SCADA system.

SCADA systems are:

- Data-oriented
- Event-driven
- Scalable and flexible

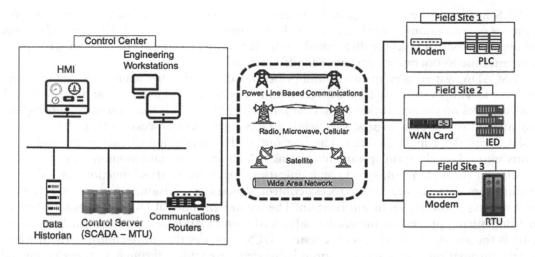

Figure 6.4 Components and general configuration of a SCADA system[28]

As shown in Figure 6.4, there are several areas within the SCADA system that would be of interest to the examiner, when looking for potential digital evidence. These areas would include:

- Wide area network (WAN) traffic
- Data Historian logs and files
- Control Server status logs
- Engineering Workstations
- Communication traffic logs
- Router traffic logs
- Remote Terminal Unit (RTU)
- Master Terminal Unit (MTU)
- Programmable Logic Controller (PLC)
- Intelligent Electronic Device (IED)
- HMI unit

Continuing on with our review of the components of SCADA systems, we next take a look at the DCS and its relationship to the broader SCADA system.

DISTRIBUTED CONTROL SYSTEM (DCS)

Distributed Control Systems are generally used to control production systems within a local area such as a factory using supervisory and regulatory control.

DCSs are used to control production systems within the same geographic location for industries such as oil refineries, water and wastewater treatment, electric power generation plants, chemical manufacturing plants, automotive production, and pharmaceutical processing facilities. These systems are usually process control or discrete part control systems. DCS are integrated as a control architecture containing a supervisory level of control overseeing multiple, integrated sub-systems that are responsible for controlling the details of a localized process.

DCS is most suited for large-scale processing or manufacturing plants wherein a large number of continuous control loops are to be monitored and controlled. The main advantage of dividing control tasks for distributed controllers is that if any part of DCS fails, the plant can continue to operate irrespective of failed section.

A SCADA system is event-driven and prioritizes data gathering, while a DCS emphasizes process-level operations. A DCS delivers data to operators, and at the same time, a SCADA concentrates on the acquisition of that data. In short, a SCADA is geared toward understanding and collecting data on processes, while a DCS emphasizes process control.

DCS has three main qualities. The first one is the distribution of various control functions into relatively small sets of subsystems, which are of semiautonomous, and are interconnected through a high-speed communication bus. Some of these functions include data acquisition, data presentation, process control, process supervision, reporting information, and storing and retrieval of information. The second attribute of DCS is the automation of manufacturing process by integrating advanced control strategies. And the third characteristic is the arranging the things as a system. DCS organizes the entire control structure as a single automation system where various subsystems are unified through a proper command structure and information flow.

Although both DCS and SCADA are monitoring and control mechanisms in industrial installations, they have different goals. There exists some commonality between DCS and SCADA in terms of hardware and its components; however, there are certain requirements by the end applications that separates a robust and cost-effective DCS from the viable SCADA system. Table 6.4 presents the key differences between a DCS and SCADA.

Table 6.4 Difference between distributed control system (DCS) and SCADA[29]

DCS	SCADA
Process-oriented	Data-gathering-oriented
Emphasizes more on control of the process and it also consists of supervisory control level and it presents the information to the operator	Concentrates more on acquisition process data and presenting it to the operators and control center
Data acquisition and control modules or controllers are usually located within a more confined area and the communication between various distributed control units carried via a local area network	Generally, covers larger geographical areas that use different communication systems which are generally less reliable than a local area network
Employs a closed loop control at process control station and at remote terminal units	There is no such closed loop control
Is process state driven where it scans the process in regular basis and displays the results to the operator, even on demand	Is event driven where it does not scan the process sequentially, but it waits for an event that cause process parameter to trigger certain actions
Does not keep a database of process parameter values as it always in connection with its data source	Maintains a database to log the parameter values which can be further retrieved for operator display and this makes the SCADA to present the last recorded values if the base station unable to get the new values from a remote location
Is used for installations within a confined area, like a single plant or factory and for complex control processes. Some of the application areas of DCS include chemical plants, power generating stations, pharmaceutical manufacturing, oil and gas industries	Is used for much larger geographical locations such as water management systems, power transmission and distribution control, transport applications and small manufacturing and process industries

DCSs are:

- Process-oriented
- Focused on central control
- Ideal for one facility
- Somewhat slower (processing times) versus the PLC/RTU SCADA environment

For the digital forensic examiner, the types of data, which may be available for examination from the DCSs, include but are not limited to:

- Process control and measurement data
- Engineering workstation, operating station or HMI, process control unit data
- Smart devices and communication system data
- Depending on the system under examination, data integrated with ERP and IT systems
- Machine set points, process, and operation variables
- Temperature, pressure limits, and equipment limitations

The final component that will be reviewed to complete our overview of operations technology is to take a quick look at the PLC and this component's role in an comprehensive SCADA environment.

PROGRAMMABLE LOGIC CONTROLLER (PLC)

SCADA and PLC are two different components of an automation process. SCADA (Supervisory Control and Data Acquisition) is a PC-based system used for monitoring process, store and retrieve process data. A PLC (Programmable Logic Controller) however, is an electronic device used for controlling process.

PLCs are used in both SCADA and DCS systems as the control components of an overall hierarchical system to provide local management of processes through feedback control as described in the sections above. PLC and DCS systems are embedded systems with their own operating systems and program languages. In the case of SCADA systems, they may provide the same functionality of RTUs. When used in DCSs, PLCs are implemented as local controllers within a supervisory control scheme. PLCs are generally used for discrete control for specific applications and generally provide regulatory control.

Almost any production line, machine function, or process can be greatly enhanced using this type of control system. However, the biggest benefit in using a PLC is the ability to change and replicate the operation or process while collecting and communicating vital information. Another advantage of a PLC system is that it is modular. That is, you can mix and match the types of Input and Output devices to best suit a particular application.

What is inside a PLC?

PLC consists of the:

- Processor (CPU)
  The central processing unit, the CPU, contains an internal program that tells the PLC how to perform the following functions:

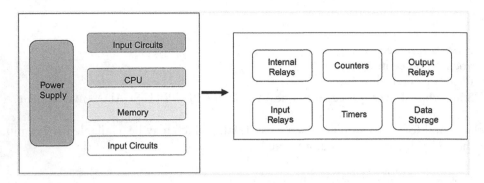

Figure 6.5 Inside the PLC

- Execute the control instructions contained in the user's programs. This program is stored in 'nonvolatile' memory, meaning that the program will not be lost if power is removed.
- Communicate with other devices, which can include I/O Devices, programming devices, networks, and even other PLCs.
- Perform housekeeping activities such as communications and internal diagnostics.
- Memory system
- Power supply
- Input/output system/circuits (The input/output (I/O) system is the section of a PLC to which all of the field devices are connected), see Figure 6.5.

For the digital forensic examiner, the types of data, which may be available for examination from the PLC, include but are not limited to:

- Control instructions contained in the user's programs stored in 'nonvolatile' memory
- Input device data, such as:

    Switches and Pushbuttons
    Sensing Devices

  - Limit Switches
  - Photoelectric Sensors
  - Proximity Sensors

    Condition Sensors
    Encoders
      o Pressure Switches
      o Level Switches
      o Temperature Switches
      o Vacuum Switches
      o Float Switches

- Output data from devices such as:
  - Valves

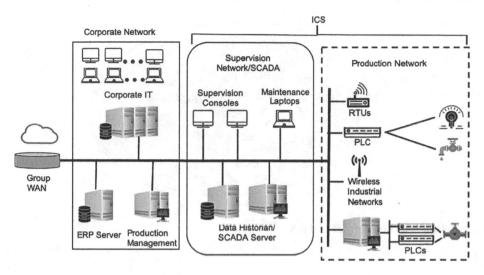

Figure 6.6 The corporate IT network and the ICS[31]

- Motor Starters
- Solenoids
- Actuators
- Horns and Alarms
- Stack lights
- Control Relays
- Counter/Totalizer
- Pumps
- Printers
- Fans[30]

As discussed in the preceding sections, Figure 6.6 highlights the interface and relationship between the corporate IT network environment and the ICS.

OT, ICS AND SCADA FUNDAMENTALS

As we wrap up this section on an overview of operations technology and ICSs, it is important to note the role that ICSs have in operating a nation's critical infrastructure, in controlling an organization's day-to-day production processes and in part, a very big part, a sustaining a global supply chain.

The discipline of operations technology while similar in some ways to its cousin information technology is, however, unique in many others. Operations technology requires a skill set that differs in both training and application than those found within information technology.

Within the ICS resides the SCADA and together they share and process information provided by the RTUs DCS and PLCs. Figure 6.7 presents a Venn diagram showing the

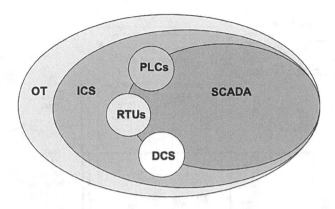

Figure 6.7 Relationship between OT and ICS

relationships among elements within operation technology while also highlighting the shared relationships within ICSs.

Incidents impacting ICS can be accidental, unplanned, localized and if fortunate, have minimal impact and interruption on business processes. However, employees, insiders, third-parties, threat actors, each have the capability and opportunity to intentionally cause operational failure and/or impact to system critical ICSs.

As discussed, there are many components within an ICS that process digital data into real-time information. Data, which may be considered vital, evidential data, should an investigation of an ICS-related incident require a forensic examination.

In the following section, we take a closer look at the differences between OT and its close cousin, IT. We will also review how OT and IT intersect with and are affected by information security (InfoSec).

CYBER FORENSICS AND OPERATIONAL TECHNOLOGY

There is an important difference between incident response (the mitigation of violations of security policies and recommended practices) and cyber (or digital) forensics (the application of science to the identification, collection, examination, and analysis, of data while preserving the integrity of the information and maintaining a strict chain of custody for the data).[32]

The goal of incident response operations is the restoration of normal operations while minimizing impacts to people, property, and the environment.

Whereas, cyber forensic techniques will enable an examiner or investigator to facilitate the recreation of events found to be illegal, unauthorized or which have disrupted operations, through the analysis of data recovered from a SCADA system. Depending on the system or systems involved, current status of these systems and prevailing conditions, the cyber forensic investigation may be conducted during or following a cyber incident, and after returning the control system to its operational state (if possible).

The results of this analysis are included in a report to management and if appropriate, may also be used as evidence admissible in a court of law. Such an examination may also prove useful in assessing the potential, future likelihood of failure or interruption to critical ICSs.

The consequences associated with cyber incidents in a control systems environment can vary and can include:

- Loss of localized or remote control over the process
- Loss of production
- Compromise of safety
- Catastrophic cascading failures that affect critical infrastructure and can extend to peer sites and other critical infrastructure sectors
- Environmental damage
- Injury or loss of human life[33]

A forensic investigation can answer several intriguing questions about an incident. For instance, consider a scenario of a SCADA system recently hit by malware, which has caused the system to malfunction. A forensic investigation can be an effective way to answer questions such as:

- Is the SCADA system still compromised by malware?
- A virus scan revealed that the Java cache[34] contains a known exploit. Was the exploit successful? What payload does it have, and has that compromised the system?
- How can the SCADA system operator clean the system after an infection, and reliably bring it back into a known good state, without having to shut down the complete system?
- An operator has installed a suspicious, untrusted application downloaded from the Internet. Did that application change components that are relevant for the stable operation of the SCADA system?[35]

OPERATIONAL TECHNOLOGY

Operational Technology (OT) refers to computing systems that are used to manage industrial operations as opposed to administrative operations such as IT. Included within OT are both hardware and software that detects or causes a change through the direct monitoring and/or control of physical devices, processes, and events in the enterprise.

OT includes ICSs and SCADA systems used in critical infrastructures such as water, oil & gas, energy, and utilities. Cyber-attacks on OT can potentially disrupt vital services, damage critical equipment, threaten human health and safety, and trigger disruptions in other sectors.

ICS are automated control systems that act upon industrial systems and processes. ICS is used as a general term that encompasses several – but not all – types of control systems. These include SCADA systems, DCS and other control systems, such as the PLCs often found in the industrial sector and critical infrastructure

Given that OT and ICSs may be a unique environment to many examiners, as well as casual readers of this text, the chapter begins with a thorough review of operational technology...the pieces, parts, components and elements, which make a SCADA system function and which in turn, may provide important examination content for the digital forensic examiner.

Cyber-connected OT devices have significantly improved automation and efficiency in the monitoring and measurement of critical functions, but these new efficiencies also introduce vulnerabilities.

Unlike the central SCADA or IT systems, OT systems are not automatically updated with service packs, new releases, and bug fixes. In reality, the OT devices are often running the same software as when they were installed 10–15 years ago at a time when physical separation form the network IT systems was considered secure.[36]

Technology Web-enabled sensing and measuring technologies have enabled the critical systems to become more reliable and automated, but have also created more vulnerabilities that differentiate OT from IT:

- Compromise of OT can disable operations, disrupt critical services to customers, and damage highly specialized equipment.
- OT must be able to survive a cyber incident while sustaining critical functions.
- Many OT systems must operate in real-time with 24/7 availability and are unable to go offline for patching or upgrades.
- OT components may be very simple devices and may not have enough computing resources to support additional cybersecurity capabilities.
- OT components may be widely dispersed and located in publicly accessible areas where they are subject to physical tampering.[37]

OPERATIONAL TECHNOLOGY AND INFORMATION TECHNOLOGY

Operational Technology (OT) $\neq$ Information Technology (IT)

The joining of physical and cybersecurity processes and the intensifying assimilation of ICSs with business networks and cloud-based applications, has resulted in the increasing frequency and sophistication of cyber-attacks on ICSs. ICSs are designed to manage physical operational processes (e.g., telecommunications, power, waste management, water control, oil and gas refining) and specialized systems such as positive train control systems (PTC).

PTC systems are specialized SCADA systems that provide positive train separation, over speed protection, and protection for roadway workers working within the limits of their authority. These systems are in contrast to business enterprise networks, which are designed to manage information.

There are many transportation, medical, building, security, and logistics systems which – though similar in many respects to traditional ICS – use different protocols, ports, and services, and are configured to operate in different modes than SCADA or DCS systems.

Table 6.5 provides a comparison of the important differences between more widely encountered and seen information technology systems and operation technology systems, typically found in the ICS environment.

Table 6.6 provides an overview comparison of security features found within the more typical and traditional information security (InfoSec) and the security typically found in ICSs.

Industrial control systems connected to business IT systems and the Internet constitute a systemic cyber risk among critical infrastructure. Cyber-connected operational technology (OT) systems improve automation and efficiency in the control of critical processes – such as generation, processing, and delivery of power, water, fuel, and chemicals – but also introduce new cyber risks.[39]

Table 6.5 Comparison – information technology systems and industrial control systems

Attribute	Information technology system	Industrial control system
Age/Component Lifetime	3–5 years	10–20 years
Architecture	Enterprise wide infrastructure and applications	Event-driven, real-time, embedded hardware and customized software
Asset Classification	Common practice, performed annually, results driven, cybersecurity and risk management focus	Performed only when required, protecting critical assets tied to budget costs
Auditability	Use of current/modern methods and tools allow for both on-site and remote data collection, analysis and audit assessment	Dependent upon age of ICS, current audit methods and automated tools may not work on legacy systems
Change Management	Regular, scheduled, timely, performed during minimal-use periods, good security procedures, often automated	Strategic and long-term scheduling required, non-trivial due to potential disruptive impact, changes must be thoroughly tested and rolled-out incrementally across the system to assure system-wide integrity
Connectivity	Corporate network, IP-based, standard protocols, continuous, primarily wired networks with some localized wireless capabilities	Control network, proprietary protocols, intermittent, long delays cause performance (safety) concerns, networks are complex and sometimes require the expertise of control engineers
Cybersecurity Testing	Use of current/modern methods	Testing has to be customized to system, modern methods inappropriate, older equipment fails/breaks
Cyber Market Maturity	Mature and maturing, advanced cyber knowledge, processes and controls	Start-up stage and limited awareness, increasing concern as IIoT grows and requires serious consideration
Forensic Examination	Most possible, data automatically retained by many processes, open access to processing environment, devices, storage, and backup built into processes	Most unlikely, data storage, backup not designed into processes, inherent design focus of applications is on continuous processing thus, automated restart features for example will overwrite potential critical digital evidential data, proprietary systems may hinder access to possible evidential data.
Incident Response	Routinely developed, deployed and tested, can be automated, data retained for analysis	More common than expected, focus on resumption of processes, response actions may destroy event data preventing root-cause analysis
Interfaces	GUI, Web browser, voice, tactical (keyboard), terminal	Electromechanical, sensors, actuators, coded displays
Malware Deployment	Common, easily deployed and updated, automated	Difficult to challenging, may unable to be deployed on legacy ICSs, ineffective
Management Key Concerns	Loss of data, operations security, privacy violations, revenue impact	Production stoppage, health, safety and environmental impact
Mobile Code	Common, easily deployed and updated, automated	Difficult to challenging, may unable to be deployed on legacy ICSs, ineffective
Operating Environment	Air-conditioned, UPS, voltage regulators, typically clean	Extreme temperatures, remote locations, vibrations, shocks, airborne particles, liquids

(Continued)

Table 6.5 (Continued) Comparison – information technology systems and industrial control systems

Attribute	Information technology system	Industrial control system
Operational Priorities	Confidentiality, integrity, availability, data integrity is essential	Availability, integrity, confidentiality, control processes cannot tolerate downtime
Patch Management	Easily defined, enterprise-wide, remote, automated	Challenging, OEM driven, long-lead times, may have direct impact on operations
Performance Requirements	Non-real-time, response must be consistent, high throughput is demanded, high delay and jitter may be acceptable, less critical emergency interaction, tightly restricted access control can be implemented to the degree necessary for security	Real-time, continuous processing, up-time reliability, fault-tolerant, fail-over... essential, response is time-critical, high delay and/or jitter is not acceptable
Physical and Environmental Security	Ranges from poor or ineffective (personnel, office systems) to excellent or hardened (critical processes, OS, telecoms)	Ranges from poor or ineffective (personnel, remote facilities) to excellent or hardened (operations center, guards, gates, guns)
Purpose	Process transactions, provide information	Controls and monitors physical processes
Recoverability	Automated, recover via reboot, may vary by application, system, contract, compliance constraints, temporal requirements	Uninterruptable, fault-tolerant, fail-over, essential and mandatory
Reliability Requirements	Responses such as rebooting are acceptable, system downtime can often be tolerated, depending on the system's operational requirements and end-user needs	Responses such as rebooting may not be acceptable because of process availability requirements, outages must be planned and scheduled days/weeks in advance
Risk Impact	Business processes, transaction processing, loss of revenues	Environmental catastrophes, destruction of equipment, production losses/stoppages, loss of life
Risk Management Issues	Manage data, data confidentiality and integrity are paramount, major risk impact is delay of business operations	Control physical world, human safety is paramount, followed by protection of the process, fault tolerance is essential, even momentary downtime may not be acceptable, major risk impacts are regulatory non-compliance, environmental impacts, loss of life, equipment, or production
Role	Support business processes and individuals	Controls industrial machinery and processes
Secure System Development	Integral to design objectives and compliance requirements	Typically, not an integral part of design consideration, focus on operational efficiency and productivity
Security Compliance	Regulatory oversight (may be limited), customer driven (competitive demand for services), contractual requirement	Specific and specified regulatory oversight (in some sectors, e.g., Electric, Nuclear) mandates compliance
System Operation	Systems are designed for use with typical, modern operating systems(OS), upgrades are straightforward	Differing and possibly proprietary operating systems, often without security capabilities built in, greater dependency on vendor due to specialized modification which may have been made to OS by vendor

(Continued)

Table 6.5 (Continued) Comparison – information technology systems and industrial control systems

Attribute	Information technology system	Industrial control system
Technology Support Lifetime	2–3 years, multiple vendors	10–20 years, same vendor
Update Capability	Straightforward, automated	Difficult, institutional knowledge oft-times required
Vendor Support	Allow for diversified support styles, through many vendors, vendor's staff keeps up-to-date on current support requirements	Service is typically only available from a sole-source provider, vendor may no longer be in business or support software/hardware depending on age of system, current vendor personnel may not have any knowledge or familiarity with the software/hardware currently installed and running in production.

Table 6.6 Comparison of IT security versus industrial control system security[38]

Security feature	Information technology	Industrial control systems
Anti-virus/Mobile Code	Common/widely used	Uncommon/Impossible to deploy
Support Technology Lifetime	3–5 years	Up to 20 years
Outsourcing	Common/widely used	Rarely used
Application of Patches	Regular/scheduled	Slow (vendor specific)
Change Management	Regular/scheduled	Rare
Time Critical Content	Generally, delays accepted	Critical due to safety
Availability	Generally, delays accepted	24 × 7 × 365 × forever
Security Awareness	Reasonably good in both private and public sector	Poor except for physical
Security Testing/Audit	Scheduled and mandated	Occasional testing for outages
Physical Security	Secure	Remote and unmanned

CYBER FORENSIC EXAMINATION OF INDUSTRIAL CONTROL SYSTEMS

Cyber forensics has been in the popular mainstream for some time, and has matured into an information-technology capability that is common among modern information security programs. Although scalable to many information technology domains, especially modern corporate architectures, developing a cyber forensic program can be a challenging task when being applied to nontraditional environments, such as control systems.

Modern IT networks, through data exchange mechanisms, data storage devices, and general computing components provide a good foundation for creating a landscape used to support effective cyber forensic examinations. However, modern control systems environments are not easily configurable to accommodate forensic programs. Nonstandard protocols, legacy architectures that can be several decades old, and irregular or extinct proprietary technologies can all combine to make the creation and operation of a cyber forensic program anything but a smooth and easy process.

Given the choice between active and passive response, in most cases, the examiner will default to a passive investigative response. This response is most appropriate for an ICS environment because the approach is both non-intrusive and there is minimal risk of disrupting ongoing critical processes or operations.

Unless the ICS has suffered an attacked or is non-operational at the time of the examination, most ICSs cannot be simply shut down or pulled offline to allow the examiner access to potential data sources. In reality, the examiner/investigator may have no recourse other than to rely on live forensics as a means of gathering any potential digital forensic evidence.

ICS DISTINCT SYSTEM ENVIRONMENTS

It is important for the examiner to have a frame of reference when approaching the planning and organization of a forensic examination of an ICS. For ease of reference, ICSs are logically grouped into three distinct classifications or system environments, based upon their technologies/architectures.

This arrangement of ICSs by technology/architecture, can assist the examiner/investigator in being better prepared for the environment in which the examination will be conducted. The level of technology present (or lack thereof) may present potential and at times, unique limitations to the identification and collection of digital evidence and to the overall forensic process, as envisioned by the examiner/investigator.

These distinct ICS classifications are:

Modern/Common Technologies: Are technologies that are critical to a control systems operation, have modern computing capabilities, and are most likely still fully supported by the vendor. These technologies will most likely run on some sort of contemporary operating system, may have some detailed information about the operations available in the open source community, and have been continuously supported since their original deployment.

Modern/Proprietary Technologies: Are technologies that are critical to a control systems operation, have been created within the last 10 years, and are still fully supported and understood primarily by the vendor (or systems integrator). In this case, the control systems technology and information about its operation are not generally available through open-source methods. Moreover, the technology and protocols associated with command and control of the operational environment may only be known to the vendor and just partially to the owner/operator.

Legacy/Proprietary Technologies: Are technologies that are critical to a control systems operation, may have been deployed more than 10 years ago, and have moderate computing capabilities (compared to modern systems). Moreover, they may or may not be supported be the vendor and are in most cases only understood (in-depth) by the vendor. The possibility that the vendor no longer has the requisite knowledge due to the age of the system further compounds this situation. As such, situations can arise when the owner of the system has key knowledge of the system (or at least how to maintain it) as the vendor no longer exists.[40]

ICS CYBER FORENSIC PROCESS

There is yet no definitive, globally accepted, court validated approach to performing a cyber forensic examination of an ICS. While there do exist definitive, globally accepted, court validated approaches to the collection, preservation, and analysis of digital evidence, the approach to acquiring this digital evidence, from an ICS, is still evolving.

ICSs present examiners/investigators with unique environments and challenges with respect to the identification, collection and processing of evidential data, especially if the examination findings are to be presented and accepted in a court of law.

It should be noted that this flowcharted process presented below, is designed to be more generic in nature. The process followed by any individual examiner will undoubtedly be customized to the meet the requirements of the particular ICS environment under examination. Additionally, prevailing laws (domestic and/or international) and the condition of the ICS environment (operational up and running or off-line, non-operational), will play a role in the examiner's approach. This may result in the addition or deletion of specific steps to this process. Figure 6.8 provides an overview of the ICS cyber forensic process.

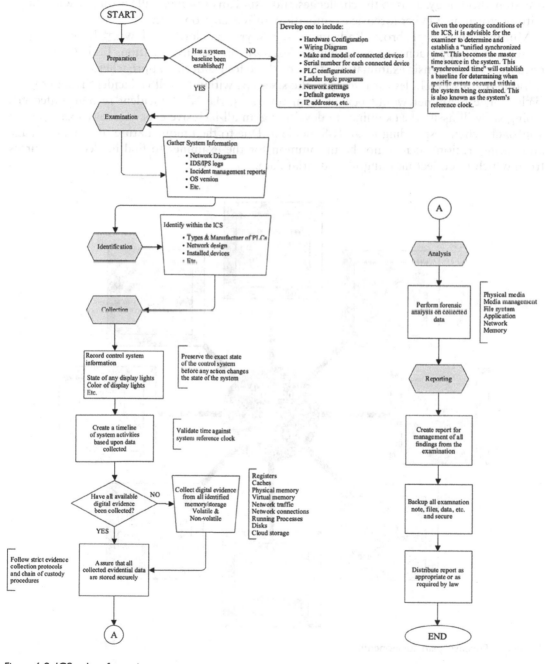

Figure 6.8 ICS cyber forensic process

FORENSIC EXAMINATION METHODOLOGIES FOR INDUSTRIAL CONTROL SYSTEM ENVIRONMENTS

In their paper 'Recommended practice: Creating Cyber Forensics Plans for Control Systems,' Technical report, for the DHS, Fabro and Cornelius identified eight major components that will contribute to creating an organizational, cyber forensic plan for ICS environments. Figure 6.9 illustrate these essential components.

The lack of a standardized SCADA forensic examination process or a single, litigation-tested methodology, adds to the challenges and issues an examiner will encounter when faced with performing a cyber forensic examination an ICS environment.

Multiple authors; Fabro,[41] Radvanovsky,[42] Spyridopoulos,[43] and Wu,[44] have each presented varied approaches to performing a SCADA forensic investigation. Table 6.7 summarizes the cyber forensic examination steps, of these four differing approaches.

Table 6.7 also includes for reference, the six steps within the NIST incident response lifecycle.[45] The examiner would benefit from reviewing the NIST incident response lifecycle. Doing so, will assist the examiner in developing an all-inclusive cyber forensic examination approach when responding to an ICS incident. Due to the unique nature of ICSs, (e.g., age and configuration), it may not be uncommon for the examiner to find inadequate sources from which to collect meaningful evidential data.

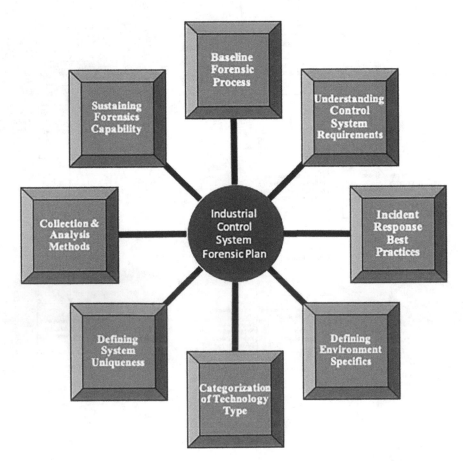

Figure 6.9 Forensic plan components

Table 6.7 Cyber forensic examination of an ICS environment – summary of varied approaches

Cyber forensic process	Fabro & Cornelius 3 steps	Radvanovsky & Brodsky 4 steps	Spyridopoulos, Tryfonas, & May 5 steps	Wu, Disso, Jones & Campos 7 steps	NIST (incident response)
Preparation				x	x
Detection					x
Examination		x	x		
Identification		x	x	x	
Preservation				x	
Collection	x	x	x	x	
Analysis	x		x	x	x
Containment					x
Eradication & Recovery					x
Documentation		x	x		
Reporting	x			x	
Post-event Activity				x	x

Developing the most comprehensive forensic process will provide the ability to accurately identify, collect, and analyze incident data. This will not only support the examination but, may uncover illegal activities and provide management with recommendations for counter-measures, designed to mitigate a reoccurrence of the incident.

CHALLENGES IN EXAMINING INDUSTRIAL CONTROL SYSTEMS

Fun Fact: There were programmable controllers before there were desktop computers. ('PC' meant 'Programmable Controller' before it meant 'Personal Computer,' then was changed to 'PLC' for 'Programmable Logic Controller' after the IBM-PC came out.)

Not-So-Fun Fact: Some of those controllers are still out there and running. They may have been programmed on dedicated hand-held terminals, with their programs stored on cassette tape, or hand-written on paper pages with pre-printed boxes for entering the data. Later, hand-helds had connections for parallel printers and the ability to add explanatory names to the input, output, and internal data points, as well as (a few) inter-line comments. Those printouts may be the only existing copy of the operating logic.

Fun Fact: There were PLCs before there were communications network standards.

Not-So-Fun Fact: Some of those controllers are still out there, running on dozens of different networking hardware layers and hundreds of software communications protocols. Names you may run into are: Modbus, CANBus, ProfiBus, FieldBus, InterBUS, CC-Link, DeviceNET, HART, CIP, Ethernet/IP, DF-1, DH-485, MELSECNet, BACnet, LON, Zigbee, SRTP, and many, many more.

Some of these protocols allow an external host to read and write anything stored in the controller's data memory, some allow an external host to read and write the controller's password or the logic program, some allow an external host to flash new firmware into the controller's operating system memory. Some protocols have undocumented functions and features which could be exploited.[46]

As discussed, Industrial Control Systems (ICSs) while similar in many respects to traditional IT systems differ in significant ways (see Tables 6.5 and 6.6). These differences pose a challenge and present potential barriers to the forensic examiner when attempting to perform a digital forensic examination of an ICS and its related operating components.

ISSUES AND CONCERNS WHEN PERFORMING AN EXAMINATION OF INDUSTRIAL CONTROL SYSTEMS

Effective cyber forensic collection in any environment requires addressing several challenges such as volatile memory, poor administrative functions, absent or inadequate logging, and general cultural limitations. Some of the additional challenges facing the examiner when reviewing an ICS environment include:

Communication Platforms

- The geographically dispersed distribution industries typically served via ICSs, utilize long-distance communication WAN and wireless/RF (radio frequency) technologies. This is compared to the more often seen LAN technologies, which examiners may be more familiar with.

  The use of long-distance communication platforms introduces several issues of concern; (a) differing security protocols and controls across differing networks; (b) the potential for data loss; (c) concerns regarding data accuracy and integrity; and (d) potential data access restrictions.[47]

  The examiner must be aware of these issues when determining if all available data have been acquired and what, if any effect potential lost data might have on the examination.
- Inherently, field devices (RTUs, PLCs, IEDs) may not have embedded activity or transaction logging. However, these devices communicate with the command and control systems within the ICS itself. These communications produce activity and transaction logs tied back to the field devices. The examiner should be certain to seek out and acquire any data that may have been captured by the activity and logging capabilities tied to a specific field device.

Continuous Processing

- Many ICS processes are continuous in nature. Unexpected outages of systems that control industrial processes are not acceptable. Control systems often cannot be easily stopped and started without affecting production.
- The use of typical forensic strategies are not acceptable solutions due to the adverse impact such practices would have on the ICS requirements for high availability, reliability, up-time, and maintainability. Indiscriminate use of cyber forensic practices in an ICS environment may cause availability and timing disruptions, which in most ICSs is unacceptable.

- Software updates on ICS cannot always be implemented on a timely basis, therefore potentially contributing to system incidents. The examiner must be aware of the version of software currently in production as this may affect the examination approach or methodology.
- Actively scanning enterprise ICSs is not as likely a consideration as it would be in a traditional enterprise IT environment. The forensic processes of actively scanning the ICSs, would raise possible safety concerns such as the introduction of additional traffic into the ICS control environment. Such actions by the examiner could interrupt critical ICS processes or operations.
- Globally, in most ICS environments, the continuation of process operations is the prime directive. As such, the immediate replacement of failed devices and subsequent rebooting procedures may contribute to the inaccessibility to these failed devices and the overwriting of possible evidential data. Either case may hamper or impede a complete forensic examination of an ICS event.
- Regardless of the age, design, or application of the ICS, the nature of the processes dictates specific protocols addressing data exchange as well as rates of data collection, transfer and the real-time overwriting of critical operating data (e.g., running processes, current connection statuses, and memory content). Such protocols must be understood by the examiner, as they will have a direct impact on the integrity and usability of any data, which may be collected for forensic analysis.

Emerging Technology

- Connection of ICSs to the Internet in order to improve performance and effectiveness have exposed these once 'closed systems,' to various threats found on the Internet. As such, ICS devices may be subject to external, unauthorized access, which can result in the attempted destruction of evidential data.
- The initial security features (such as they were) of legacy systems were designed for a disconnected infrastructure. The evolution of IIoT (Industrial Internet of Things) and Industry 4.0 place these systems at considerable risk. Subsequently, data processed or stored on these legacy ICS-connected devices may be subject to unauthorized deletion, overwrite, and destruction. Such actions could limit the amount of viable data available for forensic examination.

While not the primary subject of this text and certainly a much broader and complex topic than can be appropriately addressed here, a very brief mention of Industry 4.0 and ASCPMM is warranted. Cyber forensic investigations involving Industry 4.0/ASCPMM processes and environments, will dominate the coming decades.

- Industry 4.0
- A term coined in Germany, popularly used in Europe, and equivalent to smart manufacturing. Advanced Sensors, Control, Platforms, and Modeling for Manufacturing (ASCPMM) also known as Smart Manufacturing, represents an emerging opportunity faced broadly by the U.S. manufacturing sector. ASCPMM encompasses machine-to-plant-to-enterprise to-supply-chain aspects of sensing, instrumentation, monitoring, control, and optimization as well as hardware and software platforms for industrial automation. Advanced sensors, processors, and communication networks are used to improve manufacturing efficiency through the real-time management of energy, productivity, and costs at the level of the factory and enterprise.[48]

Smart Manufacturing, aims to reduce manufacturing costs from the perspective of real-time energy management, energy productivity, and process energy efficiency. Initiatives will create a networked data driven process platform that combines innovative modeling and simulation and advanced sensing and control. Thus, integrating efficiency intelligence in real-time across an entire production operation with primary emphasis on minimizing energy and material use; particularly relevant for energy-intensive manufacturing sectors.

Smart manufacturing is related to intelligent efficiency, as they both use Information Communication Technology (ICT) to achieve efficiency goals. Intelligent efficiency is energy efficiency achieved through sensor, control, and communication technologies, while smart manufacturing has a larger enterprise efficiency purpose with energy efficiency being a co-benefit to the improvements.

Digital Manufacturing, has as an objective, to improve product design and manufacturing processes across the board seamless integration of information technology systems across the supply chain. Digital manufacturing focuses on reducing the time and cost of manufacturing by integrating and using data from design, production, and product use; digitizing manufacturing operations to improve product, process, and enterprise performance, and tools for modeling and advanced analytics, throughout the product life cycle.

Existing Technology

- Field devices do not employ any logging mechanism. Information regarding the network communication between the field devices and the rest of the system along with activities of the field devices can be found, however, in the control center part of the SCADA system.
- The traditional configuration of many device and ICS technologies does not provide for the collection of data that could be used forensically, in the investigation of an ICS-related incident.
- The examiner should note that in some operational technology sectors (electric power for example) oversight organizations (e.g., The North American Electric Reliability Corporation [NERC]) has requirements, which state that the organization must preserve and retain and not discard or destroy any and all data or documentation pertaining to an event. The Code of Federal Regulations, 18 CFR 125.3, provides a schedule of record retention requirements for public utilities and licensees under the Federal Power Act, and includes numerous record retention requirements that exceed 3 years, including a requirement that hydro-electric plant owners keep operations and maintenance records for 25 years. See NERC Critical Infrastructure Protection (CIP) standards, which make up nearly 40 rules and almost 100 sub-requirements. The cyber forensic examiner may want to review specifically, NERC Cyber Security – Incident Reporting and Response Planning, CIP standard, CIP-008-65.

Data, which due to the very nature of the ICS process, may be quickly overwritten and therefore unstable, may become persistent, in compliance with policy or regulation, in some other data store in the system.

While these data may ultimately not prove useful in the forensic examination, the examiner should know that such retention of data requirements do exist. The examiner should therefore seek out all possible, potential retained data stores for potential examination consideration.

- ICSs are dependent upon the individual operating systems that controls them. Software changes and updates may affect the system's audit and logging tools and the correct processing reports. If so, potential evidential data may either be lost or corrupted. Data which does remain, may be suspect and deemed unusable for forensic examination purposes.
- Changes to file structures within control systems, to accommodate legacy ICSs, may result in difficulties associated with data analysis. Specifically, difficulties related to data security and authentication processes, which would be used to assure the data's integrity for forensic purposes.

Forensic Processes

- Given the very nature of ICSs, online, real-time, up-time criticality, the introduction or use of certain forensic tools used for active analysis (e.g., port scanning, and opening TCP connections) may be counterproductive and actually lead to system-wide disruption, outage, or even failure.
  The examiner should proceed with extreme caution (and seriously consider avoiding altogether) the use of an active data acquisition techniques when examining an ICS.
- The combination of legacy hardware and software configured first for operational efficiency and performance then (if at all) security, the ability to correlate and cross-reference forensic data collected from firewalls, IDS, intrusion prevention systems (IPS), etc., to individual ICS devices or system logging data may be impossible.
- Post-incident, forensic analysis, carried out by the cyber forensic examiner, is often dependent on vendor involvement. The availability, timing and commitment of the vendor to assist in the examination process by providing hardware and/or software expertise, will have a direct impact on the successful outcome of the examination.
- The data within the process and state information is deleted, removed, or overwritten at a rate that makes collection on some devices unviable or impossible (volatility of data). Within some safety systems, high-speed data recorders have been used for many years. This type of recording activity will maintain the data that is often overwritten within system components and can be used for analysis and event recreation. Although after-market solutions can be architected, they are often too cost prohibitive to implement.[49]
- It is strongly recommended, due to the volatility of ICSs, that any standard forensic tool is first operated in a test environment. It should be determined in this test environment what, if any, impact the forensic tool will have on the production environment. If such an impact cannot be identified and determined in this test environment, the forensic tool should not be used in the live production environment.
- In many ICSs, data transferring across a control system environment occurs at a very rapid pace. Individual field device information is collected and overwritten multiple of times throughout a single cycle. The ability to identify data associated with a specific incident from unrelated data, can create a problem for the examiner. This co-mingling of data must be addressed as part of the examiner's pre-examination process.

Institutional Knowledge

- Due to the nature of the ICS environment and associated components, the examiner may have to reach out to the system administrator, in order to gather knowledge on the components (e.g., HMIs, data historian, and PLC), and how they function.

- As the age of the system increases, it becomes more probable that the original vendor responsible for the development of the technology is either no longer in business, the contracts have expired, or there is simply no information about the device available. This drives demand for 'community-level' support, and as such, peer networks can become one of the few remaining support mechanisms.[50]
- The examiner may be required to possess an in-depth knowledge of the various field devices associated with the ICS, that require examination. If not possessing this knowledge, the examiner may be required to contact the system administrator and/or the equipment vendor. Depending on the nature of the incident and investigation, contacting and speaking with a system administrator (internal employee) may impede or violate examination protocol.
- When preparing to examine a modern (meaning not legacy) but, proprietary control system, interaction between the examiner and the vendor, prior to the investigation, will be required. This is to assure that the examiner has a complete working knowledge of the system(s) that will be examined and the sources of potential digital, evidential data within those systems.

Legacy Systems

- Countless ICSs employ legacy systems that are often lacking resources common on modern IT systems. Many ICSs may not have desired features including encryption capabilities, error logging, and password protection. As such, it may be difficult to accurately determine exactly who would have had access to or actually accessed components within the ICS.
- A significant number of ICSs utilize older versions of operating systems (OS) that are no longer supported by the vendor (e.g., Windows 98, 2000, XP, Vista, NP, and WindowsServer2008). Consequently, access to current information regarding the operation or function of the OS may no longer be available to the examiner.
- ICS have life expectancies that are longer than typical IT products. As such, the examiner may encounter ICS environments running hardware no longer supported by the vendor (e.g., Aydin 5215 and 5217 display generators, and Foxboro MRD 450/460 Video Controller).
- With a typical deployed lifetime extended between 10 and 15 years and often longer, ICS technology may not be compatible with the forensic tools currently used by an examiner requiring access to or connecting with, the ICS or its component pieces. According to a 2008 research study conducted by Abdul Azeez Erumban, the lifespan of industrial machinery is slightly more than 25 years.[51]
- Legacy equipment or conventional networked control devices do not retain network traffic that could provide valuable evidence in an investigation.
- Legacy HMIs run mostly on proprietary systems or operating systems that are no longer supported by the original vendor.
- Legacy field devices communicate through serial connections. This will make it impossible to capture network traffic.
- The examiner may need to request assistance from both the vendor and an experienced engineer, when examining legacy systems and proprietary field devices. The engineer should be assigned to support the examiner throughout the examination.
- Often times SCADA systems use operating systems and hardware that are not in wide usage. Due to this fact, it is important that the assessment team have access to some of these less-common platforms so that they can build their own testing environment for use during an assessment. An example might be an Alpha Server running Tru64 UNIX.[52]

- The total information sample that is resident in the investigated system is comprised of both data related to incident activity (possibly malicious) and data that is unrelated to the incident. Moreover, due to the limited memory storage of the system these data types are often indistinguishable (Data Mingling). Although it is not a unique problem to control systems, this can be attributed to inadequate labeling and is in itself a function of the control systems (vendor supplied) technology.[53]
- Legacy or systems which have reached their 'end-of-life' (no longer have vendor support), may not have the capability for detailed logging of system activity. The collection of incident-data, sufficient for a forensic examination, may be difficult or impossible to obtain.
- ICSs aging greater than 20 years present additional concerns for the examiner. Many of the field devices within the ICS are no longer manufactured. Obtaining working models of system devices for tear-down and examination purposes, in an effort to learn processing operations and to identify potential data stores, may no longer be possible.

Operations Technology Personnel

- On-site operational technology personnel may have a detailed understanding of incident response procedures; however, they may not a sense of what impact such response actions might have to real-time or latent evidential data or potential evidence sources
- As a result of a response to an incident, procedures may not exist that specify direct actions to take to preserve potential digital evidence.
- A cultural barrier within the end user community, created by a focus on the requirement for operational uptime and continuance, may override the examiner's requirement for a longer-term retention of evidence. Systems re-starts, re-boots, re-calibrations, etc., required to begin operations once an incident has passed, may contribute directly to the alteration or destruction of data, potential evidential data.

Physical ICS Design

- Control systems have been traditionally operated as isolated systems with no network connection to the rest of the world.
- Unless combined under similar operating systems most ICSs components are not capable of reading removable media such as CDs, DVDs, or USB drives, thus the examiner should not expect to find these media as possible sources of evidence.
- Unlike their IT counterparts, less specifically designed, most PLCs and RTUs will not be configured with email clients or web browsers. The examiner should verify the design configuration of all ICS components to be examined. Verifying as well any configuration change deviating from the original manufacture's design.
- The field devices that are used within ICS architectures, often have no inherent capability for detailed logging.
- On devices where, extensive logging is supported the feature is often disabled, or the devices lack sufficient capacity to store enough data. Such configuration may prohibit the device from storing or retaining sufficient data, at a quantity deemed appropriate for forensic requirements.
- The key information resources in the control systems domain will be created and deployed to handle data in such a way that the implementation of a data retention scheme is neither cost effective nor a requirement.[54]

Physical Location

- Some of the ICS components, which the examiner may desire to access may not be physically located in a centralized control room/facility.
- Distributed ICS components may be unmanned, isolated, remote substations, and require extensive transportation effort to reach.

Presiding Law and Cyber Forensic Procedures

- Evidence collection and analysis must conform to presiding County, City, State, Federal, Provincial, Country regulations, for the collection and preservation of digital evidence to assure its integrity and chain of custody, so that any findings may be used in a court of law.
- Given the very nature of ICSs many of which operate across multiple operation sites and in many cases, across multiple State, Federal, National boundaries, an incident warranting a cyber forensic examination may require the involvement of multiple agencies, legal jurisdictions, and potentially governments. Access to, collection, and eventual examination of potential evidential data, may be hampered by complicated legal systems.

Proprietary Systems and Technology

- Communication protocols and media used by ICS environments for field device control and intra-processor communication are typically different from most IT environments, and may be proprietary thus, inhibiting direct access by the examiner.
- The ability to understand device or operational log data is often a vendor-only skill. Performance of the examination may be delayed until the vendor is able to assist the examiner or until the examiner obtains a degree of proficiency and a greater degree of working knowledge regarding the ICS device or log to be examined.
- It is imperative that the examiner obtain personally (or in cooperation with the vendor) a detailed understanding of how any proprietary technology works in the operational environment. Failure to do so may affect the examiner's ability to identify, analyze, and correctly interpret any data extracted from devices or logs.
- The examiner may encounter proprietary, legacy systems where the vendor no longer retains any knowledge of the system's operation. This may be due to the age and obsolescence of the system, retirement of knowledgeable personnel and/or lack of complete and updated documentation, adequately describing the system's operations.

Security and Control

- Legacy devices that are decades old, with limited computational resources and communications bandwidth to support cybersecurity protections.
- Control and protection devices are widely distributed; some are in unmanned, remote substations or on top of poles in publicly accessible areas.
- SCADA and industrial protocols, such as Modbus/TCP, EtherNet/IP, IEC 61850, ICCP, and DNP3, are critical for communications to most control devices. Unfortunately, many of these protocols were designed without security built in and do not typically require any authentication to remotely execute commands on a control device. As such, it may be difficult for the examiner to accurately determine, who exactly acquired remote access and the opportunity to execute commands on a control device.
- SCADA systems can be part of a national critical infrastructure.
- SCADA systems are not traditionally designed to respond to security incidents and do not have automated capability to analyze and learn from what happened.

- Prior risk analysis may be lacking. The identification of critical system data may not exist or may not be reflective of current operations.
- The exact location, within the ICS and within specific field devices, where evidential data may be found, may not be documented nor known, by the system administrators.
- Embedded security features typically found in IT systems are by design, non-existent in ICSs. Access to log files (e.g., data historians) may be inadequate to prevent unauthorized access, modification, or destruction of data recorded in these devices. A general lack of security within and over ICS data recording devices may severely impact the availability, integrity, and usability of potential evidential data.
- Often times the system owner or operator will not have the requisite skill set to identify data, resulting from an incident impacting the ICS, which may have evidential value. Their primary focus is on keeping the system up, operational and running. An unchecked, immediate response to an incident, which may be appropriate, critical and required, may lead to the unintentional destruction of potentially valuable evidential data.
- According to the 2018 Fortinet report, 'Independent Study Pinpoints Significant SCADA/ICS Security Risks,' 45 percent of respondents do not use Privileged Identity Management (PIM) for Administrators, which allows organizations to monitor high-level accounts in their IT environments. Another 45 percent do not use role-based access control for employees. In the event of an incident investigation, such security failures could lead to potentially discrediting of calling into question the integrity of potentially valuable and useful evidential data.[55]
- Multiple and/or consistent ICS device failures and reboots could potentially identify deeper system problems. The examiner should be aware of such conditions and what effect these events may have on the integrity and availability of sound forensic data.

Special Considerations

- The criticality of the HMI in the control systems environment cannot be overstated. As the primary point for all command and control activity within the control systems environment, the HMI will demand special attention in a forensic investigation.
  Of concern to the investigator is the possibility that the version of the HMI software may have required initial hardening of the operating system (kernel) or use a standard 'build' that removed nonessential services and/or files from the base operating system. To that end, some of the more common features and capabilities associated with transaction monitoring, alarm and event logging, or diagnostics may be modified or absent all together. Although the core drives and resident data could be harvested for offline investigative analysis, key data stores and file structures may be so different that a vital evidence collection may be impossible. Furthermore, without in-depth understanding of how the HMI is executing the command and control function in the environment; the investigator may be unable to locate pertinent evidence that is in the HMI data stores.[56]

THE FORENSIC PROCESS

When collecting evidence, which will be used in a forensic examination, the examiner must be certain that all digital evidence meets the CARAB principles. Digital forensic evidence must be:

> Complete – gathered from all possible sources that could retain digital data, both volatile and non-volatile.

Authentic – digital forensic evidence must be traceable to its source and the examiner must be able to unequivocally connect each piece of evidence to the event under examination.

Reliable – following accepted protocols for establishing chain of custody and the security of digital evidence, there must be assurance that the data's integrity has been maintained at all times, throughout the examination.

Admissible – digital evidence acquisition methods must meet the statue of law for the collection, handling, storing, transferring, analyzing, retention, and disposition of said digital evidence.

Believable – digital evidence resulting from forensic analysis should be presented in such a manner that there is a clear and understandable linkage between the forensic evidence and the event investigated. That the evidence will be clearly understood by those who will be tasked with interpreting the evidence and taking any further action, based upon said evidence.

Collecting evidence is of utmost importance to the digital process. Collection and examination of digital evidence should proceed from the volatile to the less volatile. In order of volatility for a typical system and the preservation of digital evidence, collection should begin in this order when at all possible and feasible:

- Registers, cache
- Routing table, Address Resolution Protocol (ARP) cache, process table, kernel statistics, memory
- Temporary file systems
- Disk
- Remote logging and monitoring data that is relevant to the system in question
- Physical configuration, network topology
- Archival media[57]

IDENTIFYING POTENTIAL SOURCES OF DIGITAL EVIDENCE WITHIN AN ICS

Some SCADA systems (e.g., electricity, hydro, water, and fuel) monitor and control distribution by collecting data from and issuing commands to, geographically remote field control stations via a centralized command control location.

Sensors, actuators, and controllers (e.g., PLCs) are used to manipulate some controlled process. Data, potentially useful to an examination, may be found on these component pieces. In reality, data may be found on any component device that stores or transmits/transfers data to a centralized location within the ICS. These data are all potential examination sources.

ICS typically have a wide-range of time-sensitive operating requirements, unable to be accurately performed by human operators. To assure operational compliance, meet time-sensitivity requirements and reduce communication latency, certain ICS computations may need to be performed as close to the sensor or actuators as possible. Data recorded by both the sensors and actuators are viable candidates for forensic examination.

Due to the very nature of ICSs, real-time operations are the standard processing environment. Stopping a process or taking a system off-line to perform a cyber forensic examination (dead forensics) may be both logically and physically impossible, without incurring or creating a destructive event. Therefore, in most examination cases involving ICSs, the examiner should be prepared to perform a live forensic analysis on ICSs to be examined.

Prior to discussing possible sources of evidential data to be found within the ICS, it is best to first present a very brief discussion regarding establishing a temporal reference of the ICS-related event to be examined/investigated. This temporal reference is in establishing a valid timeline of events and the validity of all-time references associated with any collected data via the ICS's reference clock.

Reference clock system

A reference clock source that relays UTC (Coordinated Universal Time) and has little or no delay is known as a stratum-0 device. Stratum-0 servers cannot be used on the network, instead, they are directly connected to computers which then operate as primary time servers.[58]

As in any forensic investigation, an analyst must be able to establish a context of time when evaluating collected data. Unlike transactions in the corporate or modern business environment, activity and transactions within control systems environments are often required to occur in milliseconds. Combined with extensive use of volatile memory and small storage capacity, investigators looking to align incidents and consequences effectively within a control systems environment will need a very specific clock reference.

Prior to the investigation of event data or the collection of any digital forensic evidence, the investigator is advised to obtain a reference clock or timing source within the control systems domain. Fortunately, due to the way that many modern (and even some older) control systems environments are established, synchronized timing within the operations is normally addressed. Thus, the investigator may be fortunate and have access to an already pre-existing, functionally centralized time function synchronized to all elements in the control systems domain. 'Centralized time function' refers to a master time source in the system being investigated, not a centralized system in terms of geography. Also, investigators should be aware that the timing mechanism for the control domain may itself have been impacted by the cyber incident and thus should be deemed unreliable.

In addition, to compensate for the possibility of there being multiple centralized clock mechanisms for each of the control systems (and IT functions within a control systems domain), the forensic investigator is strongly advised to ascertain if more than one clocks exist. If so, it is imperative to determine if theses clocks are synchronized and which Network Time Protocol (NTP) server each system is statically set to resolve to.[59]

The following will discuss potential sources of digital evidence within the ICS, which the examiner may be able to collect in the support an investigation of an ICS-related event.

Data historian

Records all the data from the production and SCADA networks and allows exporting to the corporate IS (to the ERP for instance).

The data historian is responsible for storing and logging all of the data that the SCADA system aggregates. Through the generation of either manual or automated reports, historian data will show what happened, across the systems it monitored, over specified periods of time.

The examiner should ascertain whether there has been any external third-party or proprietary modification made to the data historian. Such modification to historian software may be made for very legitimate purposes, (e.g., provide for a more accurate and timelier picture of current production status or historical trends). However, such modifications could contribute to the potential corruption of evidence during the device's collection process.

Depending on the ICS (e.g., legacy), security controls designed to mitigate unauthorized access and modification of historian data may not exist or if they do, may not be functioning. Should the examiner find this to be the case, the examiner must take extra steps to validate the existence and proper functioning of controls are in place, to assure the integrity and accuracy of collected data, prior to those data being analyzed and relied upon as evidence.

Specific data recorded by the historian, useful to the examiner, would include but, not be limited to:

- Aggregate data
- Alarm and Event (A&E) data
- Analog readings
- Digital readings
- Client data
- DBMS logs
- Historical and real-time data
- Industrial time-series data
- Product Info (e.g., product/batch/material ID, raw material lot ID)
- Quality Info

Engineering workstations

The engineering workstation is usually a high-end very reliable computing platform designed for configuration, maintenance, and diagnostics of the control system applications and other control system equipment. The system is usually made up of redundant hard disk drives, high-speed network interface, reliable CPUs, performance graphics hardware, and applications that provide configuration and monitoring tools to perform control system application development, compilation, and distribution of system modifications.[60]

Data, which could be essential digital evidence found on ICS engineering workstations include but, are not limited to:

- Account usage
- Attached devices
- Browser usage
- Connected devices
- Logs
- PLC/HMI baseline images
- Program/File execution
- RAM data

Additional information, which may be retrieved from engineering workstations, as noted by Ramesh Singh, writing on 'Pipeline System Communication,' and a valuable source of potential evidential data include:

- Alarm and event management including alarm acknowledgment
- Asset manager
- Calibration and tuning displays
- Commands and controls to change the operating state of the system facilities such as opening or closing of valves
- Communication error displays

- Diagnostics of the system up to card level and instrumentation including system-malfunction indications
- Displays related to corporate geographical information system
- Graphical displays showing the process conditions of the system
- Intelligent cause and effect display for the station logic with 'what-if' analysis
- Production Accounting Solution (PAS) displays (PAS measurement points provide a means for recording product volume measurements and are used to track all activities related to product production)
- Process trends and analysis displays
- Programming and system-level access to the servers
- Report view and print
- Safe start up and shut down guidelines
- Safety Instrumented System (SIS) displays (A system that is composed of sensors, logic solvers, and final control elements whose purpose is to take the process to a safe state when predetermined conditions are violated)[61]
- Summaries and reports
- System maintenance and configuration changes
- Trends of selected process variables[62]

Field devices

Equipment that is connected to the field side on an ICS. Types of field devices include RTUs, PLCs, actuators, sensors, HMIs, and associated communications.

Evidential data, available from the various field devices connected throughout the ICS will vary depending on the nature of the ICS. Is the system proprietary? If so access to the data may require direct interaction with the vendor. Any changes that the vendor may have made to the device's configuration may have an impact on both the acquisition and retention of data and its integrity. Is it a legacy system? In this case, the devices may never have been originally configured to record or retain data passed between devices or have been configured to maximize production operability and not with a preservation of data in mind.

If the examiner is reviewing a proprietary ICS, it is strongly recommended that the examiner inquire as to any security-specific modifications that may have been made to the ICS that would either (a) contribute to the loss of potential evidence or (b) collect and retain potential digital evidence in other areas throughout the ICS, which the system owner may be unaware of.

Data, which could be essential digital evidence found on ICS field devices include but are not limited to:

- Applications/processes which: (a) may be currently running or (b) were running prior to ICS interruption or stoppage
- Device configurations
- Error codes
- Event date
- Event time
- Firmware installed on the ICS
- Network connections
- Open ports
- Processing IDs

Log files are tremendously important as a source of potential digital evidence for the examiner. The absence of such logs will require the examiner to identify ICS-related events and then attempt to correlate those events to a timeline of activities which took place prior to and proceeding the ICS-related event.

Human–Machine Interface (HMI)

The hardware or software through which an operator interacts with a controller. An HMI can range from a physical control panel with buttons and indicator lights to an industrial PC with a color graphics display running dedicated HMI software.

Any modifications made to the HMI that are designed to improve production monitoring and/or processing may contribute to the loss of potential evidence. Operations technologies such as automatic restart procedures, real-time allocation of memory, automated job-termination, fault tolerance configuration, data overwrite capabilities, can contribute to the loss of potentially valuable digital evidence.

Data, which could be essential digital evidence found on the HMI include but are not limited to:

- Machine hours of operation
- HMI display usage
- Operating temperatures including extreme conditions
- CPU utilization
- Storage usage
- Movement of data off local machines to central or remote repositories
- Operator interactions
- How long it takes to acknowledge an alarm
- Error codes with descriptions of the possible causes and remedies
- Log data regarding the causes of downtime or motion faults over time
- In-depth information about alarms
- Historical data
- Fault-state alarming

When it comes to any evidential data, which the examiner may obtain from the HMI, the examiner must verify and validate that appropriate internal controls are both in place and function, so that the HMI data cannot be accessed or modified. Any unauthorized access to the HMI data by a user should also be recorded. If these controls are not functioning or are non-existent, the examiner must question both the validity and integrity of the HMI data acquired.

Programmable Logic Controller (PLC) and Remote Terminal Unit (RTU)

A PLC is a solid-state control system that has a user-programmable memory for storing instructions for the purpose of implementing specific functions such as I/O control, logic, timing, counting, three mode (PID) control, communication, arithmetic, and data and file processing. As a small industrial computer, it is designed to perform the logic functions executed by electrical hardware (relays, switches, and mechanical timer/counters).

An RTU is a special purpose data acquisition and control unit designed to support DCS and SCADA remote stations. RTUs are field devices often equipped with network capabilities, which can include wired and wireless radio interfaces to communicate to the supervisory controller. Sometimes PLCs are implemented as field devices to serve as RTUs; in this case, the PLC is often referred to as an RTU.

PLCs implemented throughout the ICS also serves as a potential source for evidential data. PLCs have a user-programmable memory for storing instructions for the purpose of implementing specific functions such as I/O control, logic, timing, counting, three mode proportional-integral-derivative (PID) control, communication, arithmetic, and data and file processing. PLCs have limited capacity for long-term data storage.

The RTU function as a microprocessor-based device is connected to sensors, transmitters or process equipment for the purpose of remote telemetry and control. With the aid of appropriate sensors, the RTU monitors production processes at remote site(s) and transmits all data to a central station where it is collated and monitored. The RTU architecture comprises of a CPU, volatile memory, and nonvolatile memory for processing and storing programs and data. It communicates with other devices via either serial ports or an onboard modem with I/O interfaces.[63]

Specific data recorded by the PLC and RTU, useful to the examiner, would include but, not be limited to:

- Active processes
- Actuator data
- Data from resident user or application program written in relay-ladder logic or any other PLC-programming language
- Firmware versions
- I/O (input commands)/(output instructions)
- Ladder-logic program code
- Logs
- Program codes
- Runtime (or state) data (e.g., machine fails, alarm status)
- SD card data
- Sensor data
- Telemetry, relay data from system
- Timestamps

Master Terminal Unit (MTU)

A controller that also acts as a server that hosts the control software that communicates with lower-level control devices, such as Remote Terminal Units (RTUs) and Programmable Logic Controllers (PLCs), over an ICS network. In a SCADA system, this is often called a SCADA server, MTU, or supervisory controller.

Data, which could be essential digital evidence found on the MTU include but, are not limited to:

- Equipment functionality
- I/O data
- PLC data
- Processing environmental conditions
- RAM data
- RTU data
- Network server data
- Log files

OPC server

The OPC server provides a data connectivity interface standard used to communicate between controllers, devices, applications, and other server-based systems across the ICS.

OPC stands for OLE Process Control. Open Platform Communications (OPC) is a series of standards and specifications for industrial telecommunication. OPC specifies the communication of the real-time plant data between control devices (converts the hardware communication protocol used by a PLC into the OPC protocol for example). OPC, as an interface standard, allows Windows programs to communicate with industrial hardware devices. Factory data are collected and can be saved in a database, such as Access, SQL Server, Oracle, or MySQL.

OPC also offers plug and play connectivity from proprietary devices, and acts as an interface between various data sources like PLCs and field devices, such as sensors and actuators; applications like SCADA system, or other HMIs, RTUs, and other database servers.[64]

Data available on the DBMS collected and retained by the OPC client includes but, is not limited to:

- Real-Time Data
- Historical Data
- Event Data

The examiner should be aware that ICS components such as the OPC server (and Historian) are used and programed for specific purposes. The data retrieved from these sources may be subject to external access and therefore potentially tampering. These data may not be suitable as forensically sound digital evidence. The examiner is advised to always validate both the validity and integrity of any data collected prior to considering its use as digital evidence.

Additional sources of ICS data

Data useful to an examination/investigation can be discovered by the examiner through an extensive review of the logging functions of firewalls, routers, switches, servers, and workstations, which may be found throughout the ICS.

Depending on the environment, type, age and the overall, general configuration of the ICS, additional data sources that may be available to the examiner would include but, not be limited to:

- 8-inch floppies (legacy systems!)[65]
- 3½-inch HD floppies
- 5 ¼-inch floppies
- CD-ROMs
- DVD-ROM/RW
- Jaz© disk
- Micro SD Card
- Operation-specific handheld computers
- Personal digital assistants
- Solid state drives
- USB flash drives
- Zip© disk

The examiner should also seek out any additional sources that may provide either direct digital evidence or useful information regarding the ICS's configuration, components, operating details, modifications (changes by the system owner to vendor software, enhancements beyond the base-line delivered by the vendor), operation status, etc.

Such sources may include:

- Access point logs
- ARP Tables
- Activity logs (compliance/business reporting purposes)
- Backup archives
- Error logs
- Event logs
- Firewall logs
- Hardware configuration of the control system
- Internal status base-line documentation for each connected device
- Log files from IDS/IPS (if implemented and active)
- Make, model, and serial number of each active device in the system
- Network address, network settings such as subnet mask and default gateway, MAC address, and listening ports for devices connected to the ICS
- Operating system modification logs
- Photographic record of system status lights and 'device state' indicators
- PLC ladder logic programs
- Schematics or wiring diagrams
- Serial communication logs
- Transaction logs (compliance/business reporting purposes)

ICS status issues

The status of the ICS, up and running or shut down due to an incident/event, will present its own data collection issues for the examiner.

If the ICS is fully operational, shutting it down or stopping individual processes to gather data for forensic examination will be almost impossible. The probability of causing a catastrophic impact to processes and potential human harm will warrant a live forensic examination.

If the ICS is non-operational, acquiring evidential digital data may still prove challenging, especially if critical ICS components have been rendered inoperable or completely damaged as a result of the incident.

If this is the case, acquiring any potential evidential data require the examiner to utilize alternative acquisition approaches. Alternate approaches for acquiring data from ICS components that may have been damaged include the following:

- JTAGing (Joint Test Action Group). Using a non-destructive process that connects to a specific combination of Test Access Ports (TAPs) on a device's circuit board and instructing the processor to transfer the raw data stored on connected memory chips.
- In-System Programming (ISP). Using a non-destructive process that connects to specific points on a device, bypassing the device's processor, to directly read the device's memory.
- Chip-Off. As a last resort, a destructive process that involves physically removing the memory chip(s) from a device's circuit board and reading it on an external reader.

Once dumping memory and acquiring the data via one of the above-mentioned approaches, the data can then be analyzed to identify any areas of interest, which may be related to the incident under investigation.

ICS forensic summary

Tables 6.8 and 6.9 provide a summary of the three main categories of ICSs as defined in the literature and the typical environment within these categories, which the examiner is likely to encounter.

Table 6.8 Industrial Control systems – modern through legacy [66]

ICS category	Description	Control center components	Technology	Sources of evidential data
Modern/ Common Control System	Operate under current state-of-the-art computing capabilities. Fully supported by the vendors, which provide the hardware and software.	Engineering workstations (EWS), databases historians, HMIs, field devices (DCS, PLC, IED, MTU, RTU), SCADA, control server, I/O server, fieldbus network, firewalls, communications routers and switches	Windows OS, Linux, Unix, VxWorks, INTEGRITY-RTOS, MQX, SCADA industrial systems integrated with analytics, manufacturing execution systems (MES) platforms, TCP/IP communication, IIoT, wifi, Zigbee, WirelessHART, air-gapped systems, cloud-based, Common Industrial Protocol (CIP), small-sensor protocol MQTT v5.0	Logs (from connected devices), volatile data (from memory and registers) found typically in the engineering workstations and/or HMI computers. Data may also be found on: PLCs/RTUs/MTUs, OPC Server, routers, domain controller, switches, Historian, database server. IDS/IPS, firewalls
Modern/ Proprietary Control System	Designed, created and implemented within past 10 years. Vendor provides technical support and retains proprietary knowledge of all systems functions and operations.	Engineering workstations (EWS), database servers, historians, HMIs, field devices (DCS, PLC, IED, MTU, RTU), SCADA, control server, I/O server, fieldbus network, firewalls, communications routers and switches, WWW servers	Windows OS, Linux, Unix, VxWorks, INTEGRITY-RTOS, MQX, SCADA industrial systems integrated with analytics, manufacturing execution systems (MES) platforms, ERP systems, TCP/IP, external gateways (i.e., 3G, 4G, LTE), communication, cloud-based, Common Industrial Protocol (CIP), virtualization	Logs (from connected devices), volatile data (from memory and registers) found typically in the engineering workstations and/or HMI computers Potential evidence corruption due to vendor modification of processing and collection devices designed to achieve specific operational results.
Legacy/ Proprietary Control System	Designed, created and implemented between 10 and 20 years ago. Maintain some basic-level of computing capabilities (relative to the modern systems). May or may not still have vendor support, if vendor is still in business.	Engineering workstations (EWS), databases historians, HMIs, DCS, PLC, IED, MTU, RTU, SCADA, control server, I/O server, communications routers and switches	Windows 3.11, 95, 98, 2000, XP, Vista, NP, Windows Server2008, Modbus, Profibus, VxWorks, unsupported operating systems (DOS), field devices based on serial connections (not USB), CANbus, HART, 5-1/4" floppies, small-sensor protocol MQTT, RS-422 point-to-point hardwired connections, Siemens S5, A-B SLC500, Windows CE-based HMIs	Technologies found in systems with lifespans exceeding 10–20 years may not have the capability to retain appropriate data that would be acceptable to the forensic examiner or support extensive forensic examination.

Table 6.9 Industrial Control systems – modern through legacy[67]

ICS category	Potential discoverable data	Applicable forensic tools	Incident analysis	Notes
Modern/ Common Control System	Device date and time, current active processes and current running processes, network connections, open ports	See Appendix 6.A	Yes	Field-devices typically do not have any inherent data collection capabilities (such as local logging or audit) that could be leveraged by standard forensic methodologies. Use of live forensics, if system is currently running would be required.
Modern/ Proprietary Control System	Device date and time, current active processes and current running processes, network connections, open ports. General network and activity logs. Due to proprietary nature of ICS, if data logging from engineering workstations, Historian, field-devices, etc., has been activated.	See Appendix 6.A	Possible	Due to the proprietary environment, direct contact with vendor may be required to obtain access to the ICS to run the forensic tools. Use of passive analysis is recommended due to the lack of detailed knowledge of system operations.
Legacy/ Proprietary Control System	Most likely logging systems and mechanisms are not enabled	Due to the legacy status of most systems, the use and application of current, modern cyber forensic tools may not be practical or feasible. Cybersecurity Evaluation Tool (CSET) from DHS may be tried. See Appendix 6.A for open source tools which may have some success when used on legacy systems.	Most Likely Not	Unneeded services provided by the OS, may be allowed to run by default. OS, malware protection software, or security patches are outdated. Operational documentation, schematics, ladder diagrams, etc., which may be used to gain an understanding of ICS components may no longer exist, original manufacture may no longer be in business.

ICS DIGITAL FORENSIC EXAMINATION QUESTIONS FOR MANAGEMENT

Because of the variety of data sources, cyber forensic techniques can be used for many purposes such as discussed in this chapter, the examination, and investigation of cyber-related incidents/events which effect ICSs. With the evolution of ICSs, the rapid growth of IIoT and Factory 4.0 and the movement away from air-gapped, stand-alone ICSs to open Internet, cloud-based functionality, organizations and agencies responsible for the continued, secure operation of ICSs, will eventually require the capability to perform digital forensic examinations of these ICSs.

The following is a series of questions to assist the reader, the examiner, the investigator, management...in assessing the ICS environment in which a digital forensic examination is to be undertaken.

This list of questions is by far not exhaustive. These questions are included here as a beginning, upon which to build a more comprehensive assessment tool for both pre- and post-forensic examination of operational technology and an ICS.

Administrative

1. Are all employees, including employees who work with ICS/SCADA systems, provided security awareness training? How is this evidenced?
2. Are ICS services prioritized based on analysis of the potential impact if the services are disrupted?
3. Are these policies reviewed periodically to reflect the current threat environment, system functionality, and required level of security?
4. Does a system of internal controls and risk management procedures exist, which will enable management to identify residual risk and acceptance of that risk by management?
5. Does the organization have a plan for managing incidents?
6. Does the organization have a security policy that also applies to the ICS/SCADA systems?
7. Due to rapidly changing technology and the emergence of new threats on a daily basis, does the facility perform ongoing risk assessments?
8. Has management required all vendors to disclose any backdoors or vendor interfaces to your SCADA systems?
9. Have specific security policies been developed for the control system network and its individual components?
10. Is the incident management plan reviewed and updated?
11. Is the organization's mission, vision, values, and purpose, including the organization's place in critical infrastructure, identified, and communicated?
12. What evidence exists that senior management has expressed its commitment regarding the security of ICS/SCADA systems and acts accordingly?

Internal Controls

1. Are events analyzed to determine if they are related to other events?
2. Are events detected and reported (to include cybersecurity events related to personnel activity, network activity, the physical environment, and information)?

3. Are ongoing technical audits of SCADA devices and networks performed to assure security effectiveness?

4. Are service continuity plans developed and documented for assets required for delivery of the critical service?

5. Are the associations between assets and the critical service they support documented?

6. Are there policies and procedures for the proper labeling and handling of information assets?

7. Do system owners have a comprehensive understanding of all connections to the SCADA network, and how well these connections are protected?

8. Has a thorough risk analysis been performed to assess the risk and necessity of each connection to the SCADA network?

9. Has a vulnerability analysis and resolution strategy been developed?

10. Has each SCADA device been examined to determine whether vendor-provided security features are present? Have been disabled for ease of installation and not been reset?

11. Has strong authentication been implemented to ensure secure communications, where backdoors or vendor connections exist in SCADA systems?

12. Have all software security features been set to provide the maximum level of security?

13. Have auditing and event logs been enabled on individual ICS host-devices when possible?

14. Have protocols for communications between field devices and servers on unique or proprietary SCADA systems been identified?

15. Have proxy ARP features on routers been disabled so they can't be used by internal machines to discover routes off the ICS network without a routing table?

16. Have the following types of SCADA connections been evaluated:
    - Cloud?
    - Connections to business partners, vendors, or regulatory agencies?
    - IIoT?
    - Internal local area and WANs, including business networks?
    - Modem or dial-up connections?
    - The Internet?
    - Wireless network devices, including satellite uplinks?

17. How has the risk to the ICS, from a malicious insider fully evaluated, given that this represents one of the greatest threats to an organization?

18. Is event data logged in an incident knowledgebase or similar mechanism?

19. Is the performance of controls assessed on a scheduled basis to verify they continue to meet control objectives?

20. To assure the highest degree of security of SCADA systems, has the SCADA network been isolated from other network connections to as great a degree as possible?

21. What controls been implemented to protect communication and control networks?

22. What controls been implemented to protect data-at-rest

23. What controls been implemented to protect data-in-transit?

24. What controls exist that confirm that control systems are not using default passwords on your equipment, especially network devices such as industrial Ethernet switches, routers, wireless access points, or cellular routers?

25. When was the last audit of the SCADA system configuration, against any existing change documentation, to assure that the configurations are set correctly? What evidence exists of this audit review?
26. When was the last time that a physical security survey and inventory of access points at each facility that has a connection to the SCADA system especially, unmanned or un-guarded remote sites, performed?

Operational

1. Are access control lists enabled that can pre-register device IP or MAC addresses on the industrial network devices? Are only devices that match these access control rules allowed to use the network?
2. Are the assets that directly support the critical service inventoried (technology includes hardware, software, and external information systems)?
3. Have all unnecessary, default routes that lead back to the firewall and then to other networks, been removed? *If hosts do not need to communicate with other networks, removing the routing information will not disrupt functionality and prevents the machine from calling home if it is successfully compromised.*
4. Have router and firewall ACL rules been mirrored to reduce the chance of misconfiguration and help control versions of rules deployed?
5. How can the SCADA system operator clean the system after an infection, and reliably bring it back into a known good state, without having to shut down the complete system?
6. Is capacity management and planning performed for assets?
7. What evidence is there that security requirements are set which cover the total cycle of development, purchase, management, maintenance, and replacement of ICS/SCADA systems (hardware and software) and applying these requirements is assured?
8. When was the last time a baseline risk analysis, based on a current threat assessment used for developing a network protection strategy, conducted?

Procedural

1. Are facilities prioritized based on potential impact to the critical service, to identify those that should be the focus of protection and sustainment activities?
2. Have all data being passed from the control system to the business network been encrypted to prevent attackers from accessing and manipulating traffic between the two networks?
3. Have requirements (rules, laws, regulations, policies, etc.) for identifying event evidence for forensic purposes been identified?
4. Is there a link between the incident management process and other related processes (problem management, risk management, change management, etc.)?
5. Is there a process to ensure event evidence is handled as required by law or other obligations?
6. Whenever a change is made to the SCADA system, do procedures exist, which assure that documentation is prepared with information regarding who made the change, what change was made, when the change was made and why?

Technical

1. Are all authentication servers and access servers placed in a DMZ?
2. Are procedures in place that prevent the sharing of Active Directory, RSA ACE servers, or other trust stores between corporate and ICS networks?
3. Are separate credentials required for corporate and control network zones and are these stored in separate trust stores?
4. Are underlying causes for vulnerabilities identified (through root-cause analysis or other means) and addressed?
5. Does a policy for the use of (removable) media (such as USB sticks, hard disks, and CD-ROMs) on control systems devices (where they exist) in place? What technical measures have been taken to enforce this policy.
6. Does the facility operate a 'Red Team' whose objective is to identify potential control system and physical site attack scenarios and evaluate potential system and site vulnerabilities?
7. Has identification of control systems that serve critical functions or contain sensitive information, which require additional levels of protection been performed?
8. Has outbound traffic from the control system network been appropriately restricted? *Modern attackers utilize client-side attacks by piggybacking on existing communications. They then make the trusted device on the network call back out through the firewall, effectively bypassing security controls.*
9. Has penetration testing or vulnerability analysis been conducted on any connections to the SCADA network, to evaluate the protection posture associated with these pathways?
10. Have controls been implemented, incorporating network segregation where appropriate, to protect network integrity?
11. Have separate servers been created for authorized users that come from an external organization such as vendors or integrators? *This creates the opportunity to create vendor specific access levels and they can provide for control mechanisms that limit a number of different factors that range from time of day to traffic patterns.*
12. Have the network ARP tables been hard-coded to prevent ARP table poisoning? *ARP poisoning the most popular way to manipulate insecure protocols. While this technique is not feasible on a business network, the limited number of hosts on a control system network can be effectively protected this way.*
13. How are ICS networks isolated from any untrusted networks, especially the Internet?
14. How does the system owner assure proper configuration/patch management is applied to any control system application?
15. How has the system owner documented the network architecture that may require additional levels of protection?
16. If it is suspected that malware may be the root cause of the malfunction, how is the source identified?
17. In the case of an ICS incident-event, will the network or other support systems need to run in a restricted operational status until a forensic analysis is complete?
18. Intrusion Prevention Systems (IPS)/Intrusion Detection Systems (IDS) are the common means of identifying problems on a network. Have these tools been deployed across the ICS?

19. Is multi-factor authentication used where possible, across the ICS?
20. Is the concept of 'least privilege' used to configure enterprise-wide firewalls?
21. Is the SCADA network protection strategy based upon a defense-in-depth strategy?
22. Is there a standard set of tools and/or methods in use to detect malicious code, which may/could have been injected into the control systems?
23. Is there a standard set of tools and/or methods in use to monitor assets for unauthorized personnel, connections, devices, and software?
24. To mitigate the risk relating to the injection of malware into the data stream, does the enterprise environment use full tunnels to create a remote access solution where initial authentication and authorization must be to the corporate network first, then to the control system DMZ authorization server?
25. To the greatest degree possible, have unused services and network daemons remove or disable to reduce the risk of direct attack? *SCADA control servers built on commercial or open-source operating systems can be exposed to attack through default network services.*
26. What policies and procedures are in place to provide training to minimize the likelihood that organizational personnel will not inadvertently disclose sensitive information regarding SCADA system design, operations, or security controls?
27. What procedures exist for VLAN or physical segmentation within and between the business communication networks and control system networks?
28. What software has been implemented to detect malicious activity within the ICS? *Detection can be network or host based and requires regular monitoring of logs by experienced administrators.*

SUMMARY

This chapter has reviewed operational technology and the digital examination of ICSs. From an overview of the challenges of performing a forensic examination of an ICS, through a review of what types of data can the examiner expect to find, generally, across the ICS to where those data may be found.

As of this writing these is no standard, formally accredited or accepted approach, methodology or process that must be followed in the performance of a digital examination of an ICS-related incident/event. Several examination frameworks were reviewed and presented in this chapter for the reader's evaluation. Until an accredited examination methodology is developed and approved, accepted via legal affirmation, the examiner may elect to follow one of the frameworks presented in this chapter, combine them taking the most relevant steps from each or develop a new approach.

Inevitability, the examiner will customize the ICS examination approach based on multiple factors including the nature of the incident/event, the current operational status (fully online and running or shut-down and off-line) of the ICS, the age of the ICS (modern or legacy), and prevailing law.

As newer technologies are integrated into aging systems and ICSs continue to evolve and have a direct impact increasingly broader applications and functions, organizations and agencies alike will require a cyber forensic capability to investigate ICS-related incidents. This cyber forensic capability will be a front-line defense in keeping the nation and the world's, critical infrastructure running.

APPENDIX 6.A: CYBER FORENSIC TOOLS USEFUL IN EXAMINING INDUSTRIAL CONTROL SYSTEMS

Forensic Tool	Tool URL
AlienVault OSSIM	https://cybersecurity.att.com/products/ossim
Cell Seizure, Version 2.0.0.26685	www.paraben-forensics.com
Control Systems Cyber Security Self-Assessment Tool (CS^{2}SAT)	https://us-cert.cisa.gov/ics/Assessments
Cyber Integrity	https://cyber.pas.com/
CyberLens	www.dragos.com
dd version 1.3.4-1	http://dcfldd.sourceforge.net/
DirTools, version 0.1.2	www.ossir.org
Dumpit	www.comae.com/dumpit/
EnCase Forensic 20.2	www.guidancesoftware.com/encase-forensic
Ethereal	https://ethereal.en.softonic.com/
ETTERCAP	https://www.ettercap-project.org/
Forensic Toolkit (FTK)	https://accessdata.com/products-services/forensic-toolkit-ftk
Guardian	www.nozominetworks.com///products/guardian/
Harris STAT® Analyzer ver 3+	www.harris.com
Helix3 Pro	www.e-fense.com/helix3pro.php
IDA PRO 7.5	www.hex-rays.com/products/ida/
IEHistoryView v1.70	www.nirsoft.net/utils/iehv.html
Industrial Defender Automation Systems Manager	https://download.schneider-electric.com/files?p_enDocType=Brochure&p_File_Name=Industrial_defender_2013.pdf&p_Doc_Ref=Industrial_Def_ASM
Linux Memory Extractor (LiME)	www.digitalforensics.com/blog/linux-memory-forensic-acquisition/
Memoryze	www.fireeye.com/services/freeware/memoryze.html
Metasploit	www.metasploit.com
NESSUS	www.tenable.com/products/nessus/nessus-professional
Netcat 0.7.1	http://netcat.sourceforge.net/
NetworkMiner	www.netresec.com/?page=NetworkMiner
NIKSUN NetDetector	www.niksun.com
NMAP	https://nmap.org/
Process Explorer v16.32	https://docs.microsoft.com/en-us/sysinternals/downloads/process-explorer
Redline	www.fireeye.com/services/freeware/redline.html
SNORT	www.snort.org
Volatility Framework	www.volatilityfoundation.org
WinHex	www.x-ways.net/winhex/
Wireshark	www.wireshark.org

ACRONYMS

ACL   Access Control List
AES   Advanced Encryption Standard
AGA   American Gas Association
ANSI   American National Standards Institute

APGA	American Public Gas Association
API	American Petroleum Institute
BCP	Business Continuity Plan
CERT	Computer Emergency Response Team
CIKR	Critical Infrastructure and Key Resources
CIP	Critical Infrastructure Protection
CSSC	Control System Security Center
CVE	Common Vulnerabilities and Exposures
DCS	Distributed Control System
DHS	U.S. Department of Homeland Security
DMZ	Demilitarized Zone
DoS	Denial of Service
DOT	U.S. Department of Transportation
EMS	Energy Management System
FAT	Factory Acceptance Test
FIPS	Federal Information Processing Standards
HIDS	Host Intrusion Detection System
HMI	Human–Machine Interface
HVAC	Heating, Ventilation, and Air Conditioning
IACS	Industrial Automation and Control System
ICS	Industrial Control System(s)
ICS-CERT	CERT Industrial Control Systems – Cyber Emergency Response Team
IDPS	Intrusion Detection and Prevention Systems
IDS	Intrusion Detection System
IED	Intelligent Electronic Device
IPS	Intrusion Prevention System
IPSsec	Intrusion Prevention System Security
ISA	International Society of Automation
ISID	Industrial Security Incident Database
MES	Manufacturing Execution System
MCM	Manual Control Mechanism
MTU	Master Terminal Unit (also Master Telemetry Unit)
NIDS	Network Intrusion Detection System
NIST	National Institute of Standards and Technology
OLE	Object Linking and Embedding
OPC	OLE for Process Control
OS	Operating System
PCS	Process Control System
PLC	Programmable Logic Controller
RBAC	Role-Based Access Control
RFID	Radio Frequency Identification
RMF	Risk Management Framework
RTU	Remote Terminal Unit/Remote Telemetry Unit
SCADA	Supervisory Control and Data Acquisition
SSO	Single Sign-On
SPP-ICS	System Protection Profile for Industrial Control Systems
TIH	Toxic Inhalation Hazard
TSA	Transportation Security Administration
UPS	Uninterruptible Power Supply
VPN	Virtual Private Network

NOTES

1 Center for Strategic and International Studies (CSIS), (2020), "Significant Cyber Incidents Since 2006," https://csis-website-prod.s3.amazonaws.com/s3fs-public/200626_Cyber_Events.pdf, retrieved July 9, 2020.

2 [Gregory, D., Witty, R., Blair, R., Thielemann, K., (January 4, 2021), "Gartner Predicts 2021: Organizational Resilience," Gartner ID G00735230," www.gartner.com/doc/reprints?id=1-255A3UM3&ct=210203&st=sb, retrieved May 2021.

3 [(n.a.), (October 24, 2020), "NSA and CISA Recommend Immediate Actions to Reduce Exposure Across Operational Technologies and Control Systems, Alert (AA20-205A)," https://us-cert.cisa.gov/ncas/alerts/aa20-205a, retrieved May 2021.

4 (n.a.), (February 9, 2021), "Hacker Tampered with Chemical Processes Controls at Florida Water Treatment Plant," NewsBites, Vol. 23, Num. 011, SANS Institute, www.sans.org/newsletters/newsbites/xxiii-11, retrieved May 2021.

5 Zetter, Kim, (November 3, 2014), "An unprecedented look at Stuxnet, the world's first digital weapon," WIRED, www.wired.com/2014/11/countdownto-zero-day-stuxnet/, retrieved July 9, 2020.

6 (n.a.), (n.d.), "Platform IT, Operational Technology and Facility-Related Control Systems," Strategic Environmental Research and Development Program (SERDP), Environmental Security Technology Certification Program (ESTCP), www.serdp-estcp.org/Tools-and-Training/Installation-Energy-and-Water/Cybersecurity/Overview-of-PIT-OT-FRCS, retrieved July 9, 2020.

7 Simonovich, L., (October 2019), "Caught in the Crosshairs: Are Utilities Keeping Up with the Industrial Cyber Threat?" Ponemon Institute and Siemens, https://assets.new.siemens.com/siemens/assets/api/uuid:35089d45-e1c2-4b8b-b4e9-7ce8cae81eaa/version:1572434569/siemens-cybersecurity.pdf, retrieved July 10, 2020.

8 Tornberg, B., (June 15, 2018), "5 Types of Manufacturing Processing and How They're Different," www.e3businessconsultants.com/erp-consulting/5-types-manufacturing-processing-theyre-different, retrieved September 10, 2020.

9 Stouffer, K., Pillitteri, V., Lightman, S., Abrams, M., Hahn, A., (May 2o15), "NIST Special Publication 800-82, Revision 2, Guide to Industrial Control Systems (ICS) Security Supervisory Control and Data Acquisition (SCADA) Systems, Distributed Control Systems (DCS), and Other Control System Configurations such as Programmable Logic Controllers (PLC)", http://dx.doi.org/10.6028/NIST.SP.800-82r2, retrieved July 6, 2020.

10 Caldwell, S., June 26, 2009, "The Department of Homeland Security's (DHS) Critical Infrastructure Protection Cost-Benefit Report, www.gao.gov/new.items/d09654r.pdf, retrieved July 6, 2020.

11 Chappelow, J., (June 25, 2019), "What is the Private Sector," www.investopedia.com/terms/p/private-sector.asp#, retrieved September 10, 2020.

12 (n.a.), (2019), "American Public Power Association, 2019 Statistical Report," PublicPower.org, www.publicpower.org/system/files/documents/2019-Public-Power-Statistical-Report.pdf, pg. 23, retrieved July 9, 2020.

13 Falco, Joe, et al., IT Security for Industrial Control Systems, NIST Internal Report (NISTIR) 6859, February 2002, www.nist.gov/customcf/get_pdf.cfm?pub_id=821684, retrieved July 7, 2020.

14 (April 24, 2020), "Threat Landscape for Industrial Automation Systems: H2 2019," Kaspersky Lab Industrial Control Systems Cyber Emergency Response Team (Kaspersky Lab ICS CERT), https://ics-cert.kaspersky.com/media/KASPERSKY_H22019_ICS_REPORT_FINAL_EN.pdf, retrieved July 9, 2020.

15 Ibid.

16 McGrew, R., Vaughn, R., (February 2009), "Vulnerability Analysis of SCADA HMI Systems," Center for Infrastructure Protection, The CIP Report, volume 7 number 8, pg. 6, www.energy.gov/sites/prod/files/oeprod/DocumentsandMedia/10-The_CIP_Report_Issue79.pdf, retrieved July 16, 2020.

17 Stouffer, K., Pillitteri, V., S., Lightman, Abrams, M., Hahn, A., (My 2015), "Guide to Industrial Control Systems (ICS) Security," NIST Special Publication 800-82, Revision 2, http://dx.doi.org/10.6028/NIST.SP.800-82r2, retrieved July 15, 2020.

18 (n.a.), (n.d.), "RTU defined," SUBNET Solutions Inc., www.subnet.com/resources/dictionary/rtu.aspx, retrieved July 17, 2020.

19 Csanyi, E., (October 2, 2017), "The Basics of Hardware and Software for SCADA Systems You Should Know About," Electrical Engineering Portal, https://electrical-engineering-portal.com/hardware-software-scada-systems, retrieved July 7, 2020.

20 (n.a.), (June 28, 2011), "Master Terminal Units (MTU) in SCADA systems," http://electrical-questionsguide.blogspot.com/2011/06/master-terminal-units-mtu-in-scada.html?m=1, retrieved July 7, 2020.

21 Bailey, D., Wright, E., (2003), "Practical SCADA for Industry," IDC Technologies, https://repository.unad.edu.co/bitstream/10596/5004/1/Practical_SCADA_for_Industry-1-110.pdf, retrieved July 7, 2020.

22 Radvanovsky, R. and Brodsky, J. (2013) Handbook of SCADA/control systems security. Boca Raton, FL, USA: Taylor & Francis Group, pg. 190, www.icsdefender.ir/files/scadadefender-ir/books/ICS-SECURITY-NEW/Radvanovsky-%20Robert%20Handbook%20of%20SCADA_control%20systems%20security.pdf, retrieved July 22, 2020.

23 Bernier, K., (April 1, 2005), "Historians vs. Relational Databases," Control Engineering, www.controleng.com/articles/historians-vs-relational-databases/#:~:text=A%20plant%2Dwide%20historian%20provides,is%20built%20to%20manage%20relationships, retrieved July 7, 2020.

24 Morland, T., (November 22, 2019), "SCADA 101: Local Historian Overview," https://blog.norcalcontrols.net/local-historian-overview, retrieved July 23, 2020.

25 Comeau, S., (June 9, 2020), "Introduction and Optimization of Data Historians," www.hallam-ics.com/blog/introduction-and-optimization-of-data-historians, retrieved July 23, 2020.

26 Stouffer, K., Pillitteri, V., S., Lightman, Abrams, M., Hahn, A., (May 2015), "Guide to Industrial Control Systems (ICS) Security," NIST Special Publication 800-82, Revision 2, http://dx.doi.org/10.6028/NIST.SP.800-82r2, retrieved July 15, 2020.

27 Karim, M., Alrasheedy, A., (2015), "Compensated Mass Balance Method for Oil Pipeline Leakage Detection using SCADA, "International Journal of Computer Science and Security (IJCSS), Volume (9), Issue (6), http://citeseerx.ist.psu.edu/viewdoc/download?doi=10.1.1.734.8699&rep=rep1&type=pdf, retrieved July 17, 2020.

28 Stouffer, K., Pillitteri, V., Lightman, S., Abrams, M., Hahn, A., (May 2015), "NIST Special Publication 800-82, Revision 2, Guide to Industrial Control Systems (ICS) Security Supervisory Control and Data Acquisition (SCADA) Systems, Distributed Control Systems (DCS), and Other Control System Configurations such as Programmable Logic Controllers (PLC)", http://dx.doi.org/10.6028/NIST.SP.800-82r2, Figure re-drawn in grayscale by author, retrieved July 6, 2020.

29 Electrical Technology, (n.d.), "What is Distributed Control System (DCS)?", www.electricaltechnology.org/2016/08/distributed-control-system-dcs.html, author email exchange with Electrical Technology, used with permission, retrieved July 8, 2020.

30 Advanced Micro Controls, Inc., (n.d.), "What is a PLC?" www.amci.com/industrial-automation-resources/plc-automation-tutorials/what-plc, Advanced Micro Controls, Inc., 20 Gear Drive, Plymouth Industrial Park, Terryville, CT 06786, 877-781-5541, author email exchange with Advanced Micro Controls, Inc., used with permission, retrieved July 8, 2020.

31 Soullie, A., (October 2017), "DYODE: Do Your Own DiodE A DIY, low-cost data diode for ICS," BruCon, Wavestone, http://files.brucon.org/2017/011_Arnaud_Soullier_Wavestone.pdf, retrieved July 21, 2020, email correspondence with Arnaud Soullier, used with permission. Figure re-drawn in grayscale by author.

32 Kent, K., Chevalier, S., Grance, T., Dang, H., (August 2006) "NIST SP 800-86, Guide to Integrating Forensic Techniques into Incident Response," http://csrc.nist.gov/publications/PubsSPs.html#800-86, retrieved July 28, 2020.

33 Fabro, M., Cornelius, E. (August 2008), "Recommended practice: Creating Cyber Forensics Plans for Control Systems," Technical report, Department of Homeland Security, https://inldigitallibrary.inl.gov/sites/sti/sti/4113665.pdf, retrieved July 28, 2020.

34 A cache is a memory buffer used to temporarily store frequently accessed data. It improves performance since data does not have to be retrieved again from the original source. By storing frequently accessed or expensive-to-create objects in memory or on disk, the Java Object Cache eliminates the need to repeatedly create and load information within a Java program. The Java Object Cache retrieves content faster and greatly reduces the load on application servers.

35 Ahmed, I., Obermeier, S., Naedele, M., Richard, G., (December 2012), "SCADA System: Challenges for Forensics Investigations," IEEE Computer, Vol. 45 No. 12, pp 44–51, www.cct.lsu.edu/~golden/Papers/SCADA-IEEE-Computer-revised.pdf, retrieved August 2, 2020.

36 Fowke, B., (March 28, 2017), "Testimony before the U.S. Senate Committee on Energy and Natural Resources Subcommittee on Energy hearing to Examine Cybersecurity Threats to the U.S. Electrical Grid and Technology Advancements to Minimize the Threat," www.energy.senate.gov/public/index.cfm/files/serve?File_id=40A50EA7-75FA-4CEB-9A5A-3FE9074F4B77, retrieved July 7, 2020.

37 (n.a.), (August 2017), "Securing Cyber Assets Addressing Urgent Cyber Threats to Critical Infrastructure," The President's National Infrastructure Advisory Council, www.cisa.gov/sites/default/files/publications/niac-securing-cyber-assets-final-report-508.pdf, retrieved July 7, 2020.

38 Finco, G., Huffman, E., (September 28, 2006), "Introduction SCADA Security for Managers and Operators," www.energy.gov/sites/prod/files/oeprod/DocumentsandMedia/Introduction_to_SCADA_Security_for_Managers_and_Operators.pdf, retrieved July 7, 2020.

39 (n.a.), (August 2017), "Securing Cyber Assets Addressing Urgent Cyber Threats to Critical Infrastructure," The President's National Infrastructure Advisory Council, www.cisa.gov/sites/default/files/publications/niac-securing-cyber-assets-final-report-508.pdf, retrieved July 9, 2020.

40 Fabro, M., Cornelius, E., (August 2008), "Recommended Practice: Creating Cyber Forensics Plans for Control Systems," U.S. Department of Homeland Security (DHS), National Cyber Security Division Control Systems Security Program, https://inldigitallibrary.inl.gov/sites/sti/sti/4113665.pdf, retrieved July 22, 2020.

41 Ibid.

42 Radvanovsky, R. and Brodsky, J. (2013) Handbook of SCADA/control systems security. Boca Raton, FL: Taylor & Francis Group, pgs. 188–189, www.taylorfrancis.com/books/9780429253218/chapters/10.1201/b13869-15, retrieved July 22, 2020.

43 Spyridopoulos, T., Tryfonas, T., & May, J. H. R. (2013). Incident Analysis & Digital Forensics in SCADA and Industrial Control Systems. In System Safety Conference incorporating the Cyber Security Conference 2013, 8th IET International (pp. 1–6). Institution of Engineering and Technology (IET). https://doi.org/10.1049/cp.2013.1720, retrieved July 22, 2020.

44 Wu, T., Disso, J. F. P., Jones, K. and Campos, A. (2013), Towards a SCADA forensics architecture, in 'Proceedings of the 1st International Symposium for ICS & SCADA Cyber Security Research', p. 12, https://pdfs.semanticscholar.org/4e71/43b574d6d8bf088190802257fab20b326da3.pdf, retrieved July 22, 2020.

45 Cichonski, P., Millar, T., Grance, T., Scarfone, K., (August 2012), "Computer Security Incident Handling Guide," Special Publication (NIST SP) - 800-61 Rev 2, National Institute of Standards and Technology, https://csrc.nist.gov/publications/detail/sp/800-61/rev-2/final, retrieved July 22, 2020.

46 Salkie, P., (2016), "Securing Legacy Industrial Control Systems," 2016 NSF Cybersecurity Summit for Large Facilities and Cyberinfrastructure, https://static1.squarespace.com/static/5047a5a6e4b0dcecada15549/t/57b48ae62994ca7441105b00/1471449830854/Securing_Legacy_Industrial_Control_Systems_CTSC2016-2.pdf, retrieved August 2, 2020.

47 Permann, M., Rohde, K., (2005), "Cyber Assessment Methods for SCADA Security," The Instrumentation, Systems and Automation Society, presented at 15th Annual Joint ISA POWID/EPRI Controls and Instrumentation Conference, https://digital.library.unt.edu/ark:/67531/metadc890695/m2/1/high_res_d/911094.pdf, retrieved August 2, 2020.

48 (n.a.), (n.d.), "Advanced Sensors, Control, Platforms, and Modeling for Manufacturing (Smart Manufacturing): Technology Assessment, "www.energy.gov/sites/prod/files/2015/02/f19/QTR%20Ch8%20-%20Smart%20Manufacturing%20TA%20Feb-13-2015.pdf, retrieved September 10, 2020.

49 Fabro, M., Cornelius, E., (August 2008), "Recommended practice: Creating Cyber Forensics Plans for Control Systems," Technical report, Department of Homeland Security, https://inldigitallibrary.inl.gov/sites/sti/sti/4113665.pdf, retrieved August 2, 2020.

50 Ibid.

51 Abdul Azeez Erumban, (June 2008), "Expected Average Lifetimes: Machinery, Table 7," Review of Income and Wealth, Series 54, Number 2, pg., 255, Blackwell Publishing, 9600 Garsington Road, Oxford OX4 2DQ, UK, www.roiw.org/2008/2008-11.pdf, retrieved July 30, 2020.

52 Permann, M., Rohde, K., (2005), "Cyber Assessment Methods for SCADA Security," The Instrumentation, Systems and Automation Society, presented at 15th Annual Joint ISA POWID/EPRI Controls and Instrumentation Conference, https://digital.library.unt.edu/ark:/67531/metadc890695/m2/1/high_res_d/911094.pdf, retrieved August 2, 2020.

53 Fabro, M., Cornelius, E., (August 2008), "Recommended practice: Creating Cyber Forensics Plans for Control Systems," Technical report, Department of Homeland Security, https://inldigitallibrary.inl.gov/sites/sti/sti/4113665.pdf, retrieved August 2, 2020.

54 Ibid.

55 (n.a.), (May 7, 2018), "Independent Study Pinpoints Significant SCADA/ICS Cybersecurity Risks," Fortinet, www.fortinet.com/content/dam/fortinet/assets/white-papers/WP-Independent-Study-Pinpoints-Significant-Scada-ICS-Cybersecurity-Risks.pdf, retrieved August 4, 2020.

56 Fabro, M., Cornelius, E., (August 2008), "Recommended practice: Creating Cyber Forensics Plans for Control Systems," Technical report, Department of Homeland Security, https://inldigitallibrary.inl.gov/sites/sti/sti/4113665.pdf, retrieved August 2, 2020.

57 Brezinski, D., Killalea, T., (February 2002), "RFC 3227, Guidelines for Evidence Collection and Archiving," The Internet Society, www.faqs.org/rfcs/rfc3227.html, retrieved August 17, 2020.

58 Rutkowski, R., (September 18, 2018), "What You Should Know About Stratum System Levels," https://blog.bliley.com/what-you-should-know-about-stratum-system-levels, retrieved August 9, 2020.

59 Fabro, M., Cornelius, E., (August 2008), "Recommended Practice: Creating Cyber Forensics Plans for Control Systems," U.S. Department of Homeland Security (DHS), National Cyber Security Division Control Systems Security Program, https://inldigitallibrary.inl.gov/sites/sti/sti/4113665.pdf, retrieved August 9, 2020.

60 (n.a.), (n.d.), "Secure Architecture Design Definitions," Cybersecurity and Infrastructures Security Agency (CISA), Department of Homeland Security, https://us-cert.cisa.gov/ics/Secure-Architecture-Design-Definitions, retrieved August 10, 2020.

61 Other terms commonly used include emergency shutdown system (ESS), safety shutdown system (SSD), and safety interlock system (SIS).

62 Singh, R., (2013), "Pipeline System Communication," Arctic Pipeline Planning Design, Construction, and Equipment, Elsevier B.V., www.sciencedirect.com/topics/engineering/operator-workstation, retrieved August 9, 2020.

63 Idachaba, F., Ogunrinde, A., (2012), "Review of Remote Terminal Unit (RTU) and Gateways for Digital Oilfield Deployments," International Journal of Advanced Computer Science and Applications, Vol. 3, No. 8, www.researchgate.net/publication/267403801_Review_of_Remote_Terminal_Unit_RTU_and_Gateways_for_Digital_Oilfield_delpoyments, retrieved August 10, 2020.

64 Agarwal, T, (n.d.), "Optimum Idea about an OPC Server in Industrial Control Systems," www.elprocus.com/why-is-opc-server-needed-for-industrial-control-systems, retrieved August 10, 2020.

65 Note on 8-inch 'adapters' or 'cables for PCs' To read 8-inch floppy drives on a 'PC', you MAY be able to connect 8-inch drives to a PC's internal floppy controller (if it has one). But you need an 8-inch drive, with power supply, in a cabinet; a 50-pin cable for the 8-inch drive; a wiring adapter for the 50-pin cable to your 34-pin floppy controller; and software and software knowledge! Even with that, MOST PC floppy controllers won't read old single-density format anyway! Johnson, H., (January 10, 2019), "How to read and write old floppy disks," www.retrotechnology.com/herbs_stuff/s_drives_howto.html, retrieved August 9, 2020.

66 Tables 6.8 and 6.9 were developed by the author from multiple private-sector sources, documents, whitepapers, industry reports and from public agency documents. Two papers which contributed greatly to the overall development and compilation of these Tables and which the author wishes to specifically note are:

   Fabro, M., Cornelius, E., (August 2008), "Recommended Practice: Creating Cyber Forensics Plans for Control Systems," U.S. Department of Homeland Security (DHS), National Cyber Security Division Control Systems Security Program, https://inldigitallibrary.inl.gov/sites/sti/sti/4113665.pdf, retrieved August 9, 2020.

   Spyridopoulos, T., Tryfonas, T., & May, J. H. R. (2013), "Incident Analysis & Digital Forensics in SCADA and Industrial Control Systems," In System Safety Conference incorporating the Cyber Security Conference 2013, 8th IET International (pp. 1-6). Institution of Engineering and Technology (IET), www.researchgate.net/publication/259309151_Incident_Analysis_Digital_Forensics_in_SCADA_and_Industrial_Control_Systems, retrieved August 9, 2020.

67 Ibid.

Chapter 7

Cyber forensics and risk management

Douglas Menendez

CONTENTS

OVERVIEW OF ENTERPRISE RISK MANAGEMENT (ERM)

Introduction

This chapter introduces the reader to the concept of Enterprise Risk Management (ERM) and how ERM interconnects to both cybersecurity and cyber forensics. The chapter provides an overview of ERM, as well as addressing some of the key considerations for Cyber Risk Management (CRM). Also examined will be a look at U.S. government regulations that pertain to CRM, and some procedure on assessing cyber forensics risk. The chapter concludes with insights into how cyber forensics readiness can contribute to reducing business risk.

Basics of enterprise risk management (ERM): How to get started

Organizations exist to create value for their stakeholders. By setting objectives, developing strategies, following through and continuously improving processes, value is created.

That's the ideal situation, at least. In reality, it's not always as simple as planning and sticking to it. There is always the risk that certain events could affect the success of these plans. It is the job of management to make adequate preparations to ensure that systems are in place to continue hitting objectives, even when the beast of unforeseen circumstance rears its head.

ERM is a direct solution to these kinds of uncertainties, allowing management to oversee the continual creation of value on a complete, integrated, organization-wide level. By utilizing an effective ERM system, you can rest assured that the organization will see a consistently high success rate in terms of hitting objectives and (key performance indictors) KPIs.

Stakeholders of all kinds, from customers, suppliers, to government and regulatory bodies are all increasingly interested in how businesses are implementing ERM. A well-implemented ERM system could set the foundation for many high-quality, long-term client relationships.

Equally, not having a proper system for ERM could mean that a business is perceived as less competent, and could even result in loss of clients and damage to brand image.[1]

What is enterprise risk management (ERM)?

According to The Committee of Sponsoring Organizations of the Treadway Commission (COSO), from *Enterprise Risk Management – Integrating with Strategy and Performance*, ERM is 'The culture, capabilities, and practices, integrated with strategy-setting and performance, that organizations rely on to manage risk in creating, preserving, and realizing value.'

In addition to identifying risks, ERM also involves risk 'readiness' and prioritizing responses to make an organization as resilient as possible to active or potential risks.

ERM plans, policies, and procedures should be shared with appropriate stakeholders, investors and regulators, as risk management is a key pillar of overall corporate governance. All industries, public and private companies as well as not-for-profit and government organizations can benefit from ERM.

Another organization, the *International Standardization Organization (ISO) in ISO 31000 – Risk Management Guidelines*, defines risk management as: 'coordinated activities to direct and control an organization with regard to risk … [a] systematic application of policies, procedures and practices to the activities of communicating and consulting, establishing the context and assessing, treating, monitoring, reviewing, recording and reporting risk.'

Risk management has been around for a long time, with its roots in the insurance industry. Products such as property insurance, liability insurance, and malpractice insurance, have traditionally been how organizations and individuals would manage risk.

More recently, as risk management has become more widely adopted, ERM has developed into more of a business process management framework. With this has been the shift in focus from being reactive to risk events, to developing policies, processes and controls, that allow an organization to proactively protect itself from potentially harmful risk exposures. Some companies have even started to use their ERM program as a competitive market-place differentiator.[2]

Some of the advantages of a successful ERM program

ERM, as the name implies, is 'enterprise-wide'; a successful ERM program helps create a risk-aware culture. This allows management to identify and manage cross-enterprise risks, and focus on the most important risks to the organization. While no organization can completely eliminate vulnerability to adverse events, ERM will allow management to improve their responses when it comes to risk decisions.

In the past, most of risk management was focused on the negatives, but now ERM can be seen as a conduit for improving processes, controls, and security. Successful ERM initiatives allow organizations to align their risk appetite and strategy. This link to company growth, investment, and return, can help minimize 'surprises' and allow an organization to capitalize on opportunities as they arise.

ERM components

In the COSO publication, *Enterprise Risk Management: Integrating with Strategy and Performance* (2017 Edition), COSO proposes a framework of five components and twenty principles to assist organizations in improving their approach to managing risk to meet the demands of an evolving business environment.

The five COSO components are identified as:

1. Governance and Culture
2. Strategy and Objective-Setting
3. Performance
4. Review and Revision
5. Information, Communication, and Reporting

COSO's 20 principles, which underline the framework's five components, are shown in Table 7.1.

The five ERM components

1. Governance and culture: ERM cannot succeed unless the organization seeks to fully integrate it within the culture of their workplace. This pertains to the ethics behind worker responsibilities, codes of conduct, and the proper comprehension of risks, as well as all associated management programs and solutions.
2. Strategy and objective-setting: A fundamental part of ERM is making sure the risk management strategies align with core objectives and broader business strategies. Business objectives are the basis for planning and implementing strategies, while simultaneously serving as a launch-pad for identifying, assessing, and responding to risks.

Table 7.1 COSO ERM framework – components and principles[3]

Governance and Culture
- Exercises Board Risk Oversight
- Establishes Operating Structures
- Defines Desired Culture
- Demonstrates Commitment to Core Values
- Attracts, Develops, and Retains Capable Individuals

Strategy and Objective-Setting
- Analyzes Business Context
- Defines Risk Appetite
- Evaluates Alternative Strategies
- Formulates Business Objectives

Performance
- Identifies Risk
- Assesses Severity of Risk
- Prioritizes Risks
- Implements Risk Responses
- Develops Portfolio View

Review and Revision
- Assesses Substantial Change
- Reviews Risk and Performance
- Pursues Improvement in ERM

Information, Communication, and Reporting
- Leverages Information Systems
- Communicates Risk Information
- Reports on Risk, Culture, and Performance

3. Performance: Assessing how certain risks will impact the performance of key processes is important for risk prioritization. In this context, risks are prioritized in order of their severity. Following this, risk responses are selected based on an assessment of the potential for risk that has been identified. Results of this part of the process are typically reported to key stakeholders.
4. Review and revision: By reviewing the performance of risk management processes, organizations can determine how well the ERM program is working, including whether or not changes are needed.
5. Information, communication, and reporting: ERM is not a single checklist or a fixed set of steps; it is an ongoing process of collecting and assessing information from internal and external sources, across all parts of an organization.[4]

Summary

The concepts, core areas, and benefits of ERM have been discussed as were the five main components of an ERM, which may be implemented by any organization in any industry. The popular COSO ERM framework was also reviewed. Next, we will discuss considerations for CRM.

CONSIDERATIONS FOR CYBER RISK MANAGEMENT

Each year brings new cybersecurity threats, breaches, and previously unknown vulnerabilities in established systems. Even with unprecedented vulnerabilities such as Spectre and

Meltdown, the approach to dealing with the risks they pose is the same as ever: sound risk management with systematic processes to assess and respond to risks.

What is cyber risk management?

The International Organization for Standardization (ISO) defines 'risk' as the 'effect of uncertainty on objectives.' 'Risk management' is the ongoing process of identifying, assessing, and responding to risk. To manage risk, organizations should assess the likelihood and potential impact of an event and then determine the best approach to deal with the risks: avoid, transfer, accept, or mitigate.

A good risk management program should establish clear communications and situational awareness about risks. This allows risk decisions to be well informed, well considered, and made in the context of organizational objectives, such as opportunities to support the organization's mission or seek business rewards.

ERM essential elements

Most risk management standards, such as those from ISO, COSO, and National Institute of Standards and Technology (NIST), and have common key processes. In its best practices for an ERM program, the Government Accountability Office (GAO) identified six essential elements. See Figure 7.1.

The first element, aligning ERM to goals and objectives, sets the foundation for the program by establishing the three pillars of enterprise CRM: governance, risk appetite, and policy and procedure.

Governance should include a body of risk-decision experts and decision makers using a framework of risk management processes that ensure engagement by key stakeholders (leaders, Authorizing Officials, and Risk Committee). *Appetite for risks* should be aligned to organizational goals and objectives. *Policies and procedures* communicate risk management expectations, risk definitions, and guidance throughout the enterprise. Once the risk management program is running, the remaining five elements continuously manage risk.

Seven considerations for cyber risk management

The following seven topics are well worth considering when planning a risk management program.

1. Culture. Leaders should establish a culture of cybersecurity and risk management throughout the organization.

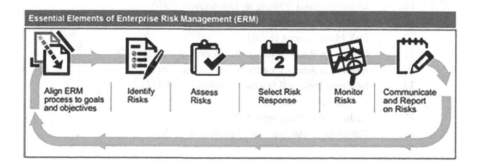

Figure 7.1 GAO six essential elements[5]

2. Information sharing. The right stakeholders must be aware of risks, particularly of cross-cutting and shared risks, and be involved in decision making.
3. Priorities. To prioritize risks and responses, you need information, such as trends over time, potential impact, time horizon for impact, and when a risk will likely materialize.
4. Resilience. Risk management must also enable continuity of critical missions during and after disruptive or destructive events, including cyber-attack. Many organizations use the CERT Resilience Management Model (CERT-RMM) to manage and improve their operational resilience. The model includes Risk Management as one of its 26 process areas (See Table 7.2.).
5. Speed. When an organization is exposed to a risk, speedy response can minimize impact. Identifying risks early helps.
6. Threat environment. Organizations should improve their intelligence into adversary capabilities while also accounting for risks from third parties and insider threats.
7. Cyber hygiene. Implementing basic cyber hygiene practices is a good starting point for CRM. Cyber hygiene focuses on basic activities to secure infrastructure, prevent attacks, and reduce risks. The Center for Internet Security (CIS) has developed a list of 20 cybersecurity controls (See Table 7.3).

Table 7.2 CERT resilience management model (CERT-RMM)[6]

1.	Asset Definition and Management (ADM)
2.	Access Management (AM)
3.	Communications (COMM)
4.	Compliance (COMP)
5.	Controls Management (CTRL)
6.	Environmental Controls (EC)
7.	Enterprise Focus (EF)
8.	External Dependencies Management (EXD)
9.	Financial Resource Management (FRM)
10.	Human Resources Management (HRM)
11.	Identity Management (ID)
12.	Incident Management and Control (IMC)
13.	Knowledge and Information Management (KIM)
14.	Measurement and Analysis (MA)
15.	Monitoring (MON)
16.	Organizational Process Definition (OPD)
17.	Organizational Process Focus (OPF)
18.	Organizational Training and Awareness (OTA)
19.	People Management (PM)
20.	Risk Management (RISK)
21.	Resilience Requirements Development (RRD)
22.	Resilience Requirements Management (RRM)
23.	Resilient Technical Solution Engineering (RTSE)
24.	Service Continuity (SC)
25.	Technology Management (TM)
26.	Vulnerability Analysis and Resolution (VAR)

Table 7.3 CIS 20 cybersecurity controls[7]

Basic CIS Controls	
1.	Inventory and Control of Hardware Assets
2.	Inventory and Control of Software Assets
3.	Continuous Vulnerability Management
4.	Controlled Use of Administrative Privileges
5.	Secure Configuration for Hardware and Software on Mobile Devices, Laptops, Workstations, and Servers
6.	Maintenance, Monitoring, and Analysis of Audit Logs
Foundational CIS Controls	
7.	Email and Web Browser Protections
8.	Malware Defenses
9.	Limitation and Control of Network Ports, Protocols, and Services
10.	Data Recovery Capabilities
11.	Secure Configuration for Network Devices, such as Firewalls, Routers, and Switches
12.	Boundary Defense
13.	Data Protection
14.	Controlled Access Based on the Need to Know
15.	Wireless Access Control
16.	Account Monitoring and Control
Organizational CIS Controls	
17.	Implement a Security Awareness and Training Program
18.	Application Software Security
19.	Incident Response and Management
20.	Penetration Tests and Red Team Exercises

The Software Engineering Institute (SEI) has identified a baseline set of 11 cyber hygiene practices (See Table 7.4).

Prepared, not bullet proof

With cyber risks continuing to grow, making good risk management decisions really matters. Cyber events will still happen to your organization, but it will be better prepared to deal with them.[9]

Cyber forensics and insider threats

One of the biggest cyber risks are insider security threats. According to Accenture, 69% of security professionals face insider security threats. In a recent Q & A interview with digital security expert, Andrew Morrison, a principal with Deloitte's Cyber Risk Services, Morrison provides some insight into how should CISOs today deploy forensics technology and expertise to stop costly data breaches before they happen.

Since the 1970s, security firms and law enforcement agencies have relied on a niche investigative discipline, digital forensics, to track down and recover stolen data from computer systems, and identify bad actors.

Table 7.4 Cyber hygiene – a baseline set of practices[8]

I.	Identify and prioritize key organizational services, products and their supporting assets.
2.	Identify, prioritize, and respond to risks to the organization's key services and products.
3.	Establish an incident response plan.
4.	Conduct cybersecurity education and awareness activities.
5.	Establish network security and monitoring.
6.	Control access based on least privilege and maintain the user access accounts.
7.	Manage technology changes and use standardized secure configurations.
8.	Implement controls to protect and recover data.
9.	Prevent and monitor malware exposures.
10.	Manage cyber risks associated with suppliers and external dependencies.
11.	Perform cyber threat and vulnerability monitoring and remediation.

In the early days, digital forensics was fairly straightforward: investigators could copy a physical hard drive to recover data. They used criminals' digital fingerprints to apprehend and prosecute them.

The practice has expanded greatly today, along with the complexity of computing and an explosion of cybercrime. The annual global market for forensics software and services is projected to grow from $3.4 billion in 2018 to $5.9 billion by 2024, according to Mordor Intelligence.

Q – How does digital forensics fit into modern cybersecurity?
It's an integral part of any cybersecurity strategy. Historically, digital forensics grew up outside of cyber. Companies used it for fraud investigations or legal discovery, duplicating evidence, preserving it in a clean state, then analyzing it and producing a report.

With digital forensics tools for behavior analysis, CISOs can do forensics in real-time rather than against static data. – Andrew Morrison

Now it's being used in incident response strategy. Although it's important to preserve evidence and make sure you're doing things right, now you can use next-generation forensics tools to restore business operations as quickly as possible. The prosecution of a cyber event is secondary to that.

Q – What are some of its new capabilities?
Today we're reinventing the wheel a little bit. We're collapsing the forensics mindset into the analytics mindset, and taking it beyond alert optimization. With alert optimization, companies look at the matrix of known threat risks, looking at existing data, developing models to analyze the threat and reporting back on what you found. That's been happening a long time.
What's new is shifting that matrix into unknown vectors and unknown risks. With digital forensics tools around behavior analysis, CISOs can try to do forensics in real-time rather than against static data, giving you the ability to react faster to threats that are moving faster than you can consume huge quantities of data.

Q – So it's shifting from an investigative to a preventive methodology?

Yes. The concept of forensics as prevention is 'cyber hunting.' Cyber hunting is the application of digital forensics to identify a compromise that has not yet detonated. Let's say someone is in your network doing something, but they haven't yet had an end result. Forensics can help reduce what we call the 'dwell time' of a vulnerability.

Once a threat has gotten through your firewall, you have a very limited window to act before it becomes a breach or something nefarious. Where most enterprises see a lot of the value in this space is capturing an adversary before damage has occurred, but after their protective measures have failed.

Q – Does it make more sense to use third-party vendors for digital forensics, or build resources in-house?

You'll probably need both. You'll need an in-house team that is doing continuous forensics, investigation, and analysis. But you also need a highly specialized skill set for deeper investigations.

In the wake of an incident, companies need a team of cyber hunters that just don't exist in the market in great quantity. It's difficult to keep them on staff, and their real worth is in the wake of an incident, to very quickly use forensics to determine the scale and severity of the damage.

Q – How can forensics help companies deal with the insider threat challenge?

Forensics can look at your entire population of employees and contractors and do predictive risk scoring about who is most likely to become an insider threat.

It's interesting how deep that can go, with people who are job-seeking, people who are demanding pay increases, people who are disgruntled or underperforming. You can add layers of monitoring using forensic tools to better scrutinize their activity. There's an entire discipline around that, but forensics is probably the most useful tool companies have to stave off insider threats.[10]

Summary

The reader was presented with an overview of CRM, and heard from a security expert about how cyber forensics can help mitigate the risk from insider threats.

Next up, a look at some of the by the U.S. government-supported CRM frameworks.

CYBER RISK MANAGEMENT AND THE U.S. GOVERNMENT

NIST risk management framework background

The Cybersecurity Framework (CSF) can help both federal agencies and private sector organizations to integrate existing risk management and compliance efforts and structure consistent communication, both across teams and with leadership.

NIST has been updating its suite of cybersecurity and privacy risk management publications to provide additional guidance on how to integrate the implementation of the CSF. NIST Interagency Report (IR) 8170 *Approaches for Federal Agencies to Use the Cybersecurity Framework* summarized eight approaches that may be useful for federal agencies and other private sector organizations. NISTIR 8170 discusses how the CSF can be valuable in managing information and information systems according to:

- The *Risk Management Framework (RMF) for Information Systems and Organizations* (RMF) (SP 800-37 Rev 2), by implementing security controls detailed in Security and Privacy Controls for Federal Information Systems and Organizations (SP 800-53 revision 4), and using the methodology outlined in Managing Information Security Risk: Organization, Mission, and Information System View (SP 800-39).

Conversely, the RMF incorporates key CSF, privacy risk management, and systems security engineering concepts. Among other things, the CSF Core can help both state and federal agencies along with private sector organizations to:

- Better-organize the risks they have accepted and the risk they are working to remediate across all systems.
- Use the reporting structure that aligns to security and privacy controls for information systems and organizations (NIST SP 800-53 rev. 4).
- Enable organizations to reconcile mission objectives with the structure of the core.

Each task in the RMF includes references to specific sections in the CSF. For example:

- Task P-2, *Risk Management Strategy*, aligns with the CSF Core [Identify Function];
- Task P-4, *Organizationally-Tailored Control Baselines and Cybersecurity Framework Profiles*, aligns with the CSF Profile construct; and
- Task R-5, *Authorization Reporting*, and Task M-5, *Security and Privacy Reporting*, support OMB reporting and risk management requirements organization-wide by using the CSF constructs of Functions, Categories, and Subcategories.[11]

Risk management framework (RMF) overview

The selection and specification of security controls for a system is accomplished as part of an organization-wide information security program that involves the ***management of organizational risk*** – that is, the risk to the organization or to individuals associated with the operation of a system. The management of organizational risk is a key element in the organization's information security program and provides an effective framework for selecting the appropriate security controls for a system – the security controls necessary to protect individuals and the operations and assets of the organization.

Risk-based approach

The RMF provides a process that integrates security and risk management activities into the system development life cycle. The risk-based approach to security control selection and specification considers effectiveness, efficiency, and constraints due to applicable laws, directives, Executive Orders, policies, standards, or regulations.

The following activities related to managing organizational risk are paramount to an effective information security program and can be applied to both new and legacy systems within the context of the system development life cycle and the Federal Enterprise Architecture (see Figure 7.2).

Prepare Step
Prepare carries out essential activities at the organization, mission and business process, and information system levels of the enterprise to help prepare the organization to manage its security and privacy risks using the RMF.

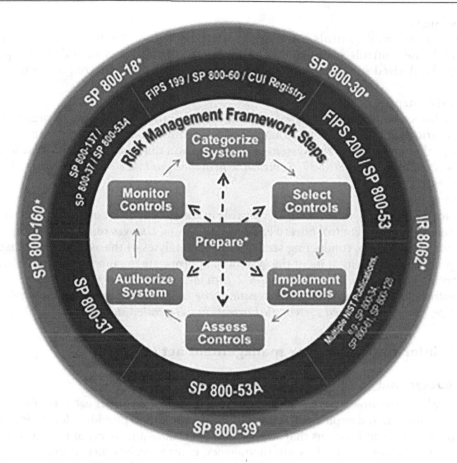

Figure 7.2 System development life cycle and the Federal Enterprise Architecture[12]

Categorize Step

Categorize the system and the information processed, stored, and transmitted by that system based on an impact analysis. *(The RMF categorize step, including consideration of legislation, policies, directives, regulations, standards, and organizational mission/business/operational requirements, facilitates the identification of security requirements. FIPS 199 provides security categorization guidance for non-national security systems. CNSS Instruction 1253 provides similar guidance for national security systems.)*

Select Step

Select an initial set of baseline security controls for the system based on the security categorization; tailoring and supplementing the security control baseline as needed based on organization assessment of risk and local conditions. *(NIST Special Publication 800-53 Revision 4 provides security control selection guidance for non-national security systems. CNSS Instruction 1253 provides similar guidance for national security systems.)*

Implement Step

Implement the security controls and document how the controls are deployed within the system and environment of operation. *(NIST Special Publication 800-53A Revision 4 provides security control assessment procedures for security controls defined in NIST Special Publication 800-53.)*

Assess Step

Assess the security controls using appropriate procedures to determine the extent to which the controls are implemented correctly, operating as intended, and producing the desired outcome with respect to meeting the security requirements for the system.

Authorize Step

Authorize system operation based upon a determination of the risk to organizational operations and assets, individuals, other organizations and the Nation resulting from the operation of the system and the decision that this risk is acceptable. *(NIST Special Publication 800-37 Revision 2 provides guidance on authorizing system to operate.)*

Monitor Step

Monitor and assess selected security controls in the system on an ongoing basis including assessing security control effectiveness, documenting changes to the system or environment of operation, conducting security impact analyses of the associated changes, and reporting the security state of the system to appropriate organizational officials. *(NIST Special Publication 800-37 Revision 2 provides guidance on monitoring the security controls in the environment of operation, the ongoing risk determination and acceptance, and the approved system authorization to operated status.)*[13]

Federal information security management act (FISMA)

FISMA background

The Federal Information Security Management Act (FISMA) requires each federal agency to develop, document, and implement an agency-wide program to provide information security for the information and systems that support the operations and assets of the agency, including those provided or managed by another agency, contractor, or other sources.

The Federal Information Security Modernization Act of 2014 amends the Federal Information Security Management Act of 2002 (FISMA) and provides several modifications that modernize Federal security practices to address evolving security concerns. These changes result in less overall reporting, strengthens the use of continuous monitoring in systems, increased focus on the agencies for compliance, and reporting that is more focused on the issues caused by security incidents.

FISMA, along with the Paperwork Reduction Act of 1995 and the Information Technology Management Reform Act of 1996 (Clinger–Cohen Act), explicitly emphasizes a risk-based policy for cost-effective security. In support of and reinforcing this legislation, the Office of Management and Budget (OMB) through Circular A-130, 'Managing Federal Information as a Strategic Resource,' requires executive agencies within the federal government to:

- Plan for security
- Ensure that appropriate officials are assigned security responsibility
- Periodically review the security controls in their systems
- Authorize system processing prior to operations and, periodically, thereafter

As a key element of the FISMA Implementation Project, NIST also developed an integrated RMF which effectively brings together all of the FISMA-related security standards and guidance to promote the development of comprehensive and balanced information security programs by agencies.[14]

Office of management and budget (OMB)

Circular No. A-130 – Managing Information as a Strategic Resource.

Overview

The Circular establishes general policy for information governance, acquisitions, records management, open data, workforce, security, and privacy. It also emphasizes the role of both privacy and security in the Federal information life cycle. Importantly, it represents a shift from viewing security and privacy requirements as compliance exercises to understanding security and privacy as crucial elements of a comprehensive, strategic, and continuous risk-based program at Federal agencies.

The Circular promotes innovation, enables appropriate information sharing, and fosters the wide-scale and rapid adoption of new technologies while strengthening protections for security and privacy.[15]

Summary

An overview of three major U.S. Government CRM frameworks, which will provide a basis for risk analysis and assessment in concert with cyber forensic procedures was examined. While these frameworks are written and intended to provide guidance for U.S. government agencies, they have applicability to public and private organizations as well. Ahead is a review of a process for assessing cyber forensic risk.

ASSESSING CYBER FORENSICS RISK

The digital forensics process is a highly technical field that is dependent on the proper implementation of specific, well-accepted protocols and procedures. Inadequate forensic tools and technical examination, as well as lack of adherence to appropriate protocols and procedures can result in evidence that does not meet legal standards of proof and admissibility. Digital forensics risk arises, for example, when personnel lack the proper tools to conduct investigations, fail to process evidentiary data properly, or do not follow accepted protocols and procedures.

To create this quantitative approach, we will borrow from the COSO Enterprise Risk Management Framework model.

Digital forensic risk management process

The process (or cycle) of digital forensics risk management has four main parts (see Figure 7.3):

- Forensic Risk Identification
- Forensic Risk Assessment
- Forensic Risk Response
- Forensic Risk Monitoring

1. Identification and documentation of digital forensic risks
   Risks are to be considered as anything that could potentially impact successful achievement of a digital forensic investigation. All risks should be clearly identified and well-documented.

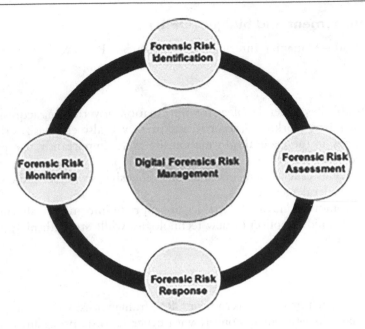

Figure 7.3 Digital forensics risk management process (cycle)

For the purposes of this methodology, we are following a previously developed, 'metric survey approach for assessing digital forensics risk,' that separated the digital forensics process into eight main risk areas:

- Protocols & Procedures
- Evidence Assessment
- Evidence Acquisition
- Evidence Examination
- Documentation & Reporting
- Digital Forensics Tools
- Legal Aspects
- Victim Relations

2. Assessment of documented digital forensic risks

Simply identifying risks is not enough; impact of the risk should be understood, as well as probability, within an estimated timeframe. Once significant risks have been adequately documented, the next task is to assess them in terms of their likelihood and estimated significance. Sometimes, it's difficult or impossible to accurately predict the probability or time-frame of certain risks. Nonetheless, this exercise should be performed to the best of the organization's ability, and across all levels.

Various methods exist for assessment of documented risks, from simple qualitative approaches to more in-depth mathematical models. The point of this task is to help management determine which digital forensic risks deserve the most immediate attention.

Sahinoglu et al., using industry best practices guidelines, such as the U.S. Department of Justice's Forensic Examination of Digital Evidence: A Guide for Law Enforcement, and collecting survey data to calculate a quantitative risk index for the digital forensics process, generated a quantitative digital forensics risk index.

Sahinoglu et al. identified eight specific vulnerabilities areas, as part of this digital forensics risk index: Protocols and Procedures, Evidence Assessment, Evidence Acquisition,

Evidence Examination, Documentation & Reporting, Digital Forensics Tools, Legal Aspects, and Victim Relations.

Questions from the Sahinoglu et al. survey were designed to elicit responses regarding the perceived risk to proper digital forensics procedures, evidence handling/examination, admissibility, and other associated issues from particular threats, as well as the counter-measures the respondents may employ to counteract those threats.

The specific digital forensic risk areas and associated digital forensic sub-processes identified by Sahinoglu and his team are summarized in Table 7.5.

Table 7.5 Digital forensics risk categories detailing vulnerabilities and threats[16]

Digital forensic risk areas	Digital forensic sub-processes
1. Protocols & Procedures	Mission Statement
	Personnel
	Administrative
	Service Request/Intake
	Case Management
	Evidence Handling/Retention
	Case Processing
	Technical Procedures Development
2. Evidence Assessment	Case Assessment
	Onsite
	Location Assessment Processing
	Search Authority
	Evaluation
3. Evidence Acquisition	Precautions
	Protection
	Preservation
4. Evidence Examination	Preparation
	Physical Extraction
	Logical Extraction
	Timeframe Analysis
	Data Handling Analysis
	Application File Analysis
	Ownership/Possession
5. Documentation & Reporting	Examiner Notes
	Examiner Report
	Findings Detail/Summation
6. Digital Forensics Tools	Hardware
	Software
	Training
	Funding
7. Legal Aspects	Jurisdiction
	Search & Seizure
	Admissibility
8. Victim Relations	Victim Rights & Support
	Court Preparation
	Media

Table 7.6 Risk rating example of the eight digital forensic risk areas

Digital forensic risk areas	Impact	Likelihood	Risk rating
	Risk = Impact × Likelihood		
1. Protocols & Procedures	1	1	1
2. Evidence Assessment	2	1	2
3. Evidence Acquisition	3	2	6
4. Evidence Examination	3	3	9
5. Documentation & Reporting	2	2	4
6. Digital Forensics Tools	3	2	6
7. Legal Aspects	1	1	1
8. Victim Relations	1	1	1

Calculating forensic risk

According to ISACA (www.isaca.org), risk is the combination of the probability of an event and its consequence. In general, this can be explained as: Risk = Impact × Likelihood.

Using the following definitions:

- Impact – based on review of the evaluation criteria, what is the impact of a failure or error upon the Digital Forensic Risk Area?
- Likelihood – based on review of the evaluation criteria, what is the likelihood of a failure or error upon the Digital Forensic Risk Area?

Along with a three-tier scoring system of High (H) = 3, Medium (M) = 2, and Low (L) = 1, the digital forensic professional will be able to assign a value to the impact and likelihood of failure or error in the specific digital forensic area, using these values then to calculate a risk rating for each individual digital forensic risk area.

Table 7.6 shows the results of using arbitrary, example values for impact and likelihood to calculate digital forensic risk, for the eight digital forensic risk areas identified by Sahinoglu et al.

This quantitative assessment of digital forensic risk will be a highly useful tool to interested parties such as investigators, administrators, and officers of the court seeking to minimize/mitigate digital forensics risk.

Minimization/mitigation of digital forensics risk, will greatly facilitate the success of digital forensics investigations, ensuring that legal standards of proof and admissibility are ultimately met.[17]

Risk assessment – heat map

A risk heat map is a tool used to present the results of a risk assessment process visually and in a meaningful and concise way.

Heat maps are a way of representing the resulting qualitative or quantitative evaluations of the probability of risk occurrence (likelihood) and impact. In this example, the impact of the digital forensic risk area failing.

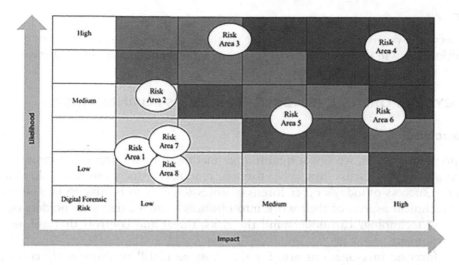

Figure 7.4 Digital forensics risk area heat map[18]

A heat map visualizing the digital forensics risk area rating calculated in Table 7.6 is shown in Figure 7.4.

So, what does this all mean to the cyber forensic examiner? After identifying and quantifying the risk areas, putting together an action plan to respond or address the high-priority risks is the next step.

There are several basic risk response strategies.

Avoidance: Eradicate the risk by eliminating the cause of the risk event. For example, an organization might decide to completely outsource their digital forensic risk process, due to a lack of funding or experience within its own ranks.

Reduction: Reduce the impact and/or likelihood of the risk event to an acceptable level. This could be accomplished by improving controls, automating procedures, or employee awareness training.

Transfer: Contractually transfer the risk and loss to a third party. This could be through the purchase of cybersecurity insurance, so that in the event of a breach, the incident response and digital forensic costs might be covered through an insurance policy. Keep in mind that many insurance policies will require proper controls and security to be documented and determined to be operating effectively.

Acceptance: Retain the risk and develop plans to cover the financial consequences. This is based on the organization's 'risk appetite,' and potentially industry-wide regulations that might levy fines or penalties in the event of a cyber event.

Risk monitoring

Remember that the risk management process is not 'one and done.' It is an ongoing process that must be constantly monitored. Risks are constantly evolving, just as organizations continually change. By monitoring risks with an ongoing process, organizations can be in the best position to respond appropriately when a risk event occurs.

Summary

In this section, we presented a quantitative method for assessing cyber forensic risk. In the next section, we will discuss how cyber forensic readiness helps to reduce business risk.

HOW CYBER FORENSIC READINESS REDUCES BUSINESS RISK

Introduction

In the previous section, we saw a quantitative method for assessing cyber forensic risk. In this section, we will discuss how cyber forensic readiness reduces business risk. One of the keys to the success of today's cyber forensic professionals is to be able to be able to 'translate' the technical aspects of their work into business language that stakeholders outside of information technology can understand the risks, threat and controls that are needed for their environment.

Digital forensic investigations are, for the most part, still predominantly conducted in response to an incident. With this reactive approach, there is extreme pressure put on the investigation team to gather and process digital evidence before it is no longer available or has been modified. Showing signs of weakness, being reactive to incidents suggests that organizations are not acting on their own initiative to identify problem areas and develop strategies for its suppression.

For investigations to truly become proactive, organizations must closely examine the time, money, and resources invested into their overall investigative capabilities. Digital forensic readiness is a process used by organizations to maximize their electronically stored information (ESI) to reduce the cost of digital forensic investigations. At the starting point, there needs to be a breakdown of risks including both internal events – those that can be controlled and take place within the boundaries of control (e.g., outages, human error) – and external events – those that cannot be controlled and take place outside the boundaries of control (e.g., floods, regulations).

Here are six practical and realistic scenarios that can be used to demonstrate a pro-active initiative to manage business risk.[19]

Scenario #1: Reducing the impact of cybercrime
Every business is facing an evolving threat environment, which increases the risk to their ongoing operations. Those organizations that implement a forensic risk methodology will have a better understanding of the threats that could potentially cripple their IT systems. This allows the organizations to implement appropriate controls to help prevent an attack and be better prepared for responding to an incident.

Scenario #2: Validating the impact of cybercrime or disputes
Despite an organization's best efforts, a breach or exfiltration of data is just a matter of when, not if. In that case, an organization needs to be prepared to quantify the impact of the event. The information pulled together during the forensic risk assessment, provides both the likelihood and impact of a forensic event.

Scenario #3: Producing evidence to support organizational disciplinary issues
While external hacking and breaches get the majority of publicity in the news, there is also a growing threat from insiders. Organizations should have a written Code of Conduct and Acceptable Use policy. These establish a professional and ethical culture for the workplace. If employees violate these guidelines, they can be subject to disciplinary actions, up to and including termination. If there is any legal recourse taken by a

terminated employee, having sound digital evidence will help protect the organization's interest.

Scenario #4: Demonstrating compliance with regulatory or legal requirements

Compliance pressure is increasing globally for all organizations in all industries. There is a difference between being in 'compliance' vs. being 'secure.' However, having documented evidence of compliance, as a result of a forensic risk assessment, is imperative to responding to regulatory authorities.

Scenario #5: Effectively managing the release of court-ordered data

Given today's litigious society, it is inevitable that an organization will face disputes that end up in court. Being able to respond the E-Discovery requests, subpoenas, and litigation hold court orders, in an efficient and effective way is imperative. From the due diligence performed during the forensic risk assessment, organizations will be prepared to maintain the admissibility of ESI and requirements described within the U.S. Federal Rules of Evidence.

Scenario #6: Supporting contractual and/or commercial agreements

With more and more organizations outsourcing IT functions to cloud providers, the risk from third-parties will likely rank high in a forensic risk assessment. By identifying and classifying the types of data that is shared with outside entities, organizations will be in a better position to protect and defend critical and sensitive information, such as PII, PHI, or credit/debit card information. It is also important for organizations to understand and monitor compliance with contractual obligations from third-party providers.

By following a reactive approach to digital forensic investigations, organizations foster a perception that they lack is initiative for managing risk. Conversely, when organizations implement strategies to proactively gather potential sources of digital evidence in support of the business risk scenarios, they showcase their ability to effectively manage risk.[20]

Now that we have presented six real-world scenarios that show the benefits of cyber forensic risk management, let's explore the advantages of cyber forensic readiness.

In the event that an organization's business is brought to a standstill by an unwanted or unforeseen event, whether natural or man-made, the business needs to recover and continue. As a result, strategies such as incident response, awareness training, disaster recovery, and business continuity planning have become basic components of organizations' operational structure.

In addition to recovery issues, an unwanted incident can also result in other issues such as insurance claims, legal matters, and regulatory issues. In the course of recovery and investigation, claims may arise against employees, third parties or even the organization, for example, pertaining to what led to the incident. Could it have been negligence, malicious intent, fraud, or sabotage?

Digital evidence becomes very important when such issues arise in an organization that uses IT infrastructure, even if the usage is minimal. Digital forensics tools and techniques are available for retrieving and analyzing digital evidence. Users of information systems leave digital footprints whenever they use the systems – be they computer systems, smartphones, mobile phones, tablets, or networks (i.e., the Internet, intranets, phone networks).[21]

What is forensic readiness?

U.K. government's National Technical Authority for Information Assurance – Communications-Electronics Security Group (CESG) Good Practice Guide No. 18, (2009) *Forensic Readiness*,

Issue 1.0, defines forensic readiness as: 'The achievement of an appropriate level of capability by an organization in order for it to be able to collect, preserve, protect and analyze digital evidence so that this evidence can be effectively used in any legal matters, in disciplinary matters, in an employment tribunal or court of law.'[22]

Besides 'readiness,' another term that is used is 'resilience.' Accenture describes resilience as a 'bend, but don't break' approach to securing the enterprise that combines the disciplines of cybersecurity, business continuity, and enterprise resilience.[23]

While data have existed for centuries in one form or another, from primitive paintings on cave walls, to digital data that reside on today's distributed server platforms. Data have four unique attributes: volume, velocity, variety, and veracity.

> **Volume** – Organizations now capture and process greater volumes of data than ever before. Only a few years ago, working with a 100-megabyte file was considered a lot of data. Today, data can be measured in zettabytes, or ZBs, which is equal to 1 trillion megabytes.
>
> **Velocity** – Beyond the vast amount of data collected, today's globalization and connectivity result in data produced at incredible and increasing speeds. IBM estimates that in 2012, 2.8 ZBs were created; by the end of 2020, the total data generated annually is forecasted to reach 40 ZBs.
>
> **Variety** – Data are being identified, captured, and stored from an increasing number of sources. From customer transactions to transmissions from outer space, the variety of data defies comprehension. Internal sources such as accounting, finance, and customer records have been complemented by the proliferation of external data sources.
>
> **Veracity** – this fourth 'V' is the most frequently overlooked attribute of data because it is often difficult to determine the quality or accuracy of data. The data must faithfully reflect the truth. In organizations that lack a strong data governance structure, records can be incomplete, entries could have errors and data might be inconsistently formatted. All of these issues can compromise analysis and produce inaccurate results.[24]

In the course of operations, organizations generate a lot of digital data and records. Such data and records can become crucial pieces of evidence in the event of an unwanted incident. Some of this digital evidence is stored and preserved as part of disaster recovery and business continuity processes as well as document-retention policies. See Table 7.7 for examples of the various types of digital evidence.

The highly volatile nature of digital evidence demands that it be treated delicately by safeguarding chain of custody. Having a forensic readiness plan in place ensures that in the event digital evidence is required, it will be readily available and in an acceptable form. This requires training of staff and having proper policies in place to ensure compliance.

Forensic readiness planning complements other organizational plans and processes, including disaster recovery, business continuity, and document-retention policies. Conventional disaster recovery and business continuity processes usually concentrate on low-frequency/high-impact events; a forensic readiness plan would, however, cover high-frequency/low-impact events as well. While the latter tend to appear insignificant, they could be the cause or source of a major disaster.

Forensic readiness planning is part of a quality information risk management approach. Risk areas have to be identified and assessed and measures must be taken to avoid and minimize the impact of such risk. Organizations with a good risk assessment and information security framework would find it easier to adopt a forensic readiness plan.

Table 7.7 Various types of digital evidence[25]

Address Books and Contact Lists	Documents and Spreadsheets Events
Audio files and voice recordings	Email messages and attachments
Backup files	Hidden and system files
Bookmarks and favorites	History files organizer items
Browser history	Log files
Calendars	Organizer items
Compressed archives	Page files and printer spooler files
Configuration and .INI files	Pictures, images, and digital photos
Cookies	Temporary files
Databases	Videos
Deleted files	Virtual systems

A forensic readiness plan should have the following goals:

- To gather admissible evidence legally without interfering with business processes
- To gather evidence targeting potential crimes and disputes that could have adverse impact on an organization
- To allow investigations to proceed at costs proportional to the incident
- To minimize interruption of operations by investigations
- To ensure that evidence impacts positively on the outcome of any legal action[26]

A forensic readiness implementation guide

A forensic readiness plan is meant to prepare an organization for an event the occurrence of which cannot be predicted. In preparation, an organization should review and analyze security – technical controls, policies, procedures, and skill sets. This can be carried out by a skilled forensic investigator, who can recommend proper amendments and action that can be taken to improve upon what is in place and ensure a good forensic readiness plan.

The goals and objectives of the organization and its risk appetite need to be identified, the security posture analyzed, employees educated and enlightened on the forensic readiness plan, and the action plan formulated to deal with identified gaps in the *status quo*. Knowing the goals, objectives, and risk appetite helps to determine what would be considered significant or relevant risk, what type of incidents should be expected, and how to respond to them. The current security level should then be reviewed for adequacy and to expose any potential loopholes. Employees need to be informed and educated regarding the forensic readiness plan to ensure their compliance. Finally, identified loopholes are mitigated by instituting appropriate measures.

The Carolina Crime Report 'Forensic Readiness Checklist,' offered the following ten points for a forensic readiness checklist:

1. Define the business scenarios that would require digital evidence. This helps to streamline where and how to concentrate evidence collection storage.
2. Identify potential evidence sources and the types of evidence.

3. Determine evidence collection requirements.
4. Establish capability for secure evidence gathering and collection in a forensically sound manner.
5. Establish a policy for proper chain of custody.
6. Ensure monitoring targets detection and deterrence of major incidents.
7. Specify the circumstances at which point the escalation of a full formal digital investigation should commence.
8. Educate and train staff on incident response and awareness to ensure that they comprehend their role in the digital evidence process and the importance and sensitivity of it.
9. Document evidence-based cases, describing the incident and its impact.
10. Ensure legal review to facilitate appropriate action in response to an incident.

A digital forensic professional performing a forensic readiness assessment should check to see that the above points can be deduced from the forensic readiness policy of an organization.[27]

Increased use and dependency on information technology for running organizations and businesses have resulted in the availability of digital footprints that can be used to unravel the what, where, how and why in the event of an unwanted incident. Digital evidence can lead to the indictment or vindication of an individual or organization. Digital evidence needs to be gathered and treated with due care, usually by applying chain-of-custody requirements, because of its high volatility.

While many organizations are currently aware of the importance and need for disaster recovery and business continuity plans, they must also recognize the need for and importance of forensic readiness planning. The tendency is to be reactive, waiting for an incident to occur then trying to handle it and carry out investigations – gathering evidence after the fact. As a result, operations become disrupted, some evidence may be altered or lost, and evidence may not be handled in an acceptable manner.

Forensic readiness greatly minimizes these problems, especially as a great deal of the evidence required is available before the incident, during the incident, and before investigations begin. As a result, time and money are saved, potential incidents are mitigated, and business continuity and compliance are ensured, with minimal disruption and interruption of operations. Forensic readiness also assists in ensuring employees' compliance with the organization's policies and regulatory requirements due to constant monitoring and review.[28]

Summary

An overview of ERM, along with some of the key considerations for CRM and how CRM interrelates to cybersecurity and cyber forensics, was addressed. Also addressed in this chapter were some of the U.S. government regulations that pertain to CRM, and a quantitative methodology for assessing cyber forensics risk. The chapter concluded with an insight into how cyber forensics readiness can assist in helping to reduce business risk.

NOTES

1 Peterson, O. (July 1, 2019), "Basics of Enterprise Risk Management (ERM): How to Get Started," www.process.st/enterprise-risk-management/, retrieved October 2020, used with permission.
2 Ibid.
3 Everson, M., Chesley, D., Martens, F., et al., (June 2017), "Enterprise Risk Management Integrating with Strategy and Performance, Executive Summary," Committee of Sponsoring Organizations of

the Treadway Commission (COSO), www.coso.org/Documents/2017-COSO-ERM-Integrating-with-Strategy-and-Performance-Executive-Summary.pdf, retrieved October 2020.

4 Ibid.

5 Mihm, J. Christopher, et al., (December 2016), "Enterprise Risk Management Selected Agencies' Experiences Illustrate Good Practices in Managing Risk," United States Government Accountability Office, GAO-17-63, page 2, www.gao.gov/assets/690/681342.pdf, retrieved December 7, 2020.

6 (n.a.), (February 2016), "CERT Resilience Management Model (CERT-RMM) Version 1.2," Carnegie Mellon University, Software Engineering Institute, https://resources.sei.cmu.edu/library/asset-view.cfm?assetid=508084, retrieved October 3, 2020.

7 (n.a.), (n.d.), "The 20 CIS Controls & Resources," Center for Internet Security www.cisecurity.org/controls/cis-controls-list, retrieved October 13, 2020.

8 Trevors, M., Wallen, C., (2017) "Cyber Hygiene: A Baseline Set of Practices," Carnegie Mellon University, Software Engineering Institute, https://resources.sei.cmu.edu/library/asset-view.cfm?assetid=508765, retrieved October 23, 2020.

9 Tobar, D., (February 9, 2018), "7 Considerations for Cyber Risk Management," Carnegie Mellon University, Software Engineering Institute, https://insights.sei.cmu.edu/insider-threat/2018/02/7-considerations-for-cyber-risk-management.html, retrieved October 10, 2020.

10 (n.a.), (July 15, 2019), "How digital forensics can prevent cybercrime", Workflow, A conversation with security expert Andrew Morrison. https://workflow.servicenow.com/security-risk/future-digital-forensics-technology. Used with permission from ServiceNow, Inc., retrieved October 30, 2020.

11 (n.a.), (March 20, 2020), "NIST Cybersecurity Framework. Risk Management Framework," www.nist.gov/cyberframework/risk-management-framework, retrieved October 30, 2020.

12 (n.a.), (December 3, 2020), "Risk Management Framework (RMF) Overview," Computer Security Resource Center, https://csrc.nist.gov/projects/risk-management/rmf-overview, retrieved December 13, 2020.

13 (n.a.), (October 13, 2020), "NIST Risk Management Framework (RMF) Overview," https://csrc.nist.gov/projects/risk-management/rmf-overview, retrieved October 30, 2020.

14 (n.a.), (October 13, 2020), "FISMA Implementation Project," NIST Information Technology Library, Computer Security Resource Center, https://csrc.nist.gov/Projects/risk-management/detailed-overview, retrieved October 29, 2020.

15 (n.a.), (n.d.), "Circular No. A-130 – Managing Information as a Strategic Resource," CIO.gov. Policies and Priorities, www.cio.gov/policies-and-priorities/circular-a-130/, retrieved October 29, 2020.

16 Sahinoglu, M., Stockton, S., Morton, S., Barclay, R., Eryilmaz, M., (June 2014), "Assessing Digital Forensics Risk: A Metric Survey Approach," Informatics Institute, AUM, Montgomery, 36124 USA; Computer Engineering, ATILIM University, Ankara, 06836 Turkey. www.researchgate.net/publication/268507819_ASSESSING_DIGITAL_FORENSICS_RISK_A_METRIC_SURVEY_APPROACH, retrieved October 2020.

17 Ibid.

18 Figure 7.4 was created and designed by authors using resources from PoweredTemplate.com, https://poweredtemplate.com/safety-risk-matrix-55083/, Image: PoweredTemplate.com, baseline, generic heatmap template downloaded from PoweredTemplate.com and used with permission.

19 Sachowski, J., (December 11, 2015), "How 'Digital Forensic Readiness' Reduces Business Risk," DARKReading. www.darkreading.com/attacks-breaches/how-digital-forensic-readiness-reduces-business-risk/a/d-id/1323508, Copyright (© 2015) From (How "Digital Forensic Readiness" Reduces Business Risk.) by (Jason Sachowski). Reproduced by permission of Taylor and Francis Group, LLC, a division of Informa plc. retrieved November 10, 2020.

20 Ibid.

21 Sule, D., (January 1, 2014), "Importance of Forensics Readiness," ISACA Journal Archives, www.isaca.org/resources/isaca-journal/past-issues/2014/importance-of-forensic-readiness, retrieved November 11, 2020, used with permission.

22 Ibid.

23 (n.a.), (n.d.), "Cyber Resilient Business," www.accenture.com/us-en/insights/cyber-security-index, retrieved December 9, 2020.

24 Stippich, W., and Preber, B., (2016), "Data Analytics, Evaluating Internal Audit's Value," https://bookstore.theiia.org/data-analytics-elevating-internal-audits-value, retrieved, December 9, 2020.

25 Gubanov, Y., (July 11, 2012), "Forensic Focus, Retrieving Digital Evidence: Methods, Techniques and Issues," www.forensicfocus.com/articles/retrieving-digital-evidence-methods-techniques-and-issues/, retrieved December 10, 2020.

26 Sule, D., (January 1, 2014), "Importance of Forensics Readiness," ISACA Journal Archives, www.isaca.org/resources/isaca-journal/past-issues/2014/importance-of-forensic-readiness, retrieved November 11, 2020, used with permission.

27 Ibid.

28 Ibid.

Chapter 8

Mobile device forensics

An introduction

Andrew Hrenak

CONTENTS

INTRODUCTION

Computer forensics versus mobile device forensics

In the course of my investigations, the comparison between computer forensic processing and that of mobile device forensic processing is often raised. Obviously, the hardware architecture of computer-based forensic processing is the primary difference between the two. Generally, computers have been designed for lengthy service, and the internal memory is replaceable. The provision for an exchangeable hard disk drive or solid-state drive within the computer chassis provides the ability to follow a reliable and consistent forensic process.

Document the hardware through photography, remove the secondary memory, prevent alteration to the memory through the use of a write blocking device, and create the forensic image. Upon completion of your verification and validation efforts, introduce the forensic image into your forensic tool of choice and find evidence! Simple enough, right?

The standardization of data and power connections provides the examiner the ability to establish physical connection to the media through whatever cable or connector is being used. Subsequently the imaging process is relatively benign, and access to the stored data is achieved.

However, the described model is dependent upon the lack, or presence, of encryption. Once encountered, encryption creates a host of alternate steps to the process. In consideration that this is not the chapter addressing computer-based forensics, we will conclude with the computer processing description at this point.

MOBILE DEVICES

In comparison, mobile devices are almost as different as their owners. The form factors can be that of a bar, a block, or a flip open style. The keypad and basic control buttons are physical buttons or virtual representations within the digitizer screen. The data port will be serial based, but is of many configurations. USB A, B, or C, or some proprietary combination thereof; as evidenced by Apple's early thirty pin data connector and current Lightning cable.

Historically, the data port had been left in the 'open' for power and data connectivity, resulting in a fairly easy access point into the operating system (OS). Depending upon the operating system, access to the data stored within the device could be gained via the data port. Trending emphasis on the security of personal information has changed the 'easy access' to data we, as forensic examiners, previously enjoyed.

Mobile device hardening

But why not just remove the memory from the mobile device as with personal computers you ask? Well, the hardware architecture of mobile devices is quite dissimilar to that found in a legacy computer. Mobile devices are built to withstand user abuse, as such they are ruggedized. The presence of seals, tape, clips, and screws creates an obstacle course of sorts just to access the interior of the device! Once inside, the *printed controller board* (PCB) has a variety of surface-mounted components soldered to it – each having a distinct purpose, and each following the economy of space rule. As the manufacturing processes improve, the size of the hardware becomes increasingly smaller. As the size decreases, the difficulty in identification of the surface-mounted device (SMD) and accessibility to them increases.

Unlike most computers, the memory module within a mobile device is soldered to the PCB. This requires specialized techniques to remove it without causing damage and potential data loss. Not to mention a varied number of adapters to be used to facilitate connection with the memory module following its removal from the PCB. There are additional factors regarding removal of the memory module that will be addressed later in this section. But for now, understand that it is undesirable to separate the memory module from the PCB.

That leaves us with data port access. As indicated, the operating system and security architecture that is resident within the mobile device may obstruct data transfer at the data port. This requires the examiner to have knowledge regarding default security features of the various operating systems as well as the nuances within each operating system's version. Once the security features are identified, paths to bypass the enabled features are required to establish connection and access to the stored data.

Mobile devices: a peek inside

In addition to the operating system-based security features, manufacturers are now using hardware that has various levels of security built into them. Each model of device will often have a different *central processing unit* (CPU) and memory module. Unless the examiner is encountering a base model mobile device, the data can be stored in an encrypted format while at rest. The data may be encrypted using *Full Disk Encryption (FDE), Full Volume Encryption* (FVE), or *File-Based Encryption* (FBE).

Thus, without knowledge of the encryption key, which is commonly based upon the pass code; the data is recovered in an encrypted format. Should this occur, an unreasonable amount of effort and time may be exhausted in attempting to recover decrypted data. Suffice it to say, having knowledge of the pass code is an essential element in modern mobile device forensics.

Historically, the examination of personal computers also required the examiner's awareness to network-based communication. More strategically, to ensure the data currently within the personal computer memory stayed in its current state. In a majority of network installations, we can isolate the personal computer by disconnecting the network cable. Should it be connecting to a wireless network, a physical switch on the computer chassis may be present to disconnect it. Or one must interact with the personal computer and disable the WiFi via configuration settings.

Isolation is also an element of consideration within mobile device forensics.

The very nature of mobile device function is to maintain continuous connection to the service network, a wireless connection point, and Bluetooth devices as directed by the user's preference. This performance results in memory that is in a state of change, almost continuously. From a forensic perspective, this is definitely not desirable. Incoming data may overwrite previously allocated, but no longer desired information. In some instances, this information may be marked for 'deletion' by the user and critical to the investigation. Even though marked for deletion, the data may still be recoverable unless overwritten.

Therefore, strict adherence to radio isolation is critical in preserving the stored data. This may be accomplished through various mechanisms such as an attenuation enclosure, removal of the *Subscriber Identification Module* (SIM), or enabling 'Airplane Mode.' The listed mechanisms have been presented in an order of greatest isolation to least reliable isolation.

PERSONAL COMPUTER FORENSICS VS MOBILE DEVICE FORENSICS

An immediate difference noted between personal computer and mobile device forensics comes in the form of how the *user-generated* data is stored. By definition, file systems store data in a hierarchal manner to assist in the storage and recovery of data for the user's interaction. In consideration of the various mobile device manufacturers, the file system formats associated with personal computers is just different than that within mobile devices.

Yes, data is stored in a reliable manner and is fully accessible to the user. It just is stored in a combination of files and applications or Apps. The Apps are made up of databases, various supporting files and *preference lists* (.plist), or extensible markup language (.xml) files.

The presence of databases within the Apps requires the forensic examiner to have capability in database interrogation. This skill set allows the examiner to locate and carve the relevant information from the database when a tool of choice simply does not support that particular App. This skill is even more important given the increasing number of Apps made available to end-users daily.

Personal computers and mobile devices both share reliance upon networks for communication and data transfer. Both can use cloud-based services for data storage, email, and third-party application functionality. However, it is with mobile device usage that globalization of data became an area of investigative resource for the forensic examiner. People, end-users, want their data accessible across all of their authenticated devices. For example, an informational presentation produced using their personal computer at work is 'shared' through cloud services.

This enables the end-user the opportunity to travel 'light' and still have access to their information at the physical location of the presentation as long as the location is accessible by WiFi or cellular service. It is then accessed through their mobile device and pulled down to complete the presentation. Another example is the photograph taken on vacation that is posted to the user's cloud service and shared among mobile devices associated to the primary account, or others given access to the account.

The predominance of this type of data sharing and storage requires an examiner to consider not only what *is* present but also what *is not* present in the mobile device. Then, consider *why* the data is not present and how to document it from the location it is stored that is beyond the mobile device itself.

Now for the ultimate difference. *It Depends!*

According to a Pew research study on mobile device usage, conducted in June, 2019. Roughly 96% of adults in the United States own a mobile device.[1] That allows for a significant number of mobile devices of various manufacture, various operating systems, and sub-versions. Of significant importance is the presence of third-party Apps within the mobile device servicing the preferences of the device owner. This is in comparison to the base line applications within a Windows-, Mac-, or Linux-based personal computer.

Adding an 'Office Suite' or other utility to facilitate data tabulation, text editing, and image editing results in storing files as files. Whereas, a mobile device stores user-generated information as files, in database tables, and as BLOBs (Binary Large Objects) within the database. Relevant information may be located in system generated files, previously identified as PLIST or XML preference files.

In my experience, I have found no two mobile devices defined as 'smart phones,' to hold the same number and type of Apps to service the needs of the user. This is especially true when a user owns and uses multiple mobile devices; one for personal use and one for work use. Comparatively, I have encountered similarly 'profiled' personal computers sharing the exact same software.

To recap, the primary differences between forensic examinations of personal computers and those of mobile devices are:

- Form Factor
- Memory
- Encryption

- Isolation
- Data storage
- Data globalization

THE MOBILE DEVICE: OPERATING SPECS

There are a number of companies that produce mobile devices, and the following list presents a few.

- Alcatel
- Apple
- Blackberry
- CoolPad (Jeyung)
- Google
- Huawei
- Kyocera
- Motorola
- Nokia
- One
- Samsung

Each of the manufacturers provides mobile device hardware to the various Service Providers within the globe. To provide a more concise understanding, our discussion moving forward will relate to the primary forms of frequency modulation in use within the United States. Those are *Code Division Multiple Access*, or CDMA, and the *Global System for Mobile communication*, or GSM.

Code Division Multiple Access, or CDMA, facilitates communication across a single channel, by multiple mobile devices simultaneously. Through the use of assigned identification codes, the messages are simultaneously transmitted. The presence of the unique code ensures the message is received by the intended recipient. This method of communication is based upon spread spectrum technology to transmit and receive messages via a single channel.

At the time of this information, the primary United States service providers using CDMA are:

- Sprint
- Verizon
- US Cellular

GSM utilizes frequency modulation based upon time to transmit and receive messages. Known as Time Division Multiple Access, or TDMA, provides multiple users the ability to transmit messages across a single channel. Each user is assigned a specific time slot within the channel on which to send and receive messages.

In addition to TDMA, the infrastructure of a GSM Network provides support for accurate and confidential message transfer. This is facilitated through an authentication key. The authentication key acts in a pre-shared key role that is recognized by the network infrastructure. Upon authentication, messaging is facilitated. The key is resident within the Subscriber Identification Module (SIM) located within the mobile device chassis.

The primary United States service providers using GSM are:

- AT&T
- T-Mobile

Mobile devices that support data transfer via cellular network service will employ a SIM for authentication purposes. The presence of the SIM does necessarily define the type of frequency, CDMA or GSM, in use. This is because SIM cards are present within any device utilizing the *Long-Term Evolution* (LTE) standard for wireless broadband communication.

Rather, the specific service provider is defined through the *Integrated Circuit Card Identifier or* ICCID. This code is comprised of a series numbers that are between 19 and 20 characters in length. The ICCID has a defined format which identifies the following features:

- Major industry identifier
- Country code or 'CC'
- Issuer Identifier or 'II'
- Account Identifier or 'SIM Number'

All of this information is presented in a linear format represented as

#

or as

MMCC III# #### #### ##C X (The '#' symbolizes the user's unique account identification)

Further defined as:

Industry Identifier (*2), Country Code (*2), Issuer Identifier (x3), Account ID (*11), Checksum (*1), X

Or commonly seen in an AT&T ICCID as being:

8901 410# #### #### ##4 X

In this example the '89' represents the Industry Identifier associated with Mobile Communications. The '01' represents the Country Code designation for the United States. This is followed by the Issuer Identifier designation for AT&T '410.' The remaining digits represent the unique account identification code for the user and the calculated checksum.

In the United States the following list of carriers and their respective Issuer Identifier codes are provided for reference.

- AT&T: 030, 150, 170, 410, 560, 680
- Verizon: 004, 005, 012, 480
- T-Mobile: 026, 160, 260, 490
- Sprint: 120

SIM cards come in various sizes. Those specifications are listed in Table 8.1.

In addition to removable SIM cards, a growing number of manufacturers are including the Embedded SIM, or eSIM, within the manufacturing process. This technology affords the end-user the ability to provision multiple carrier services within the same mobile device, without requiring the act of switching the physical SIM. Essentially, the eSIM would be assigned to the primary account and a secondary physical SIM would support a secondary account. This

Table 8.1 SIM card sizes and specifications

Form factor	Height	Width	Depth
Standard	3.37"	2.12"	.029"
Mini	0.984"	0.590"	.029"
Micro	0.590"	0.472"	.029"
Nano	0.484"	0.346"	.029"

allows the end-user an ability to switch between the two services within the graphical user interface (GUI).

The eSIM also supports online provisioning. This feature facilitates a more consistent manufacturing process and allows the end-user an ability to use the device across more than one network. It removes the need for 'unlocking' a device provisioned to one carrier network before migrating to another carrier network.

An additional advantage of the eSIM is in relation to the Internet of Things (IoT). Since eSIM technology is embedded within the PCB, it can be deployed in wearable IoT devices, thereby providing a globalization of data exchange for greater benefit to the end-user. Refer to Patrick Wilds Chapter 2 IoT and the role of cyber forensics for further information in this area.

Having knowledge in basic cellular networking principles, as well as the underlying format of frequency modulation being used; increases the effectiveness of a mobile device investigation. However, the information regarding the specific functionality and security features of CDMA or GSM network operations is quite deep. This expanse of knowledge is important to know if one were engaged in the 'Legal' practice of live communication intercepts. That practice is well beyond this discussion, so we will continue toward the mobile device itself.

One could compare a mobile device to an automobile. Both are used by consumers daily and both require power as well as a network to perform their intended purpose. More specifically, electrical power is required by both and the WiFi and cellular networks used to route communication are similar to the roads or highways facilitating vehicular travel.

MOBILE DEVICE DATA RECOVERY AND ANALYSIS

Mobile device forensics is more akin to having a very good neighborhood vehicle repair shop versus an auto dealership service center. The neighborhood shop is hosted by a relatively friendly, somewhat grumpy, albeit busy mechanic who listens to your description of the problem, identifies the cause, repairs the problem and provides preventive insight. The shop is neat, but it is a work place with parts here or there and a little dust in the corner. It also has tools, and I mean a lot of tools, old tools, new tools, as well as tools constructed for a specific repair task. The mechanic knows what is required to maintain your automobile.

In comparison, the service center is pristine, with clean painted floors and a waiting room with a barista. You pull into the service center, approach the service advisor and explain your issue. Subsequently, you are separated from your vehicle and it goes someplace within the service center, allegedly to a service technician. A period of time later, your car reappears and you are informed by the service advisor the problem is fixed, handed an invoice identifying a part replacement, and directed to pay the cashier. After leaving the dealership and traveling about fifty miles, the check engine light comes back on! After a return to the dealership you learn the technician used a replacement part having the wrong firmware.

The analogy of the two repair centers was provided, part in jest, but to illustrate a difference between being a tool centric examiner, or 'Service Technician,' and a tool agnostic forensic examiner AKA: 'mechanic.' Mobile device forensics requires the examiner to be proficient in data recovery, data analysis, and reporting. *Each area having its own depth of knowledge that must be sought out individually to be fully realized and implemented.*

Since my first mobile device examination in 2006, there has been an explosion in the capabilities of 'mobile forensic suites.' The suites provide similar functionality as is found within their personal computer counterparts such as XWays, AccessData's Forensic Tool Kit, Autopsy, and Forensic Explorer to name a few. Their strength in parsing the various data types and subsequent rendering of data into categories of use has been an invaluable time saver.

With regard to time management, they make the best use of the examiner's time freeing the examiner from routine parsing. Ideally, the processing phase is conducted while the employee is not at work, facilitating review of the data upon the employee's return. Through the development of a workflow, multiple mobile devices can be processed in an efficient manner, providing actionable information to those who seek it. In cases that merit, the mobile forensic suite allows more time to deep dive for the artifacts directly relevant to the case.

Unfortunately, the strengths of the mobile forensic suite have some consequence. Unverified reliance on a forensic suite may result in professional embarrassment as described below.

If you recall, I previously noted information regarding database interrogation and its importance in the storage of user-generated information. Also noted was the increasing number of third-party Apps used to improve the user's experience. Well, a number of years ago the commercial forensic suites were caught up in a race. At the time, third-party Apps were exploding onto the App field, each having individualized portions of encoding. Thus, the scripts within the forensic suites used for parsing App data needed to map the individual App due to the variables existing in App development. With updates to Apps came alterations in the database.

If the parsing script was not likewise updated the results of that script became unreliable to an extent. Therefore, the examiner had to identify a reliable method to reveal the data in the previously unmapped App. Or, indicate the data was not present as the mobile forensic suite did not reveal it.

To counter this complaint from examiners, the various companies released version updates at an increasing rate. Not to fix bugs or add supported devices. Rather to address improvements in App support. The mobile forensic suite market became a foot race to determine which mobile forensic suite had the greatest App support. Once released, the foot race began again until the next major release of the software.

As forensic examiners, one is taught to validate and verify. Those examiners who did not establish a practice of version validation were susceptible to embarrassment. In one example, a mobile forensic suite pushed an update out that did not decode the time stamps of native applications within a specific operating system version correctly. This resulted in a decoding error within the reported file metadata. In instances wherein chronologic reference was relevant, this defect created an unintentional mis-reporting of facts. Something a prosecuting attorney is not prone to embrace warmly. However, through validation processes the coding issues were identified and quickly addressed.

Please do not misinterpret my message. Mobile device forensic suites are a must have item. If your shop/lab can afford to have mobile device forensic suites from multiple vendors, I am all for it. These assets can be leveraged against one another to streamline the version validation process, as well as supporting the multiple processing of mobile devices simultaneously. Obviously, there are costs associated to this suggestion. Those include the initial acquisition fee, annual support fee, as well as the tool centric training necessary to operate the forensic tool suite at an optimal level of proficiency. Costly yes, however the return on investment (ROI) depending upon your business model may justify it.

Alternatively, your shop/lab can employ highly skilled 'mechanics' who can parse data via scripting, and validate tools through carving the old-fashioned way. This recommendation may be just as costly as maintaining multiple mobile device forensic suites and training due to the pay scale associated to the L33T, or shall I say, someone who is very good at what they do. An additional concern may include the retention cycle of examiners. People in this field are usually seeking self-improvement, whether through education or career path. Employee attrition is fairly common once certifications have been achieved. Simply put, a higher salary is a significant motivator to change jobs.

The acquisition and retention of examiners may lend more weight to the deployment of mobile forensic suite(s) in your shop/lab. The learning curve of the process is decreased, while production can be maintained at, or slightly below the L33T. More importantly, the mobile forensic suite won't leave you guessing, 'Now how did the script work that John wrote?'. In the absence of an employee, production would not be interrupted due to a proprietary knowledge issue.

Mobile device forensic suites

Now that we have discussed some of the issues surrounding the method of data recovery and analysis, let us venture down the road and discuss some of the mobile forensic suites that I have had the opportunity to use.

The following list of mobile device forensics tools is not all-inclusive, nor is it presented in any order of preference.

- Belcasoft Evidence Center
- Cellebrite's BlackLight
- Cellebrite's Universal Forensic Extraction Device (UFED)
- Cellebrite's Physical Analyzer
- Hancom's MD-NEXT/MD-LIVE
- Magnet Forensic's Axiom
- Micro Systemation's XRY/XAMN
- MobileEdit's Final Mobile
- Oxygen Forensic's Detective
- Paraben Forensic's E3:DS
- Susteen's DataPilot

Each of these utilities includes an extraction capability that will address the major mobile device operating systems in use today. The utilities support various levels of data recovery such as categoric, file system or physical.

Basic computer forensics defines that a physical level, or bit level, copy of the stored data is preferred. Unfortunately, not all pairings of processors and memory modules are supported for this level of recovery. Moreover, the case may dictate that the time it would take to obtain a physical level data recovery would exceed reasonableness and therefore not be a probable solution. This is commonly an issue when the search is based upon consent, or in cases involving covert operations.

To review, the physical level of data documentation is expected to start at the beginning of the memory module, or dataset, and continue until it reaches the final bit of data stored within the memory module. It is considered to be a complete and accurate representation of the data stored within the memory module at the time of capture. It may, or may not, include cyclical redundancy checksums (CRCs).

The file system level of data documentation does not start at the beginning of the memory module; rather, it starts at the beginning of the volume. Based upon the specific type of operating system, the captured data should include operating system as well as user-generated files. It should be noted, the number of system files that will be captured is based upon the type of operating system as well as the specific version of that operating system. In a comparison to personal computer forensics, this level of data documentation is similar to that being captured during a logical acquisition of an encrypted computer system such as the cases involving BitLocker.

Finally, there is the categoric, or 'bucket,' capture. I refer to it as the 'bucket' capture simply because the mobile forensic utility sweeps through the connected device and captures the buckets of data that comprise the key areas of user-generated information that is stored in a mobile device.

The buckets of data commonly include the following types of user-generated information.

- Contact List
- Call History
- Voicemail/Audio files
- Short Message System (SMS) text-based communication
- Multi Message System (MMS) communication
- Graphic Image files
- Video files
- Data files: These files include common third-party App files that may, or may not, be automatically decoded

The categoric level of data extraction is the only level that allows the examiner a choice in what specific data type is being extracted. This may be relevant should the scope of search be limited in cases involving a warranted search, or should proprietary business information be of concern.

In consideration of the aforementioned, it has been my experience that an extraction workflow should follow a systematic order to ensure the best result in the examination. The preferred ordering is:

1. Physical
2. File System
3. Categoric or bucket level

Again, each case has its own circumstances and each mobile device has its own factors that play a component in which level of data extraction should be completed. It is up to the forensic examiner to ensure the better extraction is used to fulfill the requirements of the case.

THE MOBILE DEVICE FORENSIC PROCESS

In my experience, the most influential element in mobile device forensics is the data recovery process. Quite simply, if the data has been altered or is inaccessible, the examination and analysis is greatly impacted. To reduce the likelihood of recovering altered data, there are a number of things that can be done during the seizure process to limit the alterations and provide for a greater probability of success. The following steps, or phases, are presented to provide guidance in the mobile device forensic process. Those steps are: Seize, Secure, Identify, Recovery, Analysis, and Reporting.[2]

Seize the mobile device

The process used during the seizure of the mobile device can make or break the results of the examination. Given the current emphasis on security, encryption of data at rest has become the new normal. Gone are the days when data is waiting to be harvested from the memory without some form of encryption present. Whether full volume, or file-based, encryption will inhibit your access to the data that you need to answer the question(s) posed. Therefore, it is recognized that the best practice to seize a mobile device that is powered 'On' is to leave the device powered on and maintain power to it. A mobile device powered off should be left powered off.

In the cases involving corporately owned mobile devices, the presence of a pass code should not prevent access to the stored data. The Network Administrator should be maintaining a pass word (pass code) escrow of each issued device. Thereby facilitating recovery should the employee separate from employment or an investigation into the device usage need be conducted. Additionally, a backup schedule should be maintained.

Unfortunately, in most criminal investigations the pass code is not available to the forensic examiner. The lack of a pass code can prevent accessing user-generated data in a readable format. Although there are caveats to this statement that are based upon operating system version and hardware architecture The best-case scenario for success includes knowing, or identifying, the pass code that is preventing access into the mobile device.

Since the user's experience is an important point of consideration of the manufacturer. This consideration has created chinks in the encryption armor, so to speak. Small portals may remain accessible facilitating the extraction of seemingly innocuous data that may assist in the identification of the pass code. Or, allow alternate avenues to the data itself by bypassing the operating system entirely. Thereby negating the need for a pass code all together. We will discuss these elements in greater depth but for now it is just important to remember that if the mobile device is seized powered on, leave it on and maintain power.

Identify the mobile device to the best of your ability. Observe manufacturing labels, or other identifying characters, or numeric sequences. Document what you see and even photograph the device to use as reference in the future.

This is where the embossed information on the device, or beneath the battery, becomes important. Generally speaking, a mobile device will have identification on it that assists in identifying some subscriber information which can be used in the investigative process.

In the seizure of devices utilizing a CDMA network, one will find several identifying codes. These may consist of a Mobile Equipment Identifier (MEID) or Electronic Serial Number (ESN), a serial number, a model number, and possibly a Federal Communications Commission Identification (FCC ID) number. In comparison a device using GSM one will find an MEID, a serial number, and an ICCID that is embossed upon the SIM. Within the data stored on the SIM another service network identification code is stored, that being the International Mobile Subscriber Identity (IMSI).

The MEID is a 14-digit sequence, occupying 56 bits, which identifies a hardware device to the service network. Its predecessor is the ESN which is 11 digits in length, occupying 32 bits. As ESN production neared the maximum 4 billion possible 11-digit sequences supported by ESN, the ESN format was identified to be insufficient to label the growing number of mobile devices on the market. Therefore, the MEID was set as a standard in 2005 for newly manufactured devices. There is some argument over the exact year of implementation, but whether it was 2005 or 2006, the point is there were insufficient numbering sequences to serve the needs of manufacturing. An additional provision of MEID was the ability to tag a specific mobile device as being lost or stolen.

Both the MEID and ESN are indicators that the device bearing it had been associated with a CDMA network. As with the SIM ICCID, the MEID follows a format used to define the device.

That format consists of the following three fields: a regional code, a manufacturing code, and a serial number.

When broken into the three fields the 14-digit sequence will resemble the following.

Regional Code (x 2)|Manufacturing Code (x 6)| Serial Number (x 6), depicted as:

RC,RC,MC,MC,MC,MC,MC,MC,SN,SN,SN,SN,SN,SN

The International Mobile Equipment Identity (IMEI) is a 15-digit unique identification code assigned to a specific mobile device. Dependent upon the year of manufacture, the device may depict a 17-digit code. The IMEI follows a standardized format that identifies the manufacturer, model, and serial number of the device. Similar to the MEID, the IMEI is formatted in a specific manner to depict the aforementioned information.

The Type Allocation Code (TAC) occupies the first eight (8) characters. This series identified the manufacturer of origin and the model of the device. The next six (6) characters identify the device serial number. The remaining digits are descriptive of either the check digit in the case of a 15-character format, or the operating system software version that had been installed on the mobile device in the longer 17-digit code. The remaining digits are descriptive of either the check digit in the case of a 15-character format. Or, the operating system software version that had been installed on the mobile device in the longer 17-digit code.

The IMEI will normally be embossed onto the rear chassis cover, the SIM carrier, or beneath the battery on the manufacturer's label. As indicated, the IMEI is hard coded into the device. If it is not visible on the exterior of the handset, enter the characters *#06# into the keypad and enter. This will reveal the IMEI within the digitizer screen.

Since the GSM service networks use the SIM to authenticate to the service contract, what does the IMEI provide? Simply stated, the IMEI is the fingerprint of your mobile device. It is a unique identifier for a specific device. Should your device require repair, the facility will record the IMEI to associate it to your service ticket. It can also be compared to the vehicle identification number, or VIN, of your automobile. Just as in a theft of a motor vehicle wherein the VIN is entered into a Law Enforcement database for tracking. The IMEI can be entered into the service network's database as a lost or stolen device, blacklisting it from use.

In comparison to the IMEI, the MEID is used to authenticate with CDMA service network, is used to identify the device itself, and can be used to blacklist the device.

The IMSI is a 15-digit code that identifies a specific SIM to the service network. It provides a unique code that is used to identify level(s) of service supported by the service contract. The authentication key (Ki) within the SIM provides authentication to the service, once established the IMSI is used to route the traffic to and from the specific Mobile Station International Subscriber Directory Number (MSISDN). It works to route service at the network level, and may, or may not, be known by the end-user as it is stored within the SIM. It is normally only discoverable through an analysis of the SIM or from records obtained from the service network.

The ability to relate a specific identifying code to a specific device will support an argument toward user attribution. They also afford the investigator a link to recover network-based information that may be useful in the investigative process.

Secure the mobile device

In consideration, to function as designed, a mobile device is to be in a constant state of connection; whether cellular based or network based (WiFi). This state creates an avenue for failure for the examiner as the device is continuously receiving communication, altering the memory. Isolation of the mobile device from radio signals prevents alterations to the stored data.

A constant state of connectivity is similar to a mountain stream. The data is flowing into the mobile device, just as the water is rolling over the stream bed and along the banks. This flow brings desired data to the user, just as the churning water brings fertile sediment to the fields the stream feeds into. Users are happy that their device is bountiful with information, just as the crops growing in the fields are bountiful from the stream.

This activity is acceptable until the mobile device is identified as having a role in a complaint, or other issue. Then the flow of data starts to alter the stored data, just as the stream's erosion change the stream's route over time. What once was the known route of the stream is modified forever, never to return to that previously known state.

Once data is written to the embedded memory module, it is subject to multiple layers of hardware and software protocols. The terms 'wear leveling,' 'garbage collection,' and 'vacuuming' become significant in the data documentation process. This is a direct result of the embedded memory module and under what parameters the memory functions.

Suffice it to say, preventing incoming data to a seized mobile device is paramount. Any data, no matter the amount, can cause irreparable harm to the examination process. This could be as simple as a date and time stamp change from a web site previously visited. Or, as damaging as the receipt of a 'wipe' command that restores the mobile device to a factory fresh state.

In relation to rules of evidence; the mobile device, left non-isolated, is no longer in the condition it had been in when it was seized. Defense counselors can use this error to attack the voracity of the evidence as well as attack the examiner's credibility. For those who have not experienced it, the witness stand is a lonely place. Once the argument turns from prosecuting the defendant to defending oneself, it makes for a long and unpleasant period of testimony.

To prevent these unwanted changes, disconnection from network connectivity is paramount and can be achieved in a number of ways. This is where an understanding of radio transmission and reception plays a role. Referencing our discussion regarding the relevance of identifying which frequency modulation that the target mobile device operates on, assists in determining what form(s) of isolation can be used.

In the cases of a device communicating using GSM, removal of the SIM may be sufficient to isolate the device. In the cases of CDMA, reducing radio signal attenuation must be accomplished to interrupt the radio signals. This may be in the form of a metalized cloth wrap, or by placing the device in a commercial isolation bag or chamber.

Depending upon the make, model, and version of operating system, placing the mobile device into 'Airplane Mode' may sufficiently isolate the device. However, it may not, so layering your radio wave attenuation reduction efforts is recommended.

When seizing the device, document the day, date, and time of seizure. Document your observation of the state the device is in. Meaning: is it powered on, pass code enabled, connected to WiFi, etc. Then place the device in 'Airplane Mode,' wrap it, or place it in isolating material and transport it to your shop/lab.

Identify the device

During the Seizure step you should already have documented the make, model, and identifying numeric codes. Now, it is important to identify what type of operating system is within the device, or at least what version was it first released with. In addition, knowledge of the type of central processing unit as well as the type and capacity of memory is also important to assist in gaining access into the device, as well as verification of the amount of data being extracted.

Using the make and model information in conjunction with Internet resources such as 'phonescoop.com' or 'gsmarena.com' manufacturer's specification information can be obtained regarding the target device. This resource provides necessary information that is useful in the data extraction process. Specifically, a particular acquisition method for CDMA devices based upon the central processing unit. Or that data is, or is not, stored in an encrypted state at rest based upon analysis of the hardware and operating system specifications. Thorough review of the available information prior to attempting to recover data stored within a mobile device will save the examiner time and frustration.

It may also prove to be essential in identifying an avenue for the recovery of data from 'trouble devices.' Trouble devices are those devices that demonstrate a separation from their peers operationally. They boot normally, but when connected for data documentation they do not behave in the same manner as a same make and model device. The trouble device may have been submitted in a case involving multiple devices, many being of the same make and model.

Following my preferred workflow, I obtain data from one or more devices and then run into the trouble device. For reasons beyond me, the device just does not respond as the others. However, having insight into the hardware and operating system specifications, a solution can usually be identified. Other trouble devices have some particular physical defect such as a damaged digitizer screen or damaged data port. Those defects are relatively easy to mitigate given replacement parts, time and some soldering skill.

Data recovery

Having insight into the device itself from a software, hardware and network perspective we can now move toward the data documentation step. This step is reliant upon whether your shop/lab is utilizing mobile device forensic tool suites, or by *L33Ting* it using proprietary or open source tools.

My experiences began using basic open source tools for data recovery. Initially we photographed the relevant information on the mobile device display and prepared a report reflecting our findings. An increasing number of services recorded by mobile devices quickly identified that the use of photography was inefficient.

The next step involved establishing connectivity through AT commands to 'GET' data from the device. However, the increasing number of proprietary versions of the Binary Runtime Environment for Wireless (BREW)[3] created problems in data documentation. BREW is a component of mobile devices that fall under the 'basic' and 'feature' device forms. It was associated with mobile devices having Qualcomm processors. These processors had their own operating software that managed the system interface. This allowed applications to run without having to be coded to manage system operations, as that was being managed by the chipset.

As the evolution of mobile devices progressed, the increasing number of device manufacturers, lead to a varied number of operating systems and environments. Additionally, a number of manufacturers chose to create their own proprietary versions of a data connector. In some examples, data transfer with the device was only possible through the cellular service or target action points on the PCB.

The commonly found USB 'Type A' connector was too large for the downsizing mobile device chassis. This resulted in the implementation of serial connections using mini, micro, and now 'C' USB connectors.

Due to the varied number of hardware configurations, operating systems and data port access the early mobile device examiner had his or her hands full in their data recovery efforts.

As varied as mobile devices had become, a Canadian company known as Research In Motion, or RIM, was creating a form of continuity due to the successful marketing of their Blackberry mobile devices. These devices had actual QWERTY keyboards and an increasing number of these devices proliferated the mobile device market. Thus, setting somewhat of a standard for hardware configuration. Although versions of the Blackberry OS were relatively simple to acquire, the process could be less than complete if the examiner failed to maintain an updated version of the Blackberry Management software.

The introduction by Apple of their iPhone in 2007 initiated a drastic change to mobile device forensics. We still faced a proprietary data cable, now in the form of the 30-pin data cable. However, it contained a packaged operating system that used Apps to provide services to the end-user. The IPSW, or iPhone Software file, allowed the user to replace a corrupt system with a fresh installation. The IPSW was a bundle that was, and still is, capable of receiving updates without effecting the installed Apps or the user-generated data.

Then the appearance of the Android operating system in 2008 started another concern for examiners. Managed by Google, the early concerns were relative to its being open source and functioning somewhat differently from the Apple iPhone's architecture. It does share the iOS usage of Apps to service end-user interests. However, Android did not share the strict development protocols enforced by Apple at the time. App management was treated as being more of a hands-off approach.[4] As time has progressed, the various mobile device platforms it has been installed within, has proven to be a challenging OS!

The evolution of the primary operating systems encountered today, the iOS and Android OS, has been documented and is available in various forms within the Internet. If you practice your 'Google-Fu' you will receive about 1,430,000,000 search hit returns for 'The History of the iOS.' While, 'The history of the Android OS' will return 444,000,000 search return hits. Needless to say, there is more information available to satisfy the specific area of interest you may have for either operating system. Much more than I could provide within the length of this section.

From my experience as a Digital Forensic Examiner the more important area of interest involves security. Specifically, the security features that are enabled by default within the operating system in comparison to those supported by the hardware.

I have found that in some versions of the operating systems, the software may advertise certain security functionality. However, the hardware, either the processor or flash memory, do not support that advertised functionality. In consideration of these observations I have adopted a mantra of 'least intrusive to most intrusive' in the performance of my data documentation.

An example of a 'least intrusive' method of data documentation, or data extraction as it has been coined, would consist of a connection being established with the mobile device through the data port of the device. This process may involve the loading of a client within the evidence device which extracts the stored data as determined by the scope of search. It may rely on enabling a form of 'download mode' within the device to bypass security features. In some cases, the PCB may need to be accessed to create a short in the system, creating an open pathway to the stored data. In each of the aforementioned processes a limited amount of intrusion into the device chassis is made.

In comparison the 'most intrusive' processes involve the disassembly of the mobile device to access the PCB. Upon disassembly, the data within the flash memory may be documented through processes that involve establishing a connection with various surface-mounted target access points (TAPs), surface-mounted components (SMCs), or even removal of the embedded flash memory. These processes are commonly referred to by the following nomenclature: JTAG, ISP, and chip off.

JTAG

The Joint Test Action Group (JTAG) was formed to establish a standard to be followed in the manufacturing process of integrated circuit boards. Specifically, to address the inability to test individual integrated circuits (ICs) that are installed in close proximity to one another on the PCB preventing direct connection testing processes. The need for standardized IC testing was realized and resulted in the IEEE 1149 standard being established in 1985. The IEEE 1149 standard consists of several revisions for forensic purposes we are most interested in the 1149.1 version set forth in 1990.[5]

The standard facilitates testing of the communication channels between the various ICs within the PCB. The process is also referred to as 'Boundary Scan' and uses packets of instructions that are wrapped with IC specific instructions. As the packets flow through the various ICs mounted on the PCB. A specific set of instructions for a specific IC are unwrapped and cause a response in the IC. If the instructions are not related to that specific IC, it is passed onto the next IC. This process continues through the PCB. As the instructions are received by the appropriate IC a response is received at the other end. The subsequent response indicates the IC is functioning appropriately.

On the PCB are Test Action Points (TAPS) that contain registers which respond to the wrapped chain of packets. As the instruction flows through the TAP, if the instruction relates to the specific IC being addressed, the instruction enters the IC through the Test Data In (TDI). If the instruction is not specific to the IC being addressed the instruction is sent out via the Test Data Out (TDO). To adequately manage the process the TAP uses three additional signals. The three signals are Test Clock (TCK), Test Reset (TRST), and Test Mode Select (TMS).

TMS controls the instructions while TCK controls the cadence of the overall load and the processing of the instructions. Finally, TRST simply resets the instructions.

The ability to interact with the various ICs that effect the flow of data to and from the embedded memory module provides an avenue for the forensic examiner to direct the memory to release the data. Through the use of various emulators such as Medusa Pro, RIFF Box2, Easy JTAG, and Octoplus Pro the data is copied from the memory.

The use of JTAG in data recovery is based upon a number of considerations:

- If the mobile device in question is not supported by a commercially available tool
- If the device is pass code protected
- If it is damaged either physically or logically and will not boot

When a condition such as those described is encountered, the process to a successful recovery of data is fairly straight forward. Obtain access to the memory!

Accessing mobile device memory

The first element to address is whether the device encountered is supported by JTAG and does the emulator in your shop/lab support that specific model in question? If not, researching various listservs will get you pointed in the correct direction. Once you have identified that the device is supported. The next issue is locating the correct TAPS that are necessary to facilitate communication with the memory.

The location of the TAPS is generally provided in the form of a help file within the emulator software. Next, clean the TAPS and prepare to solder wire to the designated TAPS. The loose end of the wire soldered to a TAP is then soldered to a daughter board that is used to bridge between the TAPS and the emulator connector. Test for conduction of each TAP and initiate the data recovery process through the emulator's software.

Figure 8.1 depicts the TAPS that are located within a Samsung Wave mobile device. The TAPS in this instance were located beneath the manufacturer's label, but may be located elsewhere on the PCB.

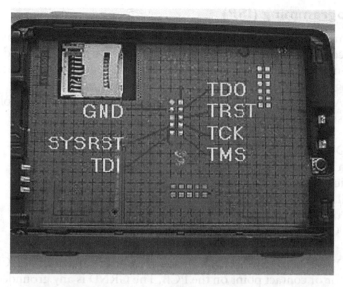

Figure 8.1 JTAG pinout Samsung Wave[6]

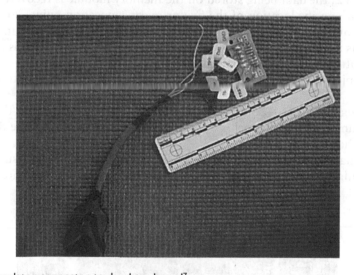

Figure 8.2 JTAG emulator connector to daughter board[7]

Figure 8.2 depicts the connecting cable to the emulator device. It has been soldered to a daughter board. Using .09 millimeter diameter enameled wire, connection is made to the JTAG TAPS on the PCB. This facilitates movement of the emulator connector while maintaining connection to the PCB, and extends the service life of the emulator connector.

Some versions of emulator software will provide verification of the extracted data through capacity measurement. From a digital forensic perspective, this measurement should be supplemented by the hashing of the recovered dataset using MD5, SHA1, or SHA256.

A duplicate, or working copy, of the hashed dataset should then be created and subsequently hashed to verify it is an accurate representation of the original recovered dataset. Upon completion of the duplication and verification process, the working copy can then be ingested in your tool of choice and searched for artifacts relevant to the complaint.

In System Programming (ISP)

In System Programming (or Processing) (ISP) involves similar principles as those described in the JTAG process. ISP involves establishing connection with the embedded flash memory through surface-mounted components (SMCs) that facilitate communication between the central processing unit (CPU) and the memory module.

As with the JTAG process, soldering wire to the various points that control communication is necessary. Those points are:

- Data 0 or D0
- VCC which equates to 2.8–3.3 volts
- VCCQ which equates to 1.8 volts
- Clock or CLK
- Command or CMD
- Ground or GRND

The VCC, VCCQ, and CMD SMDs are either a resistor or capacitor. The CLK can be either a resistor, capacitor or contact point on the PCB. The GRND is any grounded surface such as a heat shield or the main grounding plate of the PCB. Through the use of an emulator such as the Z3X, or RIFF2, the data being stored on the memory module is recovered.

To identify the specific location of the five primary connection points, knowing the location of the communication channels within the PCB is required. The manufacturer does not label them for us as D0, CLK, CMD, VCC, and VCCQ. This information is available within your emulator oftentimes referred to as a 'pinout.'

Figure 8.3 depicts an HTC mobile device being subjected to the extraction process using the Z3X box.

The connecting wire that is soldered between the daughter board and the specific SMD supplying a communication channel is 0.09 millimeters in diameter. Figure 8.4 is a close-up view depicting these connections.

If the device that you are attempting to recover data from is not directly supported by your commercial emulator, the 'pinout' will not be available through the emulator. This is where

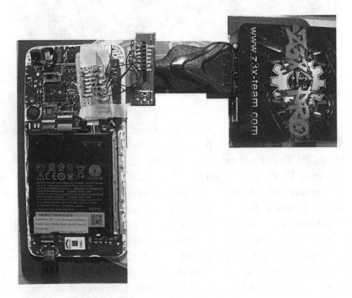

Figure 8.3 Z3X extraction[8]

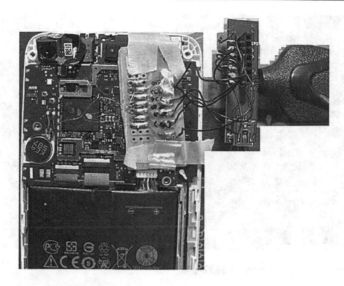

Figure 8.4 Close-up view of SMD connections[9]

your knowledge of processors and memory modules becomes important. Using the emulator interface, you can generally locate a 'generic' mode. The generic mode will allow you to enter some specific parameters concerning the type of processor or memory being addressed. If connection to the D0, CLK, CMD, VCC, VCCQ, and GRND are established, the emulator will perform its function.

That brings up the point of what to do if a pinout is not available for the device being processed. How do we locate the D0, CLK, CMD, VCC, VCCQ?

Through the use of a donor device of the same make and model, you can identify the correct contact points. This involves removing the memory module from the PCB on the donor device. Then, using knowledge gained from the memory module data sheet. Map out the applicable connection points using your voltmeter by testing for continuity between the PCB sockets beneath the removed memory module and the SMDs on the PCB.

The aforementioned reverse engineering process is not difficult, but it can be tasking due to the number of SMDs embedded upon the PCB. The following graphic depicts the PCB of a Coolpad, model 3632A that was received for data recovery. At the time, this specific make and model device was neither unsupported by commercial forensic tools, nor had it been previously 'mapped' for ISP. Therefore, the memory module was removed and the reverse engineering process was completed to identify the associated target access points to support data recovery through ISP (Figure 8.5).

If you are a member of a 'Mobile Forensic Community' many members are willing to share their pinouts. So, it is also an option to ask if anyone in the group has a pinout for that device. This correlates to the practice of sharing your work product with the community. For it has been only through the sharing of knowledge and experience that mobile device forensics is as effective as it is today.

For those who are suffering from 'elder eye' or have tremors that interfere with manual dexterity, an option is to use a VR Table (which enables forensic examiners to acquire data from Mobile Devices, GPS Devices and other electronics units) or the CODED Kit (which enables forensic examiners to acquire data from Mobile Devices without soldering). Both facilitate connection to the TAPS or ISP points through the use of pogo pins. The pogo pins are spring loaded copper pins that are attached to segmented arms that lock in place. The pogo pin can be placed in contact with the targeted IC and held in place by the locking arm.

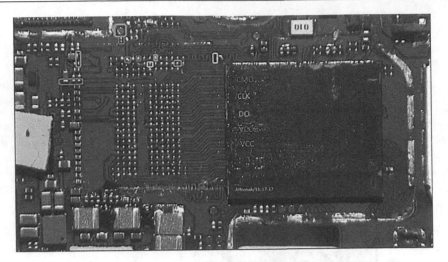

Figure 8.5 Reverse engineering of Coolpad 3632A[10]

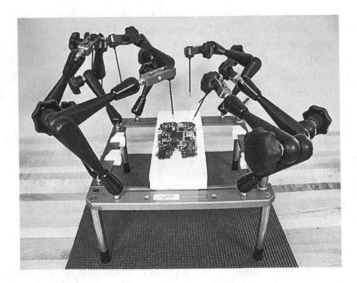

Figure 8.6 The CODED[11]

The CODED is depicted within Figure 8.6. The PCB is resting on a piece of non-conductive plastic which supports the PCB. This is necessary as pressure is applied to the PCB when the pogo pins are set onto the respective SMD. The pressure applied causes the PCB to flex, resulting in an unstable surface, leading to inconsistent communication through the pogo pins.

Connecting wires from the emulator of choice are connected to the end of the pogo pin and you are done. Well, not always as there are times the pogo pin may slip off of a SMD while you are locating another contact point. Or, you accidentally bump the table! As with soldering, time and practice with your tools lead to successfully completing the task. The VR Table and CODED Kit are good tools to have in your toolbox, but the ISP and JTAG processes can be completed through good soldering techniques just as well.

Where's my data?

So, the ISP pinouts have been located and connections to D0, CMD, CLK, VCC, VCCQ, and GRND have been established. The connections between the various points of connection on the PCB to the emulator have been tested for continuity and verified as being 'good.' Activating the emulator's 'Read' function results in the successful extraction of the stored data. Then the data is subsequently hashed, verified, and duplicated for working copy purposes. In the warm and fuzzy feeling of success, the working copy is processed with your parsing tool of choice. A period of time later the parsed data is revealed, or not.

You have just received the error message 'No Data'!

A quick review of the data using your hexadecimal editing tool reveals data is present in your working copy. However, the data is encoded in an unrecognized manner. At this point you review the steps taken to get the data from the embedded memory from your notes. Upon completion you confirm that no abnormality or deviation was noted during the process. Then what went wrong?

This is likely a result of not performing due diligence during the identification phase of the mobile device forensic process. The ISP process was successful; the stored data within the memory module was extracted. The problem is *encrypted data* was extracted. ISP is a very powerful process, but it has an Achilles heel. That weakness is encryption.

Data that is stored encrypted at rest must be decrypted during the recovery process to be parsed. ISP does not provide that functionality. The lesson is to be aware of the capabilities of the mobile device being processed before selecting a specific recovery method. This practice leads to the best possible outcome of your effort being attained.

Chip off

The most intrusive, and destructive, data recovery process in mobile device forensics is that of a chip off. The process involves removing the embedded memory module from the PCB. The primary methods used to separate the memory module from the PCB are:

* Heat Flow
* Mechanical

The Heat Flow method involves the use of heat to gradually increase the temperature of the components to reach the flow rate of the solder joining the two components. Once the flow rate has been achieved, the components are separated.

The Mechanical method involves the removal of material resulting in isolating the desired component from the undesired. In this case, separating the memory module from the PCB by grinding the PCB from the memory module.

Both of these methods require research be completed of the mobile device hardware as well as the memory module itself. The primary element of consideration during the research is that of support for encryption. If information indicating encryption is enabled, do not proceed through the chip off process.

Consistent with the ISP process, the presence of encrypted data applies to chip off as well. If one were to complete a chip off wherein the memory module contains encrypted data, the data is recovered in an unusable form. To decrypt the data, we could use some decryption tools to attempt to recover the data. However, in consideration of the level of encryption in use today, this is an unreasonable effort due to the amount of time it would take, if possible at all. In cases where 256 AES (Advanced Encryption Standard) is utilized, I personally do not have the 27 years to wait. Additionally, given proprietary encryption schemas some manufacturer's use for key generation for their File-Based Encryption, decryption efforts become even more unreasonable.

For argument's sake consider the following question and response.

Question: Shouldn't we be able to reflow the memory module back onto the PCB and try an alternate method of data recovery? After all, the device is returned to its original state.

Response: Although it is possible to reflow a memory module onto the PCB following removal; the probability of success leading to operational functionality of the mobile device is questionable. Dependent upon the method of separation used, heat flow or mechanical, it may not be even physically possible.

The takeaway is simple: Just because a process is possible, it doesn't increase the probability of success. Research the operating system as well as the hardware architecture of the device and formulate a plan that engages the variables that are present.

NOR flash memory

First things first. What is the embedded flash memory module? Essentially it is an integrated component within a system that stores information necessary for the system to meet its function. Flash memory is a form of electrically erasable programmable read-only memory (EEPROM). Being non-volatile, it can maintain the data stored within it without power. There are two basic forms of flash memory in mobile devices, those being NOR and NAND.

NOR flash memory was designed to store code and have quick random access within the memory. As such, it is well suited for execution of the code. NAND flash memory is designed to provide relatively slow random access across the memory, as compared to NOR. However, NAND is much easier to program (write-to) as well as erase as compared to NOR. This is due to the architecture that is present within the flash memory.

NOR flash memory cells are connected to the bit line in parallel; whereas NAND flash memory cells are connected in series. This results in NOR having faster access (reads) as each cell can be written to individually. Due to the ability of NAND accessing the cells in bursts of 512 bytes, it performs programming (writes) and erases at a much quicker rate than NOR. NAND is also less expensive to manufacture and has greater capacity than NOR, given comparable die size. In the implementation of mobile devices, NAND is the better choice for flash memory.

For the purpose of this discussion, I will reference the NAND flash memory that is commonly encountered in today's mobile devices. Flash memory uses electronically controlled gates, also referred to as memory cells, to store the data. The memory cells in unison make up the memory module and are managed by a controller. The controller can be located within the die of the memory module itself, or located in close proximity to it. In the chip off process of a majority of mobile devices, the controller is a component of the memory module.

NAND flash memory

Each memory cell has a positive or negative state, representing a binary '0' or '1.' These cells reflect the negative state of '1' by default. A NAND memory cell having a positive charge represents a '0' where the negatively charged gate represents a '1'. An individual cell is comprised of three components. Those being; a Control Gate, a Floating Gate, and a Drain.

The Control Gate facilitates the flow of electrons in the cell. The Floating Gate simply blocks the flow of electrons from the Control Gate that are traveling toward the Drain. A negative state within the Floating Gate closes the gate, blocking the flow of electrons between the Control Gate and the Drain. Whereas a positive state opens the Floating Gate allowing electrons to reach the drain.

To fully describe this concept, the material would consume the space of this book. A more simplified description is an analogy using barge traffic moving up the Mississippi river. A barge moving products upstream in the river will enter a 'lock' to bypass the various dams along its route. A lock is a flood canal located alongside the river bank that is perpendicular to a dam. The lock has an upstream gate and a downstream gate that is controlled by a lock operator. The lock operator controls the flooding and draining cycles of the lock and barge traffic is timed to use the lock system for one direction of travel at a time.

When the barge is traveling upstream, the upstream gate of the lock is already closed and the water level within the lock is at the downstream level. The barge enters the lock via the downstream gate which closes behind the barge. Once the gate is closed the water level within the lock is raised to that of the upstream level. Once equalized, the upstream gate opens and the barge continues upstream.

For our purposes the river represents the Control Gate, while the set of gates within the lock represents the Floating Gate. The lock itself represents the Drain. The lock operator represents the transistor monitoring threshold voltage within the Control Gate. Given these pairings, electrons are continuously flowing through each cell of the flash memory. This is because the Floating Gates are all open by default, and each represents a '1.' When data is required to be stored, a positive charge closes the Floating gate, preventing the electrons from continuing through to the Drain. Within the Control Gate the threshold voltage of the cell is monitored. If a slightly negative voltage is registered the cell represents a '1' whereas a positive voltage represents a '0.' When the memory cell is erased the positive charge is removed causing the Floating Gate to return to an open position allowing electrons to flow through to the Drain.[12]

This is a very simple description of a complex process that is continually evolving to meet performance demands. The provided example is descriptive of Single-Level Cell (SLC) memory which represents one state for each memory cell. Additional types of memory are Multi-Level Cell (MLC), and Triple-Level Cell (TLC). The MLC and TLC form factors provide for two- and three-bit addressing storage per cell, respectively. Obviously, this increases storage capacity without increasing the size of the memory module.[13]

NAND memory is similar in logic as in hard disc drives as it uses *blocks* to store *pages* containing data files. A *controller* is present to manage the *blocks* of data. Commonly, the size of a *block* is 512 bytes, but it may be larger. Data is written to the memory at the *page* level. One hundred and twenty-eight (128) *pages* make up the *block* if the *block* is 512 bytes. As the *block* fills with *pages* of data, the data continues to be written to a subsequent *page* located within another *block*.

As the data is being written to the memory, the controller manages the endurance of the overall NAND memory through a process called *wear leveling*. This process manages blocks of data across the entire memory to limit the number of erase cycles. As with anything created by man, NAND has a life cycle. By writing data across the entire flash memory before an erase cycle occurs would be the ideal method to ensure the memory cells reach their intended life span.

Unfortunately, this is not as practical as it sounds due to individual App storage requirements and operating system coding. So, wear leveling is performed by firmware or the file system to automatically remap the blocks of data to effectively make use of the larger amount of memory with the fewest erase cycles.

When a file is selected for deletion, the process of deletion occurs at the *block* level. If a file occupies a complete *block* the *pages* containing the data are returned to the state of '1,' or erased. This means that if a file marked for deletion in the file system does not occupy a complete *block*. That file, although not visible to the user at a logical level, may be recoverable at physical level.

NAND flash memory is manufactured in various forms called 'die' to suit the needs of the system, each having a varied number of pin counts. This is commonly known as the memory 'Package' which defines the number and pattern used by a specific memory module. The term ball grid array, or BGA, refers to a single type of die package. Other types are: single inline package (SIP), dual inline package (DIP) and thin small outline package (TSOP).

With your research of the mobile device submitted for examination completed, you have identified that the specific make and model of mobile device under investigation is susceptible to data recovery through the chip off process. In consideration that there are two possible avenues to follow to perform the chip off, which path do you follow? To assist in the decision making process, we will discuss each technique further.

Chip off – Heat flow technique

The heat flow technique, as defined, requires the use of increased temperature of the components to reach the flow rate of the solder. The memory module is secured to the PCB by a ball grid array (BGA) of solder balls. It is through the BGA that signals route to the memory module. Upon reaching the flow rate of the solder, the memory module is removed from the PCB. If done correctly the BGA will be raised on the memory module and you will see the empty 'sockets' on the PCB.

This process is simple in theory but still requires knowledge of the memory module and the thermal threshold the manufacturer has identified for it. Simply stated, if you overheat the memory module, the onboard controller as well as the memory cells within it become damaged and cannot be read. Therefore, it is necessary to research each memory module to ascertain the thermal threshold before initiating the heat flow process. The thermal threshold of a given die is contained within the datasheet for the specific memory module being addressed.

Having identified the thermal threshold of the memory module, the chip off process can now be initiated. Through the use of hot, forced air, or an infrared heat lamp that is set to the minimum temperature of the thermal threshold. The memory module is heated and subsequently separated from the PCB. Although not required the use of a preheater is recommended as heating the PCB prior to applying heat to the memory module will yield the highest success rate.

In the use of mechanical separation processes, the PCB is removed from connection with the memory module by removing the PCB a little bit at a time. This is commonly completed using a table-mounted end mill or a jeweler's polishing wheel. The PCB is ground away from the memory module resulting in the ability to connect to the ball grid array that connects the memory module to the corresponding sockets on the PCB. Extreme care must be used to ensure the PCB is removed evenly to avoid damaging the memory module by penetration the die package.

Once the memory module is separated from the PCB connection must be established with the communication ports of the memory module that are located within the grid array. This is most easily established through the use of test socket adapters. However, can also be completed through the use of direct connection using solder, or the VR Table, or CODED Kit.

Chip off – Mechanical technique

When using test socket adapters, the Sireda brand is very convenient when reading embedded MultiMedia Controller (eMMC), embedded MultiChip Package (eMCP), and Universal Flash Storage (UFS) forms of flash memory. The Sireda adapter makes use of the memory module's onboard controller to communicate over universal serial bus ports (USB). This facilitates the use of write blocking devices between the forensic system and the memory module. Each package type requires its own test socket adapter to effectively communicate.

Alternatively, a number of chip programmers are available that support reading the data stored on the various memory modules. Examples are the UP-828P and the XelTec SuperPro. As with the Sireda test socket adapters, the programmer will require an adapter specific to the package being processed. The more package types you want to support by chip off, the more adapters you will need. This point is made as some of the adapters are not inexpensive and the decision will have impact on ROI.

The programmers provide versatility to your shop/lab regarding the forms of memory it may read. Whereas, the test socket adapter is dedicated to eMMC, eMCP, and the UFS forms. A programmer can read those as well as many others. As previously referenced, not all flash memory has an onboard controller. In these cases, a programmer is required to establish communication with the memory, read the data, and reconstruct the data in a friendly format to be used for analysis.

A process exists that can result in the recovery of data from flash memory that involves the use of an electron microscope. The process involves shaving thin layers from the die to reveal the memory cells. The condition of each cell is recorded and another layer is shaved off and the cell conditions are noted. This process continues until all of the memory cells have been recorded. The recorded data is then reconstructed and made available for analysis.

This process is not cost effective for most businesses due to the overhead expenses encountered. It is presented to demonstrate that just because something is possible, it may not be probable.

No matter which method of data recovery is used to document the data that is stored within the flash memory, verification is a necessity. If the data was extracted using a commercial tool suite, ensure logging and hashing is enabled. When progressing through the 'least to most intrusive' processes, use proven processes to hash the data and verify subsequent working copies prior to moving onto the analysis of the data.

Analysis

The processing of the data recovered from flash memory has become significantly easier with the progression of commercial tool suites. As previously noted, each tool suite has strengths and weaknesses, but they are all capable of parsing data by identifying the individual artifact (file) format and categorizing them by use. The process of categorization is beneficial to the examiner for several reasons.

Having the ability to review containers of data that are directly relative to your investigation reduces the likelihood that the examiner, or analyst, will exceed the scope of search. In criminal law, exceeding the scope of search is a violation of the Fourth Amendment to the United States Constitution. Simply put, it is a Civil Rights Violation that could, if pursued, result in incarceration. Exceeding the scope of search may taint the evidence being presented.

For example, the investigation is relevant to illegal narcotics distribution. The scope of search includes a search of the data for communication relevant to narcotics and the distribution thereof.

Logically, the scope of search would include any form of text-based communication, as well as voice communication that is recorded within the device such as voicemails or audio recordings. Since text-based communication can occur through simple message system (SMS), multimedia messaging (MMS), or a host of third-party Apps. It is reasonable to conduct a search of all of these categories. A review of SMS revealed communication artifacts relevant to the investigation as well as photographs depicting the product. The photographs were sent as attachments with the text-based communication.

If the examiner were to report on these artifacts and secure additional authority to search the 'Image Files' category, all is fine. However, if the examiner follows the trail of the graphic image file sent via SMS to the Graphic Images category and conduct a search without obtaining authority, any artifact located beyond that which was associated to the SMS attachment will likely be excluded.

The loss of a few graphic images depicting contraband may not compromise the case entirely. However, it does expose the examiner to a critical cross examination of his or her character by the defense counsel. This can be unpleasant at best, and career ending at worst.

In Tort law, which applies to business, the intrusive search could result in civil litigation. A common outcome of litigation for the party found liable is to provide some form of compensation to the offended party. Historically, this has been in a financial form.

In both situations, the process of exceeding search authority can be detrimental to the investigation itself, personal or business revenue, and personal freedom.

The use of commercial tools or proprietary scripts can be used to segregate the data categorically. As previously mentioned, this will speed up the review process, provide substantive argument negating a claim of overreach, and afford the examiner personal protection from injury.

A significant number of commercial tools have been tested by the National Institute for Standards and Technology (NIST) through their Computer Forensics Tool Testing Program, Mobile Devices (CFTT).[14] The testing process evaluates the various commercially available tools for accuracy in reporting the data the tool is advertised to support. This process was initiated as a result of the increasing number of mobile device platforms, the evolution of their capabilities, and the relevance they hold in the investigative process. Further information regarding the CFTT and its findings can be found in Chapter 10, by Douglas Menendez 'Cyber Forensic Tools and Utilities.'

The use of commercial tool suites to conduct analysis of the extracted data has become the normal practice among mobile device forensic examiners. Due to the increasing capacity of memory within the mobile devices, the ongoing updates within the various mobile operating systems and the number of third-party Apps available. It has become unreasonable to maintain a personal library of proprietary scripts to address the continued evolution of mobile devices.

This does not infer that the forensic examiner should blindly trust the output of the forensic tool suites. Knowledge of time stamp conversion from Uniform Coordinated Time (UTC) to local time is essential. This is simply due to the observance of UTC by the service networks to document data transfer.

Since commercial forensic suites reference the local time of the forensic system on which they are installed. Notation of the time zone from which the evidence was in use can have an effect on the accuracy of the time stamps reported within the parsed data. Some commercial tool suites offer the ability to convert to either local time, or maintain the recorded time bias being observed by the device at the time of seizure. This is an important element of consideration, especially if the forensic system time is not properly managed.

Chronologic ordering of data is one of the more efficient methods within the analysis process. Locating the category of data that is within the search authority and sorting chronologically can reveal substantive actionable information in the shortest amount of time. Subsequently, artifacts that are associated to the incident in question, but previously undiscovered may be located.

Case example

Referencing the narcotics violation example presented earlier. An SMS revealed text-based communication as well as photographs of the product. Through chronologic ordering, the original message that was sent to recipient 'A' was documented. Further chronologic review reveals an incoming message from sending party 'B' having photographs attached to it. The message is dated after the initial message that was sent to subject 'A.' This message is requesting further information regarding the availability of the product depicted within photographs.

Review of the photographs revealed 'content' similarity to the original photographs sent to 'A' previously. Further analysis of the received photographs identified metadata within the photographs. The metadata inferred additional similarity in the form of date stamp(s) and camera model that associates the two sets of photographs. Finally, a hash analysis of the two sets of photographs validates that the two sets of photographs are, in fact, the same.

Through performance of the chronological sort, an additional subject associated to the original complaint was identified. Through digital forensic analytic practices applicable in both computer-based and mobile device forensics, statements of fact were established.

User attribution

Another area of significant importance in the analysis process is that of 'User Attribution.' What data within the device can be relevant to associate the user of the device to the data stored within the device? Obviously, personally identifiable information (PII) that may consist of full names, addresses, credit card numbers and their account information, with any graphic image files depicting 'selfies,' driver's license or Social Security information can be used to identify ownership of the device. However, this does not provide sufficient information to associate specific data with the owner.

User attribution is established through a methodical process. Following the investigative adage, 'Who, What, When, Where, Why and How' provides guidance in establishing user attribution.

The following list is provided for better understanding and implementation:

- Who did it?
- What was done?
- When did it happen?
- Where did it happen?
- Why did it happen?
- How did it happen?

This list is presented in chronologic ordering of occurrence, but not necessarily discovery. When a complaint is received the 'What' and 'When' phases are at least partially discovered. Therefore, the investigative process is much like a puzzle. Some answers are obvious and available relatively quickly, whereas other questions require significant effort to locate an answer for, if at all. It takes an analysis of the artifacts used to answer the aforementioned questions to reach a conclusive argument of fact.

Again, referencing the narcotics investigation example, we initially validate the complaint; or 'What happened.' Using the artifact(s) located within the device that are associated to the complaint. Gradually expand the analysis from the specific artifact(s) in an effort to identify the 'When' and 'How' the artifact became stored. From that point, establish what the operating system was processing at the time the artifact had been stored for additional clues. If the device was connected to WiFi, is the service set identifier (SSID) or basic service set identifier (BSSID) recorded? This could clarify 'Where.' Were text-based communications or the camera in use? This can be both a 'How' and a 'What.' Do either of these Apps store information that may reference a geographic location? If present this information is another form of 'Where' and possibly 'When.'

Following identification of both subject 'A' and then subject 'B' via SMS activity. Is there relational information present to assist in their positive identification? This would be another 'Who.' Expanding further, analysis of message retention duration, device backup schedule and the method of backup may prove relevant.

SQLite databases

A final note regarding the analysis process. Previously, I indicated that mobile devices have integrated the use of databases to support Apps. This practice has provided the examiner a more standardized structure to analyze which is a vast improvement over the previously encountered variety of file formats previously used to store data. As such, an important tool the modern examiner should possess is the ability to interrogate SQLite databases.

SQLite databases are self-contained and can be stored as a file within the file system. They have become a standard encountered in both native operating system and third-party App development. As such they are used to store the phonebook (Contact List), call history, SMS, MMS, Instant Messaging, as well as a variety of social networking and process specific third-party Apps.

A key to locating the SQLite databases is to run a search across the extracted data for the term 'SQLite format 3.'[15] This is the header indicator for this type of database. By completing a search for the term, all files having that header will be revealed. This would include databases associated to previously uninstalled Apps. Some forensic tool suites conduct this search as a part of their processing code; however, I am a 'verify but trust' kind of examiner and have made it a habit to be performed.

The SQLite database is made of Tables that contain Columns describing the Rows of data entered beneath the column header(s). When created, the database may rely upon additional files that afford some redundancy for the database. Those files are the Write Ahead Log (WAL) and the Journal file. The WAL is created (optionally) and used as a location to store new data that is intended for the database. Following a specific number of page changes, the data contained within the WAL file is written to the database. The Journal is created by default and stores original data. It maintains the original data to provide redundancy should the database suffer an error and require restoration.

If created, the WAL is used to store data while the operating system manages services to improve the user experience. For example, a text-based message is being created using SMS. The sender types 'This is a message' but is interrupted by a voice call. The WAL has recorded 'This is a message' and retains it without writing it to the database. Upon terminating the voice call, the sender switches back to the SMS App and is presented the entry 'This is a message.' The sender subsequently continues typing, 'This is a message depicting what a WAL' and is again interrupted. This time by a third-party App. Following the communication over the third-party App, the sender returns to the SMS App and is presented the entry, 'This is a message depicting what a WAL.' The sender returns to typing and completes the message as, 'This is a message depicting what a WAL file does.'

The SMS database is processed and only has a single message entry containing the sentence, 'This is a message depiction what a WAL file does.' However, in an analysis of the WAL file three artifacts are located. Those being the three partial messages preceding the message stored within the database. Specifically, 'This is a message' and 'This is a message depicting what a WAL.' Through comparison of the active messages within the SMS database to those recorded within the WAL file. The timeframe during which a deletion of an SMS occurred may be discoverable.[16]

So how can you determine which supporting file is being used by the SQLite database you are reviewing? Remember the header 'SQLITE format 3'? Using that as a starting point in a hexadecimal editor, navigating to the following listed Offset will identify which supporting file is in use.

- Journal File
- File Offset 18 (1 byte) = "x01" = Journaling
- File Offset 19 (1 byte) = "x01" = Journaling
- Write Ahead Log (WAL)
- File Offset 18 (1 byte) = "x02" = WAL
- File Offset 19 (1 byte) = "x02" = WAL

In reference to the importance I placed on chronologic sorting during my analysis. The method that time is recorded within the various SQLite databases is contingent upon the type of operating system in use. Understand, the value depicted is assumed to be a mathematical equivalent of Uniform Coordinated Time. But rendered according to the numerical representation supported by the operating system.

Examples of commonly encountered numeric representations are:

- UNIX Epoch which consists of a 10-digit number that represents the number of seconds since January 1, 1970 at 00:00:00 hours (01/01/1970 00:00:00)
- UNIX Epoch Milliseconds is a 13-digit number representing the number of milliseconds since January 1, 1970 at 00:00:00 hours (01/01/1970 00:00:00)
- Mac Absolute time is the number of seconds since January 1st, 1971 at 00:00:00 (01/01/2001 00:00:00). It is represented through a 9-digit number. Mac Absolute time is also stored in an 18-digit number. When encountered this represents the nanoseconds since January 1, 1971.

Each of the forensic tool suites previously listed offer SQLite Database parsing. Some include viewers to review the database tables in a native database format which also includes the option to export specific tables or content for reporting purposes. A few also include SQLite forensic tools that afford the examiner the ability to create and run queries to expedite the search processes as well as to create reports that reflect 'report friendly names' describing the content of the individual columns within documented table.

As is the case with flash memory, SQLite database form and function exceeds the length of our discussion. For further research, visit the SQLite site located at: https://sqlite.org

As with any process, training and practice is necessary to become competent. Further information can be obtained from the following sources:

- www.tutorialspoint.com/sqlite/
- www.sqlitetutorial.net/
- sandersonforensics.com

Reporting

A narrative should be prepared reflecting a case synopsis and the condition of the mobile device as received. The narrative should record the steps taken to identify the mobile device, as well as the method(s) used to document the stored data recovered from the device.

If engaged in processing contraband containing illicit images, the narrative should refer to an artifact report that contains the actual file(s) identified as being evidential to the investigation.

The artifact report resulting from a mobile device analysis should include all artifacts that depict the facts represented by the data. Using a chronologic format, prepare the report to describe the event from initiation through completion. This may include post incident information that may support 'knowledge and forethought,' such as news articles reviewed on the Internet. If user attribution can be identified, ensure it is depicted with clarity. The artifacts can be documented within an HTML format, or PDF depending upon the intended recipient. Both of these formats are supported by commercial tool suites.

SUMMARY

This chapter presented information regarding the similarities and differences between computer and mobile device forensics. We discussed the form factors of mobile device handsets, their operating systems, and memory. Mobile device identification that is used to associate a specific device with a service network was defined and a description of the cellular network topography was provided.

Next, we moved to the forensic process that follows the steps of Seize, Secure, Identify, Recovery, Analysis, and Reporting. Each of these sections provided guidance in addressing the better method to be used to complete each section of the process. Supplemental information was provided to assist in determining the better process to be followed based upon the conditions that exist in the mobile device as being relative to the investigation.

This body of work is offered to you after 15 years of digital forensic experience for law enforcement. There are suggestions made within the work relative to the data recovery process from mobile devices. The data recovery aspect of the forensic process has become increasingly difficult in consideration of improvements to security policies within the mobile device itself as well as provisions of legislative acts such as the Electronic Communications Privacy Act (ECPA) and the General Data Protection Regulation (GDPR). The legislative changes create a quasi-adversarial relationship between the manufacturing and forensic communities when attempting to identify 'truth' that may exist in a mobile device. Through knowledge of the various recovery techniques, and a realization of new techniques, the process will continue.

ACRONYMS

AES   Advanced Encryption Standard
BGA   Ball Grid Array
BLOBs   Binary Large Objects
BREW   Binary Runtime Environment for Wireless
BSSID   Basic Service Set Identifier
CDMA   Code Division Multiple Access

CFTT   Computer Forensics Tool Testing Program, Mobile Devices
CPU   Central Processing Unit
CRCs   Cyclical Redundancy Checksums
DIP   Dual Inline Package
eMCP   embedded MultiChip Package
eMMC   embedded MultiMedia Controller
ESE   Extensible Storage Engine
ESN   Electronic Serial Number
FBE   File-Based Encryption
FCC ID   Federal Communications Commission Identification
FDE   Full Disk Encryption
FVE   Full Volume Encryption
GSM   Global System for Mobile communication
GUI   Graphical User Interface
IC   Integrated Circuit
ICCID   Integrated Circuit Card Identifier
IoT   Internet of Things
IMEI   International Mobile Equipment Identity
IMSI   International Mobile Subscriber Identity
ISP   In System Programming (Processing)
IPSW   iPhone Software
JTAG   Joint Test Action Group
Ki   Authentication Key
L33T   Also 'L337.' Is really 'leet,' a corruption of 'elite' and meaning someone who is very good at what they do
LTE   Long-Term Evolution
MEID   Mobile Equipment Identifier
MLC   Multi-Level Cell
MMS   Multimedia Messaging
MSISDN   Mobile Station International Subscriber Directory Number
NIST   National Institute for Standards and Technology
OS   Operating System
PCB   Plastic Controller Board
PII   Personally Identifiable Information
PLIST   Preference Lists
SIM   Subscriber Identification Module
SIP   Single Inline Package
SLC   Single-Level Cell
SMC   Surface-Mounted Component
SMS   Simple Message System
TAC   Type Allocation Code
TAP   Target Access Point
TAPS   Test Action Points
TCK   Test Clock
TDI   Test Data In
TDO   Test Data Out
TDMA   Time Division Multiple Access
TLC   Triple-Level Cell
TMS   Test Mode Select

TRST	Test Reset
TSOP	Thin Small Outline Package
UFS	Universal Flash Storage
USB	Universal Serial Bus
UTC	Uniform Coordinated Time
VIN	Vehicle Identification Number
WAL	Write Ahead Log
WiFi	A trademarked phrase that refers to IEEE 802.11x standards
XML	Exchange Markup Language

NOTES

1 (n.a.), (June 12, 2019), "Mobile Fact Sheet," Pew Research Center, Internet & Technology, www.pewresearch.org/internet/fact-sheet/mobile/, retrieved June 10, 2020.

2 Cynthia, D., Murphy, A., (2013), "Developing Process for Mobile Device Forensics." digital-forensics.sans.org/media/mobile-device-forensic-process-v3.pdf, retrieved July 7, 2020.

3 Rouse, M., (November 16, 2010), "What Is BREW (Binary Runtime Environment for Wireless)? - Definition from WhatIs.com." SearchMobileComputing, searchmobilecomputing.techtarget.com/definition/BREW, retrieved November 20, 2020.

4 Hoog, A., (June 15, 2011), Android Forensics: Investigation, Analysis, and Mobile Security for Google Android, 1st edition, ISBN-13: 978-1597496513, Waltham, Ma: Syngress.

5 Mishra, P., (2020), "Joint Test Action Group (JTAG): Definition, Uses & Process | Study.com" Study.com," study.com/academy/lesson/joint-test-action-group-jtag-definition-uses-process.html., retrieved July 12, 2020.

6 Image, JTAG Pinout ZTE Z222, provided courtesy of the Author.

7 Image, JTAG Emulator Connector To Daughter Board, courtesy of the Author.

8 Image, Z3X Extraction, courtesy of the Author.

9 Image, Close-Up View of SMD Connections, courtesy of the Author.

10 Image, Reverse Engineering of Coolpad 3632A, courtesy of the Author.

11 Image, The CODED, courtesy of the Author.

12 Schwarz, T., (2003), "COEN 180." Www.Cse.Scu.Edu, www.cse.scu.edu/~tschwarz/coen180/LN/flash.html., retrieved June 29, 2020.

13 Painter, Z., (April 12, 2018), "The Basics of NAND Flash Memory Technology | Silicon Power." Silicon Power Blog. April 12, 2018. www.silicon-power.com/blog/index.php/guides/nand-flash-memory-technology-basics/., retrieved June 29, 2020.

14 Ayers, R. (n.d.), "Mobile Device Forensics – Tool Testing." www.nist.gov/system/files/documents/2017/05/08/mobiledeviceforensics-mfw08.pdf, retrieved June 16, 2020.

15 Sanderson, P., (May 12, 2018), SQLite Forensics, Edited by Richard Hipp, Heather Mahalik, Brett Shavers, and Eric Zimmerman. 1st ed., ISBN-13 : 978-1980293071.

16 Ibid.

Chapter 9

Forensic accounting and the use of E-discovery and cyber forensics

Richard Dippel

CONTENTS

INTRODUCTION

The use of cyber forensics has become increasingly important to the field of forensic accounting since practically all activities in the accounting profession has transitioned from paper to electronic information either on computers and/or in the cloud. Consequently, cyber forensic professionals will likely be called upon to assist forensic accountants and so it is important to understand what forensic accountants do.

A good starting point is to examine what is meant by the term forensic accounting. Forensic accounting is defined by the American Institute of Certified Public Accountants (AICPA) as follows:

> Forensic accounting services generally involve the application of specialized knowledge and investigative skills possessed by CPAs to collect, analyze, and evaluate evidential matter and to interpret and communicate findings in the courtroom, boardroom, or other legal or administrative venue. More simply, in the context of litigation, the term forensic means to be suitable for use in a court of law...[1]

In simple terms, it is the application of law to the field of accounting for the purpose of investigating, interpreting, and communicating certain findings in the legal and business arenas. To be effective, the forensic accountant must understand the many aspects of accounting along with having the skills to be an effective investigator, communicator, and witness. This includes an understanding of legal procedures and the rules of evidence. Having evidence that is not admissible in court, or does not lead to admissible evidence, is if little value in a courtroom. Therefore, considering the predominant electronic nature of accounting, the effective interaction of the forensic accountant with the cyber forensic expert in the context of applying legal procedures and the rules of evidence is critical to a successful conclusion for the client.

The application of these skills by forensic accountants requires that, unlike attorneys, forensic accountants must be independent. Attorneys must represent their client zealously. Attorneys must be advocates. Forensic accountants are not advocates. They should communicate to the client their honest opinion concerning the matter at hand. They are hired by the client to do a job and it may result in informing the client that they do not have a claim. A forensic accountant that represents a client zealously compromises his independence and so renders the expert ineffective.

A forensic accountant, who is viewed as an advocate for the client with an opinion not based on his honest analysis of a particular issue, is doing the client a disservice. However, it does not mean that the expert needs to be fair to all the parties involved. The forensic accountant is hired by a client to serve the client's interest.

As we examine forensic accounting in this chapter, we will illustrate how the forensic accountant's approach to an issue can impact how certain activities are conducted, including their interaction with the cyber forensic expert. A cyber forensic expert must assist the forensic accountant in a manner that the findings can be used in a legal proceeding and/or in a presentation to the client.

The forensic accountant has the expertise to guide the cyber forensic expert in their efforts and vice versa. Keeping this in mind, we are going to look at two primary areas involving the experts that are critical to a successful outcome for the client in a legal proceeding: Discovery of information and presentation of evidence involving Electronically Stored Information (ESI).

DISCOVERY

Before focusing on Electronically Stored Information (ESI), it would be helpful to review the discovery process in general. Discovery is utilized by parties to a legal proceeding in order to discover facts, to be used as a substitute for testimony, and to be utilized to cross exam a witness.

It is important to note that there are no overall set of rules that govern discovery in all courts in the United States. The rules vary based on whether it is a criminal or civil matter. Discovery in civil cases is adversarial versus criminal cases where, along with the adversarial nature of the process, the government has a non-adversarial role. The government is charged with ensuring that justice is done.

The discovery rules also vary based on whether the proceeding is in a federal court versus a state court. Many of the differences are due to the nature of discovery. The initial question is whether the question involves how a court operates (procedural) versus that involving the legal issue in dispute (substantive). The courts, with respect to substantive law issues, should have more of a consistent approach. However, discovery rules are considered matters of procedure and each court determines its own procedure. Experienced practitioners, though, know that from a practical standpoint, the outcome of any particular issue can also be impacted by the particular judge who is presiding over the case.

In addition, the discovery rules can vary from state to state. So when involved with a particular court, it is important to know the discovery rules for that state and the local court rules that may involve issues of discovery. When investigating a dispute that is not currently a lawsuit, it is important to anticipate the impact of both the particular state court's rules and the impact of the federal rules on the process since, at that point in time, you would not know with certainly what venue the issue would be litigated in.

Is there a significant impact on a case due to the variations in the federal and state and state to state discovery procedural rules? One example is the application of attorney–client privilege, accountant-client privilege and work product. The privileges and work product doctrine apply to both criminal and civil matters and can vary from state to state and federal versus state.

Attorney–client privilege protects from disclosure the communications between a client and the client's attorney. There are a few exceptions, but it is a fairly universally applied throughout the various state and federal court systems. Therefore, its application in the various courts would be less problematic.

On the other hand, the accountant-client privilege can vary greatly in it application in a state. Some states have a strong accountant-client privilege that is similar to the attorney-client privilege and other states have a weak privilege. Except for certain matters involving the IRS, there is no federal accountant-client privilege. Therefore, depending on your venue, reliance on this privilege can be problematic.

Information can also be protected from disclosure by utilizing the work product doctrine. Work product involves the protection from disclosure of materials 'prepared in anticipation of litigation.' The attorney does not have to be involved in order to assert the work product doctrine. It protects from disclosure the thoughts, theories and strategies of the case developed by attorneys and litigation support staff. It is also a doctrine that varies from state to state and from state to federal courts.

Why does this matter to the forensic accountant? One of the objectives of the expert should be to look after the client's interest by protecting information from disclosure to the other parties. Unless required by the law, there is no professional obligation to disclose information to parties adverse to the client's interest. The use of the privileges and the work product doctrine can be used to help the client without impairing the independence of the expert. The expert may want to involve the attorney for the client to perhaps take advantage of the attorney–client privilege. The expert should also be careful with respect to whether something should be recorded. What is recorded may be in the context of thoughts and strategy that would make it clear to a court that such was 'prepared in anticipation of litigation' and should be protected from disclosure.

One example of how different courts deal with work product is the treatment of draft reports created by a forensic accountant. Draft reports would be those reports that the expert develops and then refines into the final report. The changes, the deletions, the additions, and any other changes could provide a party an insight into how the opposing party intends to proceed or perhaps it could show an admission against interest. Under the federal rules, draft reports are protected from disclosure except upon 'substantial need.' However, individual state may not have such protection and an expert may be required to produce such drafts. Such drafts can then be used against the party.

So, does using the rules to minimize disclosure make the expert an advocate and not independent? No.

The expert is representing the client and protecting their privacy. Please note that the privileges and work product are not automatic and that the expert has to operate in a manner that they can be utilized to protect from disclosure. Accordingly, if the expert wants to effectively serve the client, then the use of such tools cannot be ignored. Both the forensic accountant and the cyber forensic expert should be aware of the application of these tools so they can be utilized in serving the client.

Due to the technical and specialized knowledge pertaining to accounting and aspects of cyber forensics, the use of an expert in this phase of the legal proceeding is vital. How you word or respond to a discovery request in an area that is so technical and specialized requires an expert to guide the attorney for the client and those who support the attorney. If you do not know what and how to ask it, then you risk being surprised in the trial if the adverse party uses such information against the client.

If despite your best efforts you are required to disclose information, the use of protective orders can protect the interest of the clients despite the disclosure of information. Protective orders protect confidential and sensitive information by providing sanctions in a court order if such information is released in violation of such orders. Should the party be required to disclose certain information, then it may be necessary to protect the information from being disclosed to individuals who are not parties to the legal proceeding.

Rule 26 of the Federal Rules of Civil Procedure defines the parameters of a protective order and includes the protection by 'requiring that a trade secret or other confidential research, development, or commercial information not be revealed or be revealed only in a specified way...'[2].

CRIMINAL DISCOVERY

Discovery in a criminal case, unlike in a civil case, must also include information that exonerates a defendant. Such a disclosure does not require a specific request from the defendant. This information includes oral and written statements, documents, objects, physical or mental examinations, and scientific tests. Discovery in the federal courts is governed by federal rules and court decisions.

Some of the significant rules and decisions are as follows:

- Rule 16 of the Federal Rules of Criminal Procedure.
- Brady v. Maryland, 373 US 83 (1963) – Must disclose exculpatory evidence.
- Jenks Act, 18 USC Section 3500 – Requires the disclosure of statements of witnesses after they have testified on direct examination.

There are certain tools that can be utilized by the prosecutor for discovery in a criminal matter such as: Search warrants, depositions (Please see below the section on Civil Discovery), examinations, and tests. Search warrants are issued upon presentation of probable cause to a court. There are also a number of exceptions to the requirement of needing a search warrant. (Please note that use of such warrants is beyond the scope of this chapter and is governed extensively by constitutional law.)

Persons also can be detained for questioning and certain tests can be run involving the person and related items. The ability to detain or run certain tests could require permission from a court and involve issues of state or federal law and/or constitutional law. Some of these tools can be used for discovery before someone is formally charged with a crime. The defendant also has certain disclosure obligations under Rule 16 of the Federal Rules of Criminal Procedure.

If the defendant requests disclosure of certain information and the government complies:

then the defendant must permit the government, upon request, to inspect and to copy or photograph books, papers, documents, data, photographs, tangible objects, buildings or places, or copies or portions of any of these items if:

(i)   the item is within the defendant's possession, custody, or control; and
(ii)   the defendant intends to use the item in the defendant's case-in-chief at trial.[3]

If the defendant requests physical or mental examination and of any scientific test or experiment and the government complies,

> the defendant must permit the government, upon request, to inspect and to copy or photograph the results or reports of any physical or mental examination and of any scientific test or experiment if:

(i)   the item is within the defendant's possession, custody, or control; and
(ii)   the defendant intends to use the item in the defendant's case-in-chief at trial, or intends to call the witness who prepared the report and the report relates to the witness's testimony.[4]

With respect to expert witnesses, the defendant must, at the government's request, give to the government a written summary of testimony that the defendant is going to use if the defendant has requested such from the government concerning their expert or if the defendant has given notice to present expert testimony of the defendant's mental condition.[5]

CIVIL DISCOVERY

In regard to discovery in a federal civil case, Rule 26 (b)(1) of the Federal Rules of Civil Procedure states that

> 'unless otherwise limited by court order, the scope of discovery is as follows: Parties may obtain discovery regarding any nonprivileged matter that is relevant to any party's claim or defense and proportional to the needs of the case, considering the importance of the issues at stake in the action, the amount in controversy, the parties' relative access to relevant information, the parties' resources, the importance of the discovery in resolving the issues, and whether the burden or expense of the proposed discovery outweighs its likely benefit. Information within this scope of discovery need not be admissible in evidence to be discoverable.'

Thus, it is important to realize that information need not be admissible in evidence to be discoverable. It only needs to be calculated to lead to admissible evidence.

As previously stated, the wording of requests is important in both criminal and civil cases, but since in a civil case, discovery is an adversarial process, if information is not requested by a party to the lawsuit then, unless the court requires it, you do not have to produce such information. It is the responsibility of the individual parties to make the requests.

The forensic accountant and cyber forensic expert in a civil matter are even more of an important component in this process. Assistance in identifying certain information that is necessary to the success of the case makes the role of the expert vital in this process. Depending on the case, the knowledge of certain processes such as the use of depreciation, fair value adjustments, revenue recognition, and other processes must be identified and explained not only to the court but possibly to the attorneys and other parties assisting the client.

In a civil matter, the tools of discovery include interrogatories, requests for production, requests for admission, and depositions. Interrogatories are questions that are sent to only the parties to the lawsuit for a response. Normally they are used to obtain information that someone may have to research an answer. Such information may include the identification of the party's insurance company or the name of their auditors. Requests for production are also sent only to the parties to the lawsuit. These requests generally ask for documents and other items, such as a computer hard drive that can be inspected. Request for admission are sent only to the parties to the lawsuit and ask a party to admit or deny a statement of fact. These requests can be useful in discovering information and for establishing facts that will then not have to be proved in court.

The last tool of discovery discussed here is depositions. In a deposition, the attorneys for all the parties are invited and are normally present and a witness is asked a series of questions not only in the presence of the attorneys but in front of a stenographer. The stenographer then records what is said by all the parties. The person testifying does not have to be a party and the person being deposed can be required to bring in materials to the deposition. This is the most expensive means of discovery, but it is also the tool with the greatest reach since you can examine persons who are not the named parties to the lawsuit.

Without having to employ one of the tools of discovery, certain information from an expert employed to testify must be disclosed. Rule 26(a)(2) of the Federal Rules of Civil Procedure requires the disclosure of certain information, including a report, if the expert is employed to provide testimony. If the witness is a non-testifying expert, then Rule 26(b)(4)(D) of the Federal Rules of Civil Procedure, under most circumstances, bars the discovery of the opinions and facts known to such expert.

LIMITATIONS ON CIVIL INVESTIGATIONS VERSUS CRIMINAL INVESTIGATIONS

There are certain limitations on investigations in a civil case versus a criminal case. The tools of discovery in a civil case are unavailable until a lawsuit is filed. Therefore, investigations are more limited than in a criminal matter where such tools as search warrants and interrogations are available before anyone is charged with a crime. Since the investigative process is limited in a civil matter before a lawsuit is filed, an expert in a civil case could be helpful in obtaining information that can be especially useful in evaluating the value of a claim, determining whether to proceed, and for preparing a lawsuit should the client decide to file one.

ELECTRONICALLY STORED INFORMATION (ESI)

From the standpoint of forensic accountants and cyber forensic experts, we can now focus on Electronically Stored Information (ESI) in the context of the discovery process. ESI can be originally created or it could be converted from written or typewritten documents.

ESI can now be email, social media, cell phone data, digital audio or video recordings, global databases, apps, global positioning data, data stored in a household appliance, onboard computers in a car, or any of the thousands of digital records produced by an average person on an average day.[6]

THE E-DISCOVERY PROCESS

In the context of illustrating the discovery of ESI, here is an Electronic Discovery Reference Model (Figure 9.1):

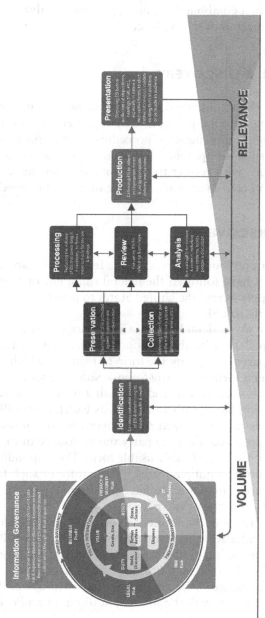

Figure 9.1 Electronic Discovery Reference Model[7]

This process is necessary for discovery to be effective in obtaining useful information and to enable it to be admitted as evidence. The initial focus of the forensic accountant would be on the 'Identify' aspect of the relevant information and the cyber forensic experts and analysts must be directed to what would be relevant with respect to aspects of accounting issues. The expert can then go about the collect, preserve, analyze, and the remaining activities resulting in the presentation of the evidence.

CRIMINAL E-DISCOVERY

There is a lack of guidance both in the federal or state courts with respect to the discovery of ESI involving criminal procedure. Civil procedure, however, is more developed with respect to ESI. Though applying civil procedure to the criminal area for guidance in this area can be helpful, such an approach has limitations due to the different policy and constitutional considerations with respect to the criminal versus the civil procedures. So, we will focus on the process in civil cases while keeping in mind the aspects that are similar to criminal cases.

CIVIL E-DISCOVERY

As has been previously referenced, the federal rules are different from the state rules though certain states have followed the federal rules. For our purposes, we will focus on the federal rules that apply across the United States to provide us with insight on how these rules generally will apply to ESI.

Rule 34 of the Federal Rules of Civil Procedure specifies that ESI can be requested directly or after translation by the responding party into a reasonably usable form and that the requesting party can specify the form or forms in which the ESI is to be produced. The expert should prefer a form where the forensic software tools can be utilized. There are certain audit software tools such as ACL and IDEA that the forensic accountant should be familiar with in their profession along with such tools as Excel, Power BI, and R.

These tools can be utilized to discover relevant information if it is in a usable form. If a form is not specified, then the party must produce in form or forms in which it is ordinarily maintained or in a reasonably usable form. The responding party only has to produce the ESI in one and not in multiple forms.[8] The nature of such form could be a point of contention where the expert may need to have the discovery in a particular form so the expert can use certain tools to perform an analysis.

Can such requests be limited? Note that the responding party can object to the form requested. Under Rule 26 of the Federal Rules of Civil Procedure, ESI need not be provided if it results in 'undue burden or cost.' However, if the requestor shows good cause then the court could order that such ESI be produced. This is an area where the court has discretion and the outcome would be dependent on the attitude of the judge. It is important then for the attorney to consult with the forensic accountant and cyber forensic expert to determine the appropriate form to request.[9]

An obstacle that a requesting party may encounter is the destruction or alteration of ESI since by its nature, it can be easily destroyed or altered. Rule 37 of the Federal Rules of Civil Procedure covers the situation where ESI, which should have been preserved in the anticipation of litigation, is lost because a party fails to take reasonable steps to preserve such ESI.

Under those circumstances, if there is prejudice, then the court may order what is necessary to fix the prejudice. If the party intended to deprive the other party of ESI, then the court can assume that the ESI was unfavorable to the party and can instruct a jury that it was unfavorable or dismiss the action or enter a default judgment.[10]

EVIDENTIARY ISSUES

Like with discovery issues, evidentiary issues can vary depending on whether it is a federal court or state court proceeding. Since each court can follow its own procedures, then the court must determine certain evidentiary issues in accordance with such procedures. Evidentiary issues can also vary depending on whether it is a criminal or civil proceeding, for example, there is greater potential use of character evidence in a criminal matter. The Federal Rules of Evidence apply to both civil and criminal cases, however, some sections apply in a criminal case and other sections apply to a civil case.

Admission of ESI

The admission by the court of ESI requires the authentication of such evidence. Authentication is a rule that requires that the evidence is what it is purported to be. Other issues that pertain to the admissibility of evidence are hearsay and relevance. These issues can be critical to the success of the case since the inability to get the ESI in evidence can result in the inability of the party to move forward with their claim or defense.

Authentication of ESI evidence

> To satisfy the requirement of authenticating or identifying an item of evidence, the proponent must produce evidence sufficient to support a finding that the item is what the proponent claims it is.[11]

To be admissible, the evidence must be authentic. To prove that ESI is authenticate may require testimony from someone in custody of such evidence, a witness who has first-hand knowledge of the facts and how it was obtained from the computer or testimony from someone concerning the process that generated ESI. One of the relevant issues for authentication is that of chain of custody. The participants must show that there is no break in the chain of custody when introducing evidence. For example, assume the expert obtains a hard drive from the computer at a company under investigation. The expert would then have to show that there was no break in the custody of such hard drive.

If there is no break in the chain of custody, the court can rely on the assertion that the ESI was from the company in question and that it has not been altered by some third party who temporarily gains possession of such drive. Such expert then does an analysis of the ESI and presents the ESI and the analysis in court.

Under Rule 902 of the Federal Rules of Evidence certain ESI is self-authenticated so eliminating the need for some of the steps of authentication. The ESI 'generated by an electronic process or system that produces an accurate result, as shown by a certification of a qualified person' as required by the rule is self-authenticated. In addition, 'data copied from an electronic device, storage medium, or file, if authenticated by a process of digital identification, as shown by a certification of a qualified person' is also self-authenticated. The qualified person could be the cyber forensic expert. Chain of custody considerations is also relevant with respect to self-authentication.[12]

Relevance and hearsay

Even if evidence is authenticated, it may still be inadmissible due to it being irrelevant or consisting of hearsay. Relevance is an important factor for admissibility of evidence. Evidence is relevant in that

(a) 'it has any tendency to make a fact more or less probable than it would be without the evidence; and

(b) the fact is of consequence in determining the action.'[13]

Hearsay can also render evidence inadmissible. Hearsay is an out of court statement offered to prove the truth of what is asserted in the statement. However, courts have held that records generated by the system are not considered 'statements' of a person and so such records are not hearsay. In addition, if certain business records are considered statements by a court, they could still be admitted as a business records exception to the hearsay rule.[14]

DAUBERT

One of the hurdles in court that a forensic accountant or any expert must overcome is for the court to qualify them as an expert. An expert can testify in court concerning their opinion of the matter at hand and an expert can conclude about the matter at hand. If the forensic accountant is not qualified as an expert then the accountant is only a fact witness and can only testify about the facts that they have personal knowledge and cannot render an opinion about an accounting issue.

The main court decision concerning the admissibility of expert testimony is a Supreme Court decision known as Daubert v. Merrell Dow Pharmaceuticals, 509 U.S. 579 (1993)[15]. Daubert established a test for qualifying an expert and such test is contained in Rule 702 of the Federal Rules of Evidence (FRE).

The court under FRE 702 will allow expert opinion if:

1. 'Such expert qualifies as an expert based on knowledge, skill, experience, training or education in order to testify in the form of an opinion;
2. Expert will help the trier of fact understand the evidence or to determine a fact in issue;
3. Testimony is based on sufficient facts or data;
4. Testimony is the product of reliable principles and methods;
5. Expert has reliably applied the principles and methods to the facts of the case.'[16]

With respect to the forensic accountant, the expert must satisfy all the criteria, however, in most situations the requirement that the 'testimony is the product of reliable principles and methods' would be less of a challenge when addressing accounting issues. This requirement appears to be more applicable to those cases based on science and not accounting rules.

While many issues are decided by the jury as a fact finder, this rule establishes the trial judge as the gatekeeper with respect to testimony by an expert.

As with the issues involving discovery and evidence, each state has its own rules with respect to expert testimony. However, the state rules may be similar to the FRE and the Daubert decision.

Since the federal rules require disclosure of an expert including the submission of the report, the process by which an expert's testimony may be challenged would be in a pretrial Motion in Limine where the motion is argued before the trial. With respect to the states, there may be a Motion in Limine or it is also possible that the challenges could be asserted during the trial. In any event, the failure to qualify as an expert in most cases would destroy the usefulness of the expert as a testifying expert.

Since many times the success of a case is dependent on the court qualifying a party's witness as an expert, it is important for the attorney to have assurance than an individual can be successfully qualified as an expert. Therefore, it can be helpful to review such expert's successes in court. One of the tools for accomplishing this is known as the Daubert Tracker. The Daubert Tracker can be found at www.dauberttracker.com/casereport.cfm.

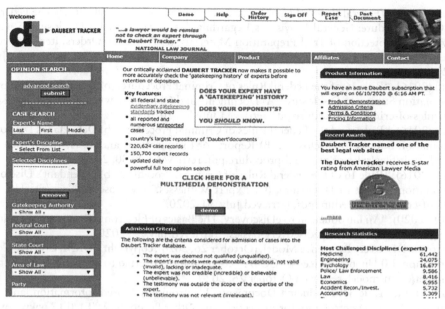

Figure 9.2 Screen image – Daubert Tracker[17]

This tracker provides information on whether certain experts were successful as an expert in court. An example, of a page from the Daubert Tracker is provided in Figure 9.2.

Another such tool is the Expert Witness Profiler, which may be found at www.expertwitnessprofiler.com. Where available, the Expert Witness Profiler also provides access to expert witness transcripts, briefs (including memoranda in support of or in opposition to motions to exclude testimony), and other relevant supporting documents.

The Profile is an indispensable tool for the attorney to use before retaining an expert or before opposing an expert in deposition or trial. The report is also useful to the expert witness to learn what attorneys know about him or her to eliminate surprises during the trial proceeding.[18]

One of the benefits of the Expert Witness Profiler is that it accesses more sources of information than the Daubert Tracker; however, it is also more expensive. For persons preparing a case, such tools can provide guidance and reassurance that they can rely on their expert or experts in the matter. They can also be used to provide insight on the expert testifying for the opposition.

CONCLUSION

This chapter serves to acquaint the reader with forensic accounting and to review how the forensic accountant must navigate the process of discovery and evidentiary issues with a particular focus on ESI. Since so much of the accounting processes and records are electronic, the interaction of the forensic accountant and cyber forensic expert is necessary to the success of their clients.

For additional, recommended reading resources, which accompany this chapter and are provided as a downloadable eResource, readers are encouraged to visit the Publisher's website at https://routledge.com/9780367524180.

NOTES

1 (n.a.), (n.d.), "Forensic Accounting," AICPA, www.aicpa.org/interestareas/forensicandvaluation/resources/forensic-accounting.html, retrieved July 29, 2020.

2 (n.a.), (2019), "Federal Rules of Civil Procedure, § Rule 26 et seq., Rule 26 Of The Federal Rules Of Civil Procedure: General Provisions Regarding Discovery; Duty Of Disclosure, 5. Claims of Privilege or Protection of Trial Preparation Materials, (c) Protective Orders. Item 7," www.ilnd. uscourts.gov/_assets/_documents/_forms/_legal/frcpweb/FRC00029.HTM, retrieved July 29, 2020.

3 (n.a.) (2016), "Rule 16 of the Federal Rules of Criminal Procedure, (b) Defendant's Disclosure, (1) Information Subject to Disclosure, (A) Documents and Objects," www.uscourts.gov/sites/default/ files/rules-of-criminal-procedure.pdf, retrieved July 30, 2020.

4 (n.a.) (2016), "Rule 16 of the Federal Rules of Criminal Procedure, (b) Defendant's Disclosure, (1) Information Subject to Disclosure, (B) Reports of Examinations and Tests," www.uscourts.gov/ sites/default/files/rules-of-criminal-procedure.pdf, retrieved July 30, 2020.

5 (n.a.) (2016), "Rule 16 of the Federal Rules of Criminal Procedure, (b) Defendant's Disclosure, (1) Information Subject to Disclosure, (C) Expert Witnesses," www.uscourts.gov/sites/default/files/ rules-of-criminal-procedure.pdf, retrieved July 30, 2020.

6 (n.a.), (2020), "An Introduction to eDiscovery: The Basics of Electronic Discovery, www.logikcull. com/guide/introduction-to-ediscovery-basics," retrieved July 29, 2020.

7 Electronic Discovery Reference Model, (October 26, 2020), copyright EDRM, Creative Commons Attribution 3.0 Unported License (CC BY 3.0), graphic converted to grayscale by author, https:// edrm.net/edrm-model, retrieved October 26, 2020.

8 (n.a.), (n.d.), "Rule 34. Producing Documents, Electronically Stored Information, and Tangible Things, or Entering onto Land, for Inspection and Other Purposes, 2, (E), (ii)," www.law.cornell. edu/rules/frcp/rule_34, retrieved July 29, 2020.

9 (n.a.), (2019), "Rule 26. Duty to Disclose; General Provisions Governing Discovery, b(2)(B)" www.uscourts.gov/sites/default/files/federal_rules_of_civil_procedure_dec_1_2019_0.pdf, retrieved July 29, 2020.

10 Rule 37 of the Federal Rules of Civil Procedure, Failure to Make Disclosures or to Cooperate in Discovery; Sanctions, (e) Failure To Preserve Electronically Stored Information, (2)(A)(B), www. uscourts.gov/sites/default/files/federal_rules_of_civil_procedure_dec_1_2019_0.pdf, retrieved July 29, 2020.

11 (n.a.), (n.d.), "Rule 901 of the Federal Rules of Evidence, Authenticating or Identifying Evidence," www.law.cornell.edu/rules/fre/rule_901, retrieved July 29, 2020.

12 (n.a.), (n.d.), "Rule 902 of the Federal Rules of Evidence (13) Certified Records Generated by an Electronic Process or System" www.utcourts.gov/resources/rules/ure/0902.htm, retrieved July 29, 2020.

13 (n.a.), (December 1, 2019), "Federal Rules of Evidence Rule 401, Test for Relevant Evidence, Article IV. Relevance and Its Limits (a)(b), www.uscourts.gov/sites/default/files/evidence-rules-procedure-dec2017_0.pdf, retrieved, July 2020.

14 (n.a.), (December 1, 2019), "Federal Rules of Evidence, Rule 803, Exceptions to the Rule Against Hearsay—Regardless of Whether the Declarant Is Available as a Witness. www.uscourts.gov/sites/ default/files/evidence-rules-procedure-dec2017_0.pdf, retrieved July 29, 2020.

15 Readers interested in additional, detailed information regarding this landmark case are referred to Zonana, H., (October 21, 1993), "Daubert V. Merrell Dow Pharmaceuticals: A New Standard for Scientific Evidence in the Courts?" http://jaapl.org/content/jaapl/22/3/309.full.pdf

16 (n.a.), (December 1, 2019), "Federal Rules of Evidence, Rule 702, Testimony by Expert Witnesses, www.uscourts.gov/sites/default/files/evidence-rules-procedure-dec2017_0.pdf, retrieved July 21, 2020.

17 Levin, M. (n.d.), Daubert Tracker, www.dauberttracker.com/casereport.cfm, used with permission, phone conversation and email exchange with M. Levin, image converted to grayscale by author, retrieved July 29, 2020.

18 (n.a.), (n.d.), "Expert Witness Profile" https://expertwitnessprofiler.com/product/detail/1, retrieved July 22, 2020.

Chapter 10

Cyber forensic tools and utilities

Douglas Menendez

CONTENTS

Today's cyber forensic investigator has literally hundreds of specific and unique application software packages and hardware devices that could qualify as cyber forensic tools. Added to that, the hundreds of utilities available for the same task, the job of identifying and examining the best, and most utilized tools are daunting and overwhelming to say the least.

The information contained in this chapter is intended to be used as a reference, and not as an endorsement, of the included providers, vendors, and information resources. Reference herein to any specific commercial product, process, or service by trade name, trademark, service mark, manufacturer, or otherwise does not constitute or imply endorsement, recommendation, or favoring by the authors or the publisher, nor does it imply that the products mentioned are necessarily the best available for the purpose.

Websites included in this chapter are intended to provide current and accurate information; however, it is impossible for anyone (read authors, publisher, etc.) to warrant that the information contained on the sites is accurate or timely.

Relying on information contained on these sites is done at one's own risk. Use of such information is voluntary, and reliance on it should only be undertaken after and independent review of its accuracy, completeness, efficacy, and timeliness. As such, users of this information are advised and encouraged to confirm specific claims for product performance as necessary and appropriate.

It is worth noting that no single text, guideline, or reference book can adequately and definitively state which cyber forensic tool should be used when and under which circumstances and conditions. It is the responsibility of the cyber forensic investigator to (a) have a thorough understanding of the environment and case specifics of the investigation to be performed and (b) to assess and know the specific limitations of each tool before placing unfretted reliance on any single pieces of software or hardware.

Failing to heed these precautions, and to assess one's skill and abilities in utilizing the tools reviewed here, is both unethical and places everyone involved in the investigation at risk.

Good! Now that, that has been said, this chapter will help the reader sort through this exhaustive list and provide a succinct overview of the host of cyber forensic tools available for the 21st-century cyber forensics investigator.[1]

Computer Forensics, like most other areas of Information Technology, continues to expand at a rapid pace. Likewise, the number and type of computer forensic tools and utilities continue to grow and become more specialized.

One of the best sources for information on computer forensic tools and techniques is the catalog maintained by National Institute of Standards and Technology (NIST), through the U.S. Department of Commerce. This forensic tools catalog can be found on the NIST website at www.nist.gov.

NIST COMPUTER FORENSICS TOOLS AND TECHNIQUES CATALOG

The primary goal of the NIST Forensics Tools & Techniques Catalog is to provide an easily searchable catalog of forensic tools and techniques. This enables practitioners to find tools and techniques that meet their specific technical needs. The Catalog provides the ability to search by technical parameters based on specific digital forensics functions, such as disk imaging or deleted file recovery.

The reader/user of this catalog is advised that the information in the catalog is provided by the developer. Any mention of commercial or non-commercial products is for information only and does not imply that a product has been tested.

A secondary goal of the Forensics Tool Catalog is to provide a picture of the digital forensics tool landscape, showing where there are gaps, i.e., functions for which there are no tools or techniques (see Figure 10.1).

The NIST forensics tools website is divided into three major sections. The first section provides an overall search feature to find tools and techniques (see Figure 10.2).

Figure 10.3 provides an example of a search under the 'Deleted File Recovery' Tool Category.

An example of one of the 18 'Deleted File Recovery' Tools found in the NIST catalog is represented in Figure 10.4.

The second major section of the NIST cyber forensics tools catalog includes a page for developers to input information about their tools and techniques (see Figure 10.5).

The third major section of the NIST Catalog provides for a description of the functions and technical parameters of each tools listed in the catalog. This is the Tools and Techniques Taxonomy (see Figure 10.6).

Table 10.1 summarizes the 37 different NIST Forensic Tool Functionalities.

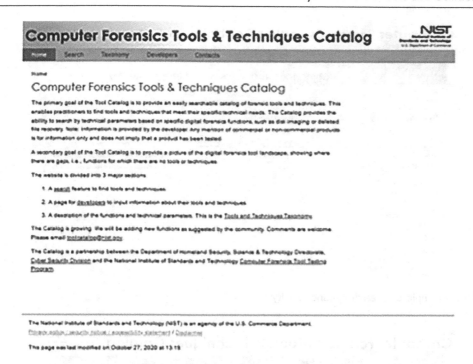

Figure 10.1 Computer Forensics Tools & Techniques Catalog[2]

Figure 10.2 Computer Forensics Tools & Techniques Catalog search feature[3]

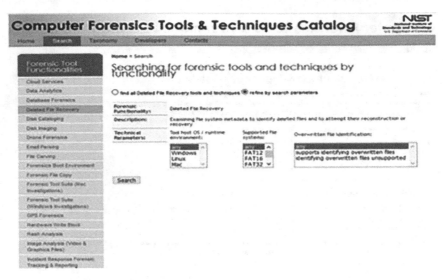

Figure 10.3 Example of a search by functionality[4]

Figure 10.4 Results from a search of the NIST Catalog by functionality[5]

In addition to serving as a 'clearinghouse' for cyber forensic tools, NIST also offers the Computer Forensics Tool Testing Program (CFTT).

NIST – COMPUTER FORENSICS TOOL TESTING PROGRAM (CFTT)

There is a critical need in the law enforcement community to ensure the reliability of computer forensic tools. The goal of the Computer Forensic Tool Testing (CFTT) project at the National Institute of Standards and Technology (NIST) is to establish a methodology for

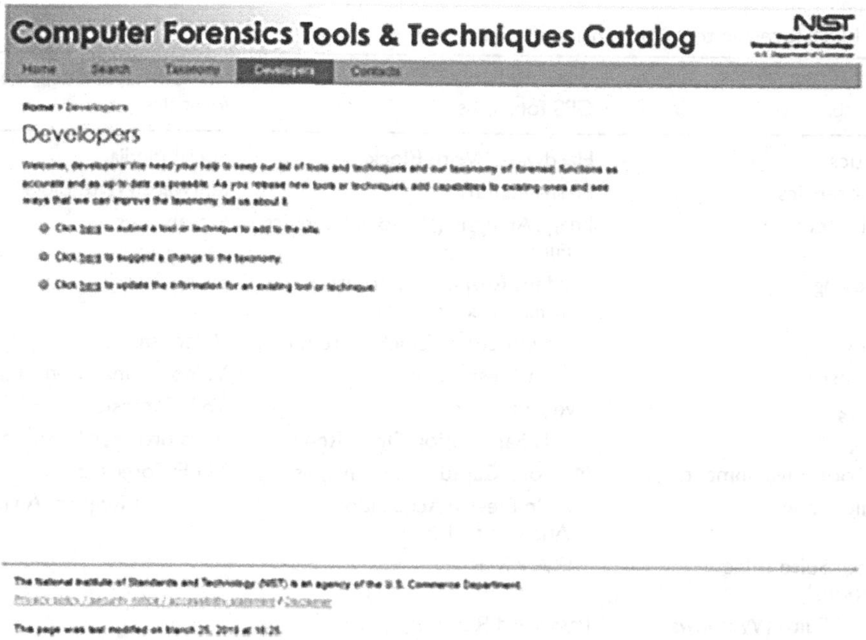

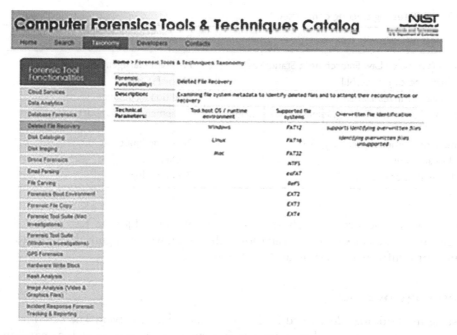

Figure 10.5 NIST Cyber Forensics Tools & Techniques Catalog developer's portal[6]

Figure 10.6 Description of a Tool's Functions and Technical Parameters[7]

testing computer forensic software tools by development of general tool specifications, test procedures, test criteria, test sets, and test hardware.

The results provide the information necessary for toolmakers to improve tools, for users to make informed choices about acquiring and using computer forensics tools, and for interested parties to understand the tools capabilities. A capability is required to ensure that

Table 10.1 NIST forensic tool functionalities[8]

Cloud services	GPS forensics	Remote capabilities / remote forensics
Data Analytics	Hardware Write Block	Social Media
Database Forensics	Hash Analysis	Software Write Block
Deleted File Recovery	Image Analysis (Video & Graphics Files)	Steganalysis
Disk Cataloging	Incident Response Forensic Tracking & Reporting	String Search
Disk Imaging	Infotainment & Vehicle Forensics	Video Analytics
Drone Forensics	Instant Messenger	Video Format Conversion
Email Parsing	Live Response	VoIP Forensics
File Carving	Media Sanitization/Drive Re-use	Web Browser Forensics
Forensics Boot Environment	Memory Capture and Analysis	Wi-Fi Forensics
Forensic File Copy	Mobile Device Acquisition, Analysis and Triage	Windows Registry Analysis
Forensic Tool Suite (Mac Investigations)	P2P Analysis	
Forensic Tool Suite (Windows Investigations)	Password Recovery	

Table 10.2 CFTT testing steering committee[11]

Project sponsors (aka steering committee)	Providing
NIST/OLES (Office of Law Enforcement Standards)	Program Management
National Institute of Justice, NIJ	Major Funding
Federal Bureau of Investigation, FBI	Additional Funding
Department of Defense, DCCI (Defense Cyber Crime Institute)	Equipment and Support
Department of Homeland Security, DHS	Technical Input
States & Local Agencies	Technical Input
Internal Revenue Service, IRS	Technical Input

forensic software tools consistently produce accurate and objective test results. The NIST approach for testing computer forensic tools is based on well-recognized international methodologies for conformance testing and quality testing.[9]

Methodology overview

The testing methodology developed by NIST is functionality driven. The activities of forensic investigations are separated into discrete functions or categories, such as hard disk write protection, disk imaging, string searching, etc. A test methodology is then developed for each category.

The CFTT testing process is directed by a steering committee composed of representatives of the law enforcement community (see Table 10.2). Currently the steering committee selects tool categories for investigation and tools within a category for actual testing by CFTT staff. A vendor may request testing of a tool; however, the steering committee makes the decision about which tools to test.[10]

Methodology process

1. **Specification Development Process**

After a tool category and at least one tool is selected by the steering committee, the development process is as follows:

1. NIST and law enforcement staff develops requirements, assertions, and test cases document (called the tool category specification).
2. The tool category specification is posted to the web for peer review by members of the computer forensics community and for public comment by other interested parties.
3. Relevant comments and feedback are incorporated into the specification.
4. A test environment is designed for the tool category.

2. Tool test process

After a category specification has been developed and a tool selected, the test process is as follows:

1. NIST acquires the tool to be tested.
2. NIST reviews the tool documentation.
3. NIST selects relevant test cases depending on features supported by the tool.
4. NIST develops test strategy.
5. NIST executes tests.
6. NIST produces test report.
7. Steering Committee reviews test report.
8. Vendor reviews test report.
9. NIST posts support software to web.
10. DHS posts test report to web.[12]

CFTT TECHNICAL INFORMATION

The CFTT provides Technical Information for the following tool categories:

1. Disk Imaging
2. Forensic Media Preparation
3. Write Block (Software)
4. Write Block (Hardware)
5. Deleted File Recovery
6. Mobile Devices
7. Forensic File Carving
8. String Search
9. MS Windows Registry Tools
10. Download Raw Test Files
11. Archived Documents[13]

As an example, the format under each tool category is:

- Test Specifications
- Test Support Software
- Test Setup Documents
- DHS Reports – All Test Results[14]

Figure 10.7 NIST CFTT reports portal[15]

All DHS reports for these technical areas can be found at: www.dhs.gov/science-and-technology/nist-cftt-reports (see Figure 10.7).

CFTT RAW TEST FILES

Overview

The CFTT allows access to the raw log files generated during CFTT tool testing. The raw files provide interested parties with the ability to examine the data used to create a CFTT tool test report. The usual procedure to access the raw files is as follows:

- Select the report of interest (bzipped tar file) containing all the test case run directories and any setup directory. Reports are categorized by the functionality, currently disk imaging, write blocking, drive erasing, file carving, and deleted file recovery.
- The bz file can be opened on either a Mac or a Linux system. For help understanding the setup or analysis logs, see the documentation provided with the test support software. All software and documentation are available at www.cftt.nist.gov.[16]

Federated testing project

The Federated Testing project is an expansion of the Computer Forensics Tool Testing (CFTT) Program designed to help digital forensics investigators to test the tools that they use in their labs and to enable sharing of tool test results within the digital forensics community.

Federated Testing Version 5 provides testing for disk imaging, forensic media preparation, forensic string search, hardware write blocking, and mobile forensics data extraction.

Shared test suites

CFTT has developed test suites that will help examiners test forensic tools. The test suites are packaged together in a live Linux .iso file. The process for testing a forensic tool, using the Federated Testing test suites is as follows.

1. Download the latest Federated Testing live Linux .iso file (see the Downloads section below) and use it to create either a bootable flash drive or a bootable DVD.
2. Insert the bootable flash drive or DVD into your forensic workstation and boot to it (you may need to change your computer's boot options to select your flash drive or DVD drive as your boot device). NOTE: to test Hardware Write Blocking and Disk Imaging tools you must boot a computer using a Federated Testing flash drive or DVD; when testing other types of tools however, e.g., a Mobile Forensics Data Extraction tool, one may consider booting a virtual machine in lieu of a computer.
3. Use the user interface (Firefox Web browser) to select the type of tool you want to test. The user interface will tell you what items you will need to have on hand to get started.
4. Use the interface to generate the test cases for testing your tool and follow the instructions to run each test.
5. Use the interface to generate a test report for your tool.
6. (Optional) Submit the test report and the log files created during testing to CFTT to share with the digital forensics community! See the Sharing Test Results section below for instructions on how to share your test results.

CFTT's approach to tool testing is to test a tool based on the functionalities it supports. Currently, you can use the Federated Testing .iso to test disk imaging, forensic media preparation, forensic string search, hardware write blocking, and mobile forensics data extraction tools, but CFTT will add new test suites in future releases to allow you to test more forensic functionalities and more types of tools, e.g., deleted file recovery, forensic file carving, etc.[17]

Shared test reports

A primary goal of the Federated Testing project is to produce tool test results that can be shared throughout the digital forensics community. The Federated Testing test suites allow any lab, agency or individual to test their tools using the same test methodology CFTT uses.

The final step of this process is to generate a test report for the tool. The test suites generate that test report for you in a common format that makes it easy for you and others to understand how the tool was tested and what the test results are. If someone has already tested a tool for the features you use in your lab, you can take advantage of their results in your evaluation of the tool.[18]

Sharing test results

Email your test reports produced using CFTT's Federated Testing test suites and a zipped copy of the testing log files to: cftt@nist.gov to share your results with the digital forensics community. CFTT staff will review your logs and the test results documented in the test reports before sharing the reports with the community. Shared test reports from Federated Testing will be publicly available through this website.

Downloads

Anyone can download version 5 of CFTT's Federated Testing live Linux .iso file (contains test suites for testing disk imaging, forensic media preparation, forensic string search, hardware write blocking, and mobile forensics data extraction tools). The ISO file sha1 value is b4162 a68ff3b2d902dfddd4f256273fe1b5015a4.

If you are testing a forensic string search tool, you will need to also download the string search test suite's companion data set from the Federated Testing Test Data Sets section.[19]

CFReDS

NIST continues to develop Computer Forensic Reference Data Sets (CFReDS) for digital evidence. These reference data sets (CFReDS) provide an investigator with documented sets of simulated digital evidence for examination.

Since CFReDS would have documented contents, such as target search strings seeded in known locations of CFReDS, investigators could compare the results of searches for the target strings with the known placement of the strings. Investigators could use CFReDS in several ways including validating the software tools used in their investigations, equipment check out, training investigators, and proficiency testing of investigators as part of laboratory accreditation.

The CFReDS site is a repository of images. Some images are produced by NIST, often from the CFTT (tool testing) project, and some are contributed by other organizations.[20]

In addition to test images, the CFReDS site contains resources to aid in creating your own test images. These creation aids will be in the form of interesting data files, useful software tools and procedures for specific tasks.[21]

Listed below are some useful links provided by NIST.[22]

- DHS Test Reports: www.dhs.gov/science-and-technology/nist-cftt-reports
- Computer Forensics Tool Catalog: https://toolcatalog.nist.gov/
- National Software Reference Library (NSRL): www.nist.gov/itl/ssd/software-quality-group/national-software-reference-library-nsrl
- DFIR Review: https://dfir.pubpub.org/
- SWGDE (Scientific Working Group on Digital Evidence): www.swgde.org/
- NW3C (National White-Collar Crime Center): www.nw3c.org/
- Digital Corpora: https://digitalcorpora.org/

CYBER FORENSIC TOOLS AND UTILITIES

From an article in SC Magazine, the forensic tool world is evolving; long-time products get new user interfaces and new underlying capabilities. Some tools have merged multiple capabilities into a single coherent product.

When we look back on the state of forensics over the years, we see a convergence of single point solutions to single point problems. We saw separation between computer, network and software forensics. Now, the playground is much larger and, instead of fragmenting tool capabilities even more, developers have continued to converge them. This is necessary because forensic science has merged into a landscape where the bits cannot be distinguished from each other.

Of course, this makes sense since the entire notion of threat hunting, incident response and threat, and event detection all depend in their own ways on forensic techniques. Virtually every competent SIEM, IDS/IPS and advanced firewall has the capability to collect forensic data. When we look at next-generation threat detection, analysis, and intelligence tools, they are almost universally built around forensic approaches to data collection and analysis.

When we collect forensic evidence in a legal environment – as evidence of a crime, tort, or contract dispute – we need to be able to account for each place the evidence has been handled since its collection (chain of custody), its source and 'life story' (provenance), as well as the logical sequence of evidentiary events that the forensic evidence represents.

Table 10.3 Top 10 cyber forensics tools as of December 2020[24]

Rank	Forensic tool	Tool's URL
1.	Sleuth Kit	www.sleuthkit.org
	Autopsy	www.autopsy.com
2.	CAINE	www.caine-live.net
3.	SANS SIFT	https://digital-forensics.sans.org/community/downloads
4.	Xways	www.x-ways.net/forensics
5.	Xplico	www.xplico.org
6.	Volatility	www.volatilityfoundation.org
7.	Magnet	www.magnetforensics.com
8.	Encase	www.guidancesoftware.com
9.	Wireshark	www.wireshark.org
10.	Access Data FTK	www.accessdata.com
	Pro Discover Forensics	www.prodiscover.com

However, when we use cyber forensics for intelligence analysis we are not quite as concerned with chain of custody. Unfortunately, we have seen a breed of cyber forensic tools – particularly in the network arena – that are weak in chain of custody. That does not, by any means, obviate their use in a forensic investigation. It just changes their contribution a bit. Now, instead of becoming part of the evidence chain, these data become investigative leads. Good investigators know, though, that if there is enough corroborating evidence we can – sometimes – get away with a weak or non-existent chain of custody.[23]

The top 10

Based on a number of Google searches, this chapter's author has identified the following cyber forensics tools based on the number of times these tools appeared in Google search results. The top 10 cyber forensics tools along with the tool's URL reference have been provided in Table 10.3.

Cyber forensics tools top 10 overviews

The following is a brief overview of each of the top 10 cyber forensics tools. The reader desiring more information or a deeper product description and functionality assessment is directed to the individual tool developer's website.[25]

The Sleuth Kit (TSK)
www.sleuthkit.org
Version: Sleuth Kit 4.10.1[26]

About:
  The Sleuth Kit® is a collection of command line tools and a C library that allows you to analyze disk images and recover files from them. It is used behind the scenes in Autopsy and many other open source and commercial forensics tools.

Features:

The volume system (media management) tool allows you to examine the layout of disks and other media. TSK supports DOS partitions, BSD partitions (disk labels), Mac partitions, Sun slices (Volume Table of Contents), and GPT disks. With these tools, you can identify where partitions are located and extract them so they can be analyzed with file system analysis tools.

Autopsy
www.autopsy.com
Version: Autopsy 4.17.0[27]

About:

Autopsy® is an easy to use, GUI-based program that allows you to efficiently analyze hard drives and smart phones. It has a plug-in architecture that allows you to find add-on modules or develop custom modules in Java or Python.

Features:

Autopsy was designed to be an end-to-end platform with modules that come with it out of the box and others that are available from third-parties. Some of the software's modules provide cyber forensic features such as:

- Timeline Analysis – Advanced graphical event viewing interface (video tutorial included).
- Hash Filtering – Flag known bad files and ignore known good files.
- Keyword Search – Indexed keyword search to find files that mention relevant terms.
- Web Artifacts – Extract history bookmarks and cookies from Firefox Chrome and IE.
- Data Carving – Recover deleted files from unallocated space using PhotoRec.
- Multimedia – Extract EXIF from pictures and watch videos.
- Indicators of Compromise – Scan a computer using STIX.

SANS SIFT[28]
https://digital-forensics.sans.org/community/downloads
Version: 18.04

About:

The SIFT Workstation is a group of free open-source incident response and forensic tools designed to perform detailed digital forensic examinations in a variety of settings. It can match any current incident response and forensic tool suite. SIFT demonstrates that advanced incident response capabilities and deep dive digital forensic techniques to intrusions can be accomplished using cutting-edge open-source tools that are freely available and frequently updated.

Features:

- Ubuntu LTS 16.04 Base
- 64-bit base system
- Better memory utilization
- Auto-DFIR package update and customizations
- Latest forensic tools and techniques
- VM Appliance ready to tackle forensics
- Cross compatibility between Linux and Windows

- Option to install stand-alone system via SIFT-CLI installer
- Expanded File System Support

CAINE[29]
www.caine-live.net
Version: 11.0

About:

CAINE (Computer Aided INvestigative Environment) is a GNU/Linux live distribution created as a Digital Forensics project. CAINE offers a complete forensic environment that is organized to integrate existing software tools as software modules and to provide a friendly graphical interface.

Features:

- An interoperable environment that supports the digital investigator during the four phases of the digital investigation.
- A user-friendly graphical interface.
- User-friendly tools.
- All devices are blocked in read-only mode, by default.
- New tools, new OSINT, autopsy 4.13 onboard, apfs ready,btrfs forensic tool, nvme ssd drivers ready.
- Ssh server disabled by default.
- Scrcpy – screen your android device.
- Autopsy 4.13 + additional plugins by McKinnon.
- X11vnc server – to control CAINE remotely.
- Hashcat.
- New scripts (forensics tools – analysis menu).
- Automactc – a forensics tool for mac.
- Bitlocker – volatility plugin.
- Autotimeliner – automatically extract forensic timeline from volatile memory dumps.
- Firmwalker – firmware analyzer.
- CDQR – cold disk quick response tool.

X-Ways[30]
www.x-ways.net/forensics
Version: 20.0

About:

X-Ways Forensics is an advanced work environment for computer forensic examiners. Runs under Windows XP/2003/Vista/2008/7/8/8.1/2012/10/2016, 32 Bit/64 Bit, standard/PE/FE. X-Ways Forensics is fully portable and runs off a USB stick on any given Windows system without installation. Downloads and installs within seconds (just a few MB in size, not GB). X-Ways Forensics is based on the WinHex hex and disk editor and part of an efficient workflow model where computer forensic examiners share data and collaborate with investigators that use X-Ways Investigator.

Features:

X-Ways Forensics comprises all the general and specialist features known from WinHex, such as…

- Disk cloning and imaging
- Ability to read partitioning and file system structures inside raw (.dd) image files, ISO, VHD, VHDX, VDI, and VMDK images
- Complete access to disks, RAIDs, and images more than 2 TB in size (more than 2^{32} sectors) with sector sizes up to 8 KB
- Built-in interpretation of JBOD, RAID 0, RAID 5, RAID 5EE, and RAID 6 systems, Linux software RAIDs, Windows dynamic disks, and LVM2
- Automatic identification of lost/deleted partitions
- Native support for FAT12, FAT16, FAT32, exFAT, TFAT, NTFS, Ext2, Ext3, Ext4, Next3®, CDFS/ISO9660/Joliet, UDF
- Superimposition of sectors, e.g. with corrected partition tables or file system data structures to parse file systems completely despite data corruption, without altering the original disk or image
- Access to logical memory of running processes
- Various data recovery techniques, lightning fast and powerful file carving
- Well maintained file header signature database based on GREP notation
- Data interpreter, knowing 20 variable types
- Viewing and editing binary data structures using templates
- Hard disk cleansing to produce forensically sterile media
- Gathering slack space, free space, inter-partition space, and generic text from drives and images
- File and directory catalog creation for all computer media

Xplico[31]
www.xplico.org
Version: 1.2.2

About:
The goal of Xplico is to extract application data from a file capture of Internet network traffic. For example, from a pcap file Xplico extracts each email (POP, IMAP, and SMTP protocols), all HTTP contents, each VoIP call (SIP), FTP, TFTP, and so on. Xplico isn't a network protocol analyzer. Xplico is an open source Network Forensic Analysis Tool (NFAT). Xplico is released under the GNU General Public License.

Features:

- Protocols supported: HTTP, SIP, IMAP, POP, SMTP, TCP, UDP, IPv6, …;
- Port Independent Protocol Identification (PIPI) for each application protocol;
- Multithreading;
- Output data and information in SQLite database or Mysql database and/or files;
- At each data reassembled by Xplico is associated a XML file that uniquely identifies the flows and the pcap containing the data reassembled;
- Real-time elaboration (depends on the number of flows, the types of protocols and by the performance of computer -RAM, CPU, HD access time, …-);
- TCP reassembly with ACK verification for any packet or soft ACK verification;
- Reverse DNS lookup from DNS packages contained in the inputs files (pcap), not from external DNS server;
- No size limit on data entry or the number of files entrance (the only limit is HD size);
- IPv4 and IPv6 support;
- Modularity. Each Xplico component is modular. The input interface, the protocol decoder (Dissector), and the output interface (dispatcher) are all modules;

- The ability to easily create any kind of dispatcher with which to organize the data extracted in the most appropriate and useful to you.

Volatility[32]
www.volatilityfoundation.org
Version: 2.6

About:

Volatility development is now supported by The Volatility Foundation, an independent 501(c) (3) non-profit organization. The foundation was established to promote the use of Volatility and memory analysis within the forensics community, to defend the project's intellectual property (trademarks, licenses, etc.) and longevity, and, finally, to help advance innovative memory analysis research. Along these lines, the foundation was also formed to help protect the rights of the developers who sacrifice their time and resources to make the world's most advanced memory forensics platform free and open source.

Features:

- Enhanced support for Windows 10 (including 14393.447).
- Added new profiles for recently patched Windows 7, Windows 8, and Server 2012.
- Optimized page table enumeration and scanning algorithms, especially on 64-bit Windows 10.
- Added support for carving Internet Explorer 10 history records.
- Added support for memory dumps from the most recent VirtualBox version.
- Updated the svcscan plugin to show FailureCommand (the command that runs when a service fails to start multiple times).
- Add APIs to paged address spaces (x86 and x64) to allow easy lookups of PTE flags (i.e. writeable, no-exec, supervisor, copy-on-write).
- Add support for tagging Mac memory ranges as heaps, stacks, etc.
- Add plugins for checking Mac file operation pointers, C++ classes in the kernel, IOKit interest handlers, timers set by kernel drivers, and enumeration of processes that filter file system events.
- Add support for KASLR Linux kernels.

Magnet Axiom[33]
www.magnetforensics.com
Version: 4.8

About:

A digital forensics solution tailored to meet the needs of organizations that perform remote acquisitions as well as collect and analyze evidence from cloud storage and communication services, computers and mobile devices.

Features:

- AXIOM Cyber acquires and analyzes data from corporate cloud storage services like AWS S3, EC2, and Azure in addition to other cloud sources including Office 365, G Suite, Box, Dropbox, Slack, and iCloud.
- AXIOM Cyber provides the most comprehensive and powerful recovery, search, analysis, and reporting tools for Macs and PCs. Powerful and intuitive Analytics features in AXIOM Cyber like Timeline, Connections, and Magnet.AI allow you to immediately

focus on the most relevant data, enabling you to work your case faster and easily present your findings to HR, Legal, and other stakeholders.
- Comprehensive parsing and carving techniques find more artifacts like browser history, chats, emails, and documents. Easily visualize and present evidence by showing emails and chats in their original format that are often needed for HR investigations like employee misconduct or harassment cases.

OpenText™ EnCase™ Forensic[34]
www.guidancesoftware.com
Version: 20.4

About:

EnCase Forensic is one of the heavyweights of the forensic software market, having been around for many years. The software offers efficient data acquisition and encryption support. It streamlines the entire investigation process, from triage to collection, investigation and reporting. It can quickly search and rank probable evidence in a range of devices.

Features:

- Enhanced indexing engine
- Easy reporting
- Extensibility
- Workflow automation
- Updated encryption support
- Apple File System (APFS) support
- Volume shadow copy capabilities
- Apple T2 Security Bypass

Wireshark[35]
www.wireshark.org
Version: 3.4.0

About:

Wireshark is an open source multi-platform network protocol analyzer. It allows you to examine data from a live network or from a capture file on disk. You can interactively browse the capture data, delving down into just the level of packet detail you need. Wireshark has several powerful features, including a rich display filter language and the ability to view the reconstructed stream of a TCP session. It also supports hundreds of protocols and media types.[36]

Features:

Wireshark has a rich feature set which includes the following:

- Deep inspection of hundreds of protocols, with more being added all the time
- Live capture and offline analysis
- Standard three-pane packet browser
- Multi-platform: Runs on Windows, Linux, macOS, Solaris, FreeBSD, NetBSD, and many others
- Captured network data can be browsed via a GUI, or via the TTY-mode TShark utility
- The most powerful display filters in the industry

- Rich VoIP analysis
- Read/write many different capture file formats: tcpdump (libpcap), Pcap NG, Catapult DCT2000, Cisco Secure IDS iplog, Microsoft Network Monitor, Network General Sniffer® (compressed and uncompressed), Sniffer® Pro, and NetXray®, Network Instruments Observer, NetScreen snoop, Novell LANalyzer, RADCOM WAN/LAN Analyzer, Shomiti/Finisar Surveyor, Tektronix K12xx, Visual Networks Visual UpTime, WildPackets EtherPeek/TokenPeek/AiroPeek, and many others
- Capture files compressed with gzip can be decompressed on the fly
- Live data can be read from Ethernet, IEEE 802.11, PPP/HDLC, ATM, Bluetooth, USB, Token Ring, Frame Relay, FDDI, and others (depending on your platform)
- Decryption support for many protocols, including IPsec, ISAKMP, Kerberos, SNMPv3, SSL/TLS, WEP, and WPA/WPA2
- Coloring rules can be applied to the packet list for quick, intuitive analysis
- Output can be exported to XML, PostScript®, CSV, or plain text

AccessData Forensic Toolkit (FTK)[37]
www.accessdata.com
Version: 7.4

About:

FTK is one of the mainstays in the digital forensic tool marketplace. It allows users to create images, process, and analyze a wide range of data types from forensic images to email archives and mobile devices, create custom scripts, review data offline, and scale within distributed processing and the cloud.

Features:

- QView™ integration introduces a simple, intuitive, and customizable review interface. Utilize multi-case functionality such as tagging, searching, labeling, and bookmarking across multiple cases. Enjoy easy mobile chat application and multimedia review, along with similar face and image detection all backed by a unified database. And, a panels-driven interface means that you can customize the view to your liking.
- Export your data into a portable case for offline review and sync back labels, bookmarks, comments, and notes to the original case. Reviewers will also appreciate the ability to view the data in a near-native format.
- Similar face and object detection allow investigators to quickly locate all images of a person or object across the case without having to train the system, which can use up valuable time and resources. Also, upload an image from outside the case and compare it to pictures within the current case without ingesting it.
- Get a head start on your investigation with URL detection and parsing capabilities across devices without regard to browser, neatly organized under one section to easily review the data and connect the dots in your investigation.
- FTK will ingest and support updated versions of LX01 and E01 images.
- Automatically import and expand a nested forensic image with image within an image support.
- Import and parse AFF4 images created from Mac® computers (generated by third-party solutions like MacQuisition by BlackBag).
- Parse XFS file systems when investigating and collecting from RHEL Linux environments.
- Leverage the power of your forensic environment with optimized support for unified database for the AWS/Amazon RDS configuration. Host your FTK database in AWS to upload, process, and review for unmatched speed and scalability.
- Cut down on OCR time by up to 30% with our efficient OCR engine.

- Locate, manage, and filter mobile data more easily with a dedicated mobile tab. Use the message application filter to quickly isolate data from message applications like WhatsApp or Facebook.
- View all associated EXIF data, including location, make, and model of the device used to capture the images or video.
- Collect, process, and analyze data sets containing Apple file systems that are encrypted, compressed, or deleted.
- Decrypt a computer drive encrypted by the latest version of McAfee Drive Encryption and new L01 export support which eases the workflow of users when data must be used within multiple tools.
- Custom processing options help establish enterprise-wide processing standards, creating consistency for your investigations and reducing the possibility of missed data.
- The easy-to-use GUI provides a faster learning experience.
- Visualization technology that displays your data in timelines, cluster graphs, pie charts, geolocation, and more helps you get a clearer picture of events.

ProDiscover Forensics[38]
www.prodiscover.com
Version: 8.2.0.5

About:
ProDiscover Forensics:

- Is a powerful computer security tool that enables law enforcement professionals to find all the data on a computer disk while protecting evidence and creating evidentiary quality reports for use in legal proceedings.
- Provides a host of features to capture and analyze disks.
- Supports a wide variety of Windows and Linux file systems.
- Ensures that both the capturing and analysis processes are performed by applying forensically sound methods.
- Is integrated with a full text search engine, set of embedded viewers, and hash comparison methods, all together providing an easy-to-use and yet powerful toolkit to forensic investigators.

Features:

- Preview and image disks.
- Preview and search suspect files to find evidence quickly and without altering any data or metadata.
- Automatically creates and records MD5, SHA1, and SHA256 hashes of evidence files to prove data integrity.
- Creates bit-stream copy of entire suspect disk, including hidden HPA section, to keep the original evidence safe.
- Maintains multi-tool compatibility by reading and writing images in the pervasive UNIX .dd format.
- Examine any or all of the following file systems:
    - Windows: FAT12, FAT16, FAT 32, and all NTFS file systems including Dynamic Disk and Software RAID.
    - Mac OS X: HFS, HFS+.
    - Linux: EXT2, EXT3 and EXT4.
    - Solaris: UFS.

- Integrated graphics thumbnail viewer and registry viewer; Outlook email viewer; History viewer; Registry viewer; and Event Log viewer.
- Extract Clusters/Files into Logical File Collections.
- File/Cluster Cross Reference.
- Import/Export .dd format images.
- Add comments to evidence of interest.
- Disk Wipe Capability.
- Extracts EXIF information from JPEG files to identify file creators.
- Linux boot disk provided to image systems without removing hard disk drive.
- Automated report generation in XML format saves time, improves accuracy and compatibility.
- GUI interface and integrated help function assure quick start and ease of use.
- Designed to NIST Disk Imaging Tool Specification 3.1.6 to ensure high quality.
- Support for VMware to run a captured image.

FORENSICS TOOLS – INTERVIEWS WITH THE EXPERTS

Immediately following the chapter summary is the transcript of two interviews conducted by the author with cyber forensic investigators, Greg Chatten and Andrew Hrenak. Andrew is also the author of Chapter 8 on Mobile Forensics. These interviews were conducted in the third quarter 2020, adhering to all COVID-19 protocols.

SUMMARY

In this chapter, we introduced the NIST Computer Forensics Tools & Techniques Catalog and highlighted the vast amount of information that is available from this free U.S. government website. We also looked at the NIST Computer Forensics Tool Testing Program (CFTT) that provides a documented and independent testing process for a wide range of digital forensic tools.

We finished up the chapter with an overview of a 'Top 10' list of popular digital forensic software products. Also, we learned from two forensic experts, about their experience with the forensic tools they have used throughout their careers.

APPENDIX 10.A: INTERVIEWS WITH GREG CHATTEN AND ANDREW HRENAK[39]

Interview #1 with greg chatten

AUTHOR Greg, please tell the readers a little bit about yourself and background.

GREG CHATTEN I am currently the president of Forensic Computer Service, Inc. (Now a division of Universal Data Forensics, LLC). Universal Data Forensics processes all types of computer data from PCs to vehicle monitoring systems and provides wide variety of data forensic services with leading hardware and software used internationally by military, law enforcement and professionals in the private sector.

I began my career in 1981, in programming, and began my career-long use of computer forensic software at that time. I have worked in data forensic services and have provided expert testimony, with over 39 years of IT experience. I am a qualified trial expert witness in data forensics and cellular phone mapping in EDMO, WDMO, KS, NDIA, SDIA, CDIL, SDIL, and numerous Counties in MO, IL, GA, IA, KY.

I have conducted cyber forensics investigations for civil and criminal litigation, internal matters such as employee theft of data and monitoring of data networks, also investigations for domestic matters such as divorce and child custody.

AUTHOR Greg, when and how did your role as a cyber forensics' investigator evolve?

GREG CHATTEN My IT experience started in the late 1970s/early 1980s working with mainframe computers when there was a need to recover data, for a variety of reasons, like crashed hard-drives. As IT changed and expanded, so did the forensic world. From desktops, to laptops, to mobile phones, to the Internet of Things, my role as a digital forensic examiner evolved as well. As an investigator, I have worked with corporate clients, law enforcement and law firms, with an opportunity to work on several important and high-profile cases.

AUTHOR Greg, tell us, what are your 'go-to' cyber forensic examination tools?

GREG CHATTEN Some of the typical tools that I use are file imaging tools. One of the first and most important steps is to ensure potential electronic evidence cannot be compromised by or through any steps of the investigation process. Using hardware and software like Tableau writeblocker, helps to assure the preservation of electronic evidence. I am also a strong advocate that any tools used by a forensics examiner be listed on the NIST website. (Author's note: The role of NIST and the NIST Cyber Forensic Catalog was discussed earlier in the chapter. NIST performs tests on these cyber forensic tools but, does not endorse the forensic tools).

AUTHOR Greg, do you follow a standard procedure or process when performing a cyber forensic investigation?

GREG CHATTEN Once I have acquired an exact image of the data, the selection, and use of an individual or specific tool will greatly depend on the nature of the case and type of analysis required. While not exclusively, I will often employ the use of several forensics tolls such as; Encase (guidancesoftware.com), FTK (accessdata.com), Magnet Axiom (magnetforensics.com), and Cellebrite (cellebrite.com).

For investigations that will involve technology and products from Apple, I use Paladin-Pro Linux Forensic Suite (sumuri.com), which boots from a USB drive. I have also used open source cyber forensic tools in my investigations and I have also, over the years, developed several propriety tools.

AUTHOR Greg, what have been some of your biggest challenges in performing a cyber forensic investigation?

GREG CHATTEN The biggest challenge is the volume of data that must be examined! Some of the more recent sources of key evidence, from recent investigations, have been data from cellphone mapping, vehicle telematics, and the vast amount of video recordings that are available from Ring Doorbells and sophisticated video surveillance systems. In these types of cyber forensic analysis, I use tools from DME Forensics (dmeforensics.com).

Another challenge frequently encountered is with encrypted data. While there are some tools available, trying to crack encryption can be difficult if not nearly impossible. Twenty years ago, people didn't worry too much about encryption or the method of encryption used was easily hacked. In today's world it could take years of constant processing to crack an encryption key. Many well branded cell phones can still be decrypted but others, such as Apple branded phones and devices, remain the hardest to crack.

AUTHOR Thank you Greg for providing readers the benefit of your years or experience and first-hand cyber forensic investigative experience.

Interview #2 with andrew hrenak

AUTHOR Andrew, please tell the readers a little bit about yourself and background.

ANDREW HRENAK I am a 31-year police force veteran detective and digital forensic examiner, with extensive knowledge and hands-on examination experience in the fields of computer and mobile device forensics.

I am currently the Operations Supervisor of the Regional Computer Crimes Education and Enforcement Group (RCCEEG) [www.rcceeg.org) St. Louis, Missouri. RCCEEG is an organization of law enforcement officers, prosecutors, and computer professionals in the St. Louis area and surrounding counties; dedicated to providing manpower, technical, and legal assistance in computer crime education and investigation.

AUTHOR Andrew, from your perspective and in your opinion, what is most essential about the cyber forensics process?

ANDREW HRENAK Being part of a police digital forensics unit, it is important to understand the technical elements of digital forensics, the importance of maintaining 'chain of custody' and the difference between incriminating and exculpatory evidence.

It is important to distinguish between criminal and civil investigations. In criminal investigations 'Beyond Reasonable Doubt' is required for determination of guilt. To gather evidence, it is often required for obtaining a search warrant to recover the evidence according to the Fourth Amendment of the United States Constitution. For criminal cases, it is necessary to establish probable cause to apply for and be issued a search warrant. The facts and circumstances of the investigation provide a basis on which the search is authorized. This limits the scope of the search for material relevant to the investigation. The scope of search is used to guide the examiner in their review of the seized data and to report on the facts of the investigation.

Knowledge of these elements and the ability to 'connect the dots' of evidence to determine fact are elemental in the forensic process.

AUTHOR What cyber forensic tools do you routinely use as part of your investigative process at RCCEEG?

ANDREW HRENAK Forensic tools can be separated as 'dead-box' forensics – analyzing data at rest vs. 'live-box' forensics, looking at volatile data on a running system. Secondarily, mobile devices and Internet of Things (IoT) devices are supported using a differing set of tools.

Here in RCCEEG, we use a variety of cyber forensic tools, each with their own specialty and each providing use with the capability to identify, retrieve, and analyze data. Some of the cyber forensics tools which we use include:

- AccessData FTK is a forensic suite for computer analysis
- Autopsy is a forensic suite for computer analysis
- Cellebrite's UFED series and Physical Analyzer has mobile and IoT extraction and analysis capabilities
- EnCase is a forensic suite for computer analysis
- Magnet Forensic's Axiom a forensic suite servicing both computer and mobile forensics
- X-Ways is an advanced forensic environment for computer forensic examiners.
- Oxygen Forensics for mobile device analysis
- Paraben E3 has a variety of forensic tools and patented faraday protection products.
- Susteen is another vendor of mobile forensic tools.
- Elcomsoft has desktop and mobile integrated forensic solutions.
- Passware is a tool used for password recovery.
- Sanderson's Forensic Toolkit for SQLite (databases)

When there is a need to analyze cases involving illicit graphic images, GriffEye is a versatile software platform, providing a variety of investigative tools. DVR examiner from DME Forensics is a software solution for the recovery of video and metadata from DVR surveillance systems.

There are many other open source tools that are available which we may use, depending on the need and case requirements.

There are many digital forensic tools available, and it is good to have several to choose from, even when evaluating the same evidence. These tools provide the cyber forensic examiner the ability to cross-reference and substantiate her/his findings and conclusions, which is a very important part of any investigation.

AUTHOR: Your expertise with mobile forensics must keep you in high demand and very busy at RCCEEG. What mobile forensic tools do you routinely use as part of your investigative process at RCCEEG?

ANDREW HRENAK: Mobile devices have really expanded in the consumer world, and as such, so have these devices' role in holding key evidence in many crimes. Cellebrite and their line of products have become a leader in this area. Their UFED series of extraction tools coupled with Physical Analyzer helps decode digital data comprehensively and quickly. Adding the capabilities contained in Magnet Forensic's Axiom, the two are leveraged against the data to reveal the greatest amount of relevant information in the shortest period of time. Although we do not use it currently, I have experience with BlackBag's BlackLight, MSAB's XRY along with Paraben's DS:E3. These utilities have data extraction and decoding capability for smartphones and other digital devices.

AUTHOR: When it comes to performing a cyber forensic examination of these mobile devices, how do you select the most appropriate tool for the various environments in which these devices operate?

ANDREW HRENAK: In general, the make, model, and type of operating system define the selection process. Additional consideration is given to the condition of the mobile device. If it is damaged or pass code protected, low-level forms of data recovery may be deployed to extract the data and then utilize a commercial tool for analysis.

AUTHOR: From your prospective as a cyber forensics subject matter expert, what in your opinion are some of the daily challenges cyber forensics investigators face daily?

ANDREW HRENAK: With the explosion of the Internet of Things (IoT), there are more and more devices on the market (thermostats, doorbell cameras, etc.) that can contain digital evidence. Many times, an investigator is called upon to document the stored data to reveal relevance to the investigation. This can lead to the reverse engineering process of a product to understand what data it contains.

If you really want to get into the nitty-gritty, it requires time. When performing an investigation these devices become components in the investigation and must be addressed quickly to provide actionable information. Often times a forensic examiner must act quickly and having knowledge of JTAG fundamentals is very useful and important. JTAG fundamentals, which involves the details of circuit boards and chip design is very useful in the forensic examination of these devices. [Author's note, JTAG {Joint (European) Test Access Group} is a common hardware interface that provides computers with a way to communicate directly with the chips on a board. It was originally developed to address the increasing difficulty of testing printed circuit boards (PCBs)].

In other areas of data explosion, Berla's iVe is a tool for quick, intuitive acquisition and decoding of user data from vehicles' infotainment and telematics systems at both the logical and physical levels.

AUTHOR : Thank you, Andrew, for providing readers the benefit of your knowledge and field-level experience in cyber forensic investigations.

NOTES

1 Marcella, A., Menendez, D., (2008) *Cyber Forensics II: A Field Manual for Collecting, Examining, and Preserving Evidence of Computer Crimes*, Second Edition, Taylor & Francis Group, ISBN 0-84938-328-5, retrieved November 10, 2020, used with permission.

2 (n.a.), (October 27, 2020), Computer Forensics Tools & Techniques Catalog, National Institute of Standards and Technology, https://toolcatalog.nist.gov/index.php, retrieved November 30, 2020.

3 (n.a.), (October 27, 2020), Computer Forensics Tools & Techniques Catalog, National Institute of Standards and Technology, https://toolcatalog.nist.gov/search/index.php, retrieved November 30, 2020.

4 (n.a.), (October 29, 2019), Computer Forensics Tools & Techniques Catalog, National Institute of Standards and Technology, https://toolcatalog.nist.gov/search/index.php?ff_id=2, retrieved November 30, 2020.

5 (n.a.), (October 29, 2019), Computer Forensics Tools & Techniques Catalog, National Institute of Standards and Technology, https://toolcatalog.nist.gov/search/index.php?all_tools= refine&ff_id=2&1%5B%5D=any&2%5B%5D=any&3%5B%5D=any, retrieved November 30, 2020.

6 (n.a.), (March 25, 2019), Computer Forensics Tools & Techniques Catalog, National Institute of Standards and Technology, https://toolcatalog.nist.gov/developers/index.php, retrieved November 30, 2020.

7 (n.a.), (March 25, 2019), Computer Forensics Tools & Techniques Catalog, National Institute of Standards and Technology, https://toolcatalog.nist.gov/taxonomy/index.php, retrieved November 30, 2020.

8 (n.a.), (March 25, 2019), Computer Forensics Tools & Techniques Catalog, National Institute of Standards and Technology, https://toolcatalog.nist.gov/taxonomy/index.php, retrieved November 30, 2020.

9 (n.a.), (November 15, 2019), Computer Forensics Tool Testing Program (CFTT), National Institute of Standards and Technology, Information Technology Library/Software and Systems Division, Software Quality Group, www.nist.gov/itl/ssd/software-quality-group/computer-forensics-tool-testing-program-cftt, retrieved November 1, 2020.

10 (n.a.), (February 22, 2018), Methodology Overview, National Institute of Standards and Technology, Information Technology Library/Software and Systems Division, Software Quality Group, www.nist.gov/itl/ssd/software-quality-group/computer-forensics-tool-testing-program-cftt/cftt-general-0, retrieved November 1, 2020.

11 (Lyle, J.), (May 8, 2017), Computer Forensics Tool Testing at NIST, Computer Forensics Show Presentation, www.nist.gov/document/cftt-pres-computer-forensic-tool-testing-nist-feb-2008, retrieved November 1, 2020.

12 (n.a.), (February 22, 2018), Methodology Overview, National Institute of Standards and Technology, Information Technology Library/Software and Systems Division, Software Quality Group, www.nist.gov/itl/ssd/software-quality-group/computer-forensics-tool-testing-program-cftt/cftt-general-0, retrieved November 1, 2020.

13 (n.a.), (April 2, 2018), CFTT Technical Information, National Institute of Standards and Technology, Information Technology Library/Software and Systems Division, Software Quality Group, www.nist.gov/itl/ssd/software-quality-group/computer-forensics-tool-testing-program-cftt/cftt-technical, retrieved November 2, 2020.

14 (n.a.), (August 28, 2020), Disk Imaging, National Institute of Standards and Technology, Information Technology Library/Software and Systems Division, Software Quality Group, www.nist.gov/itl/ssd/software-quality-group/computer-forensics-tool-testing-program-cftt/cftt-technical/disk, retrieved November 3, 2020.

15 (n.a.), (n.d.), National Institute of Standards and Technology (NIST) Computer Forensic Tool Testing (CFTT) Reports, Homeland Security, Science and Technology, www.dhs.gov/science-and-technology/nist-cftt-reports, retrieved November 3, 2020.

16 (n.a.), (February 22, 2018), CFTT Raw Test Files, National Institute of Standards and Technology, Information Technology Library/Software and Systems Division, Software Quality Group, www.nist.gov/itl/ssd/software-quality-group/computer-forensics-tool-testing-program-cftt/cftt-technical/cftt-raw, retrieved November 3, 2020.

17 (n.a.), (June 2, 2020), "Federated Testing Project," NIST Information Technology Laboratory / Software and Systems Division, www.nist.gov/itl/ssd/software-quality-group/computer-forensics-tool-testing-program-cftt/federated-testing, retrieved November 3, 2020.

18 Ibid.

19 Ibid.

20 (n.a.), (October 7, 2019), The CFReDS Project, www.cfreds.nist.gov/, retrieved November 3, 2020.

21 (n.a.), (November 15, 2019), CFReDS, National Institute of Standards and Technology, Computer Forensics Tool Testing Program (CFTT), www.nist.gov/itl/ssd/software-quality-group/computer-forensics-tool-testing-program-cftt/cfreds, retrieved November 3, 2020.

22 (n.a.), (April 22, 2019), Useful Links, National Institute of Standards and Technology, Information Technology Library/Software and Systems Division, Software Quality Group, www.nist.gov/itl/ssd/software-quality-group/computer-forensics-tool-testing-program-cftt/useful-links, retrieved November 4, 2020.

23 (n.a.), (October 3, 2016), SC Magazine Product Group Tests, Cyberforensics, Group Summary, www.scmagazine.com/group-test/cyberforensics/, retrieved December 1, 2020.

24 Data for this table was collected and developed by the author, from publicly available Internet sites.

25 The inclusion of cyber forensics tools by brand, product, name, manufacture or website, is not an endorsement of that forensics brand, product, software, tool or developer by the author and have been included herein as reference, research or as an example, of which there may be many within the specified cyber forensics tool marketspace.

26 (n.a.), (n.d.), Sleuthkit, http://sleuthkit.org/sleuthkit/, retrieved December 1, 2020.

27 (n.a.), (n.d.), Autopsy, http://sleuthkit.org/autopsy/, retrieved December 1, 2020.

28 (n.a.), (n.d), SANS SIFT Workstation, https://digital-forensics.sans.org/community/downloads, retrieved December 1, 2020.

29 (n.a.), (n.d.) CAINE, www.caine-live.net/, retrieved December 1, 2020.

30 (n.a.), (n.d.), X-Ways, www.x-ways.net/forensics/index-m.html, retrieved December 1, 2020.

31 (n.a.), (n.d.) Xplico, www.xplico.org/about, retrieved December 1, 2020.

32 (n.a.), (n.d.), Volatility, www.volatilityfoundation.org/26, retrieved December 1, 2020.

33 (n.a.), (n.d.), Magnet Axiom, www.magnetforensics.com/products/magnet-axiom/, retrieved December 1, 2020.

34 (n.a.), (n.d.), Opentext EnCase, www.guidancesoftware.com/document/product-brief/encase-forensic-product-overview, retrieved December 1, 2020.

35 (n.a.), (n.d.), Wireshark, www.wireshark.org/, retrieved December 1, 2020.

36 (n.a.), (n.d.), SECTOOLS.ORG, https://sectools.org/tool/wireshark/#comments, retrieved December 19, 2020.

37 (n.a.), (n.d.), AccessData Forensic Toolkit (FTK), https://accessdata.com/products-services/forensic-toolkit-ftk, retrieved December 1, 2020.

38 (n.a.), (n.d.), ProDiscover Forensics, www.prodiscover.com/products-services, retrieved December 1, 2020.

39 The author nor any interview participant has received payment from nor endorses any forensic tool identified or discussed during the interviews conducted. The interview participant's comments and responses are provided here in their role as cyber forensic professionals and remain their opinions and not that of the author, publisher, or of their employers. Comments provided by the interview participants are provided solely as additional value-added information for the reader.

Index

Page numbers in **bold** indicate tables, page numbers in *italic* indicate figures.